Material Flow Systems in Manufacturing

Material Flow Systems in Manufacturing

Edited by

J.M.A. Tanchoco

School of Industrial Engineering
Purdue University, West Lafayette, USA

CHAPMAN & HALL

London · Glasgow · Weinheim · New York · Tokyo · Melbourne · Madras

Published by Chapman & Hall, 2–6 Boundary Row, London SE1 8HN, UK

Chapman & Hall, 2–6 Boundary Row, London SE1 8HN, UK

Blackie Academic & Professional, Wester Cleddens Road, Bishopbriggs, Glasgow G64 2NZ, UK

Chapman & Hall GmbH, Pappelallee 3, 69469 Weinheim, Germany

Chapman & Hall USA, One Penn Plaza, 41st Floor, New York NY 10119, USA

Chapman & Hall Japan, ITP-Japan, Kyowa Building, 3F, 2-2-1 Hirakawacho, Chiyoda-ku, Tokyo 102, Japan

Chapman & Hall Australia, Thomas Nelson Australia, 102 Dodds Street, South Melbourne, Victoria 3205, Australia

Chapman & Hall India, R. Seshadri, 32 Second Main Road, CIT East, Madras 600 035, India

First edition 1994

Typeset in 10/12 Times by Thomson Press (India) Ltd, New Delhi
Printed in Great Britain at the University Press, Cambridge

ISBN 0 412 49180 X

A catalogue record for this book is available from the British Library

Library of Congress Catalog Card Number: 94-71061

♾ Printed on permanent acid-free text paper, manufactured in accordance with ANSI/NISO Z39.48-1992 and ANSI/NISO Z39.48-1984 (Permanence of Paper).

Contents

Contributors

Colin B. Besant
Department of Mechanical Engineering
Imperial College of Science, Technology & Medicine
London, SW7 2BX, UK

Cyrus C.K. Chang
Department of Industrial Engineering
National Tsing Hua University
Hsinchu, Taiwan
Republic of China

Christine G. Co
School of Industrial Engineering
Purdue University
West Lafayette, IN 47907-1287, USA

Jocelyn Drolet
Section de Genie Industriel
Universite du Quebec
Trois-Rivieres, Quebec
Canada

Pius J. Egbelu
Department of Industrial and Management Systems Engineering
The Pennsylvania State University
University Park, PA 16802, USA

Ying-Shin Ho
School of Industrial Engineering
Purdue University
West Lafayette, IN 47907-1287, USA

Suresh K. Khator
Department of Industrial and Management Systems Engineering
University of South Florida
Tampa, FL 33620, USA

Chang Wan Kim
Samsung Data Systems Co., Ltd.
Seoul, Korea

Kap Hwan Kim
Department of Industrial Engineering
Pusan National University
Pusan 609-735, Korea

Lawrence C. Leung
Department of Decision Science and Managerial Economics,
Chinese University of Hong Kong
Shatin, New Territories
Hong Kong

James T. Lin
Department of Industrial Engineering
National Tsing Hua University
Hsinchu, Taiwan 30043
Republic of China

Charles J. Malmborg
Department of Decision Sciences and Engineering Systems
Rensselaer Polytechnic Institute
Troy, NY 12180-3590, USA

Colin L. Moodie
School of Industrial Engineering
Purdue University
West Lafayette, IN 47907-1287, USA

James S. Noble
Department of Industrial Engineering
University of Missouri-Columbia
Columbia, MI 65211, USA

Bernhard F. Rembold
Mercedes-Benz, AG
Stuttgart, Germany

David Sinriech
Faculty of Industrial Engineering and Management
Technion-Israel Institute of Technology
Technion City, Haifa
Israel

J.M.A. Tanchoco
School of Industrial Engineering
Purdue University
West Lafayette, IN 47907-1287, USA

Graeme M.H. Warren
School of Industrial Engineering
Purdue University
West Lafayette, IN 47907-1287, USA

Preface

This book contains a collection of contributions related to the design and control of material flow systems in manufacturing. Material flow systems in manufacturing covers a broad spectrum of topics directly affecting issues related to facilities design, material handling and production planning and control. In selecting the papers to include in this book, the scope was limited to the design and operational control aspects related to the physical movement of parts, tools, containers and material handling devices. Recent developments in this area naturally led to concentration on flow systems involving cellular manufacturing, and automated transport equipment such as automated guided vehicles. However, the concepts discussed have general applicability to a wide range of manufacturing flow problems.

The book is organized in five major sections:

1. design integration and justification;
2. cell design and material handling considerations;
3. alternative material flow paths;
4. operational control problems; and
5. tooling requirements and transport equipment.

In the section on *design integration and justification*, Bernhard Rembold and J.M.A. Tanchoco address the problem of creating a seamless integrated design environment for material flow systems. A central object-based material flow model is created, and model evaluation and improvements are accomplished through a variety of analytical and simulation tools. In James Noble and J.M.A. Tanchoco, the problem of design concurrency is discussed. Issues related to the integration of economics within a single design framework are addressed. The objective is to not only come up with an efficient material handling system, but to design one that is economically justifiable.

Under the heading of *cell design and material handling considerations*, Colin Moodie *et al.* consider the problem of manufacturing cell design. Beyond just the classical problem of determining which machine goes to each cell, the physical arrangement of machines within a cell is addressed. The concept of virtual cells and their material handling requirements are discussed. In the chapter by Pius Egbelu, the impact of unit load sizing and containerization is explored. A common view in scheduling research is that each job is moved as a single unit load. In real factory situations, parts are moved in containers and there are usually several containers required to move the job. Accounting for these multiple entities in material flow analysis provides useful insights. Next, Charles Malmborg describes a modeling framework for addressing the tradeoff between a capacity of the material handling system and the level of work-in-progress. A decision support system based on several analytical models is described and illustrated.

The chapters on *alternative material flow paths* cover several new ideas on material flow topologies. The chapter by Kap Kim and J.M.A. Tanchoco gives a specific method to ensure that conventional factory flow networks are reachable and optimal. Network connectivity is essential if carriers in a factory transport system are required to serve all pickup and delivery stations. We use the term 'optimal' with caution since there are many important considerations which are not accounted for in the analytical model, e.g. the movement of empty carriers, real-time dynamics, etc. In J.M.A. Tanchoco and David Sinriech, the concept of single-loop flow paths is discussed. The single-loop topology can be found in many implementations of fixed path material handling systems such as automated guided vehicles, monorail systems, etc. This configuration eliminates many of the complexities associated with conventional material flow systems, since there are no contention points to deal with and the routing rule is very simple. The procedure described in this paper simultaneously determines the single-loop flow path and the locations of pickup and delivery stations. An extension is proposed for breaking up a single-loop flow path into several segments with bidirectional flows. In a follow-up chapter, David Sinriech and J.M.A. Tanchoco describe a new material flow topology based on the idea of splitting and segmenting an entire facility into several zones. Unlike other zoning strategies proposed previously which require connectivity, the SFT (Segmented Flow Topology) design procedure first splits the network into several components. The resulting split network is then analysed and the flow patterns for each zone determined. The determination of the locations of pickup and delivery stations is simultaneously performed.

In the section on *operational control issues*, Chang Kim and J.M.A. Tanchoco address the problem of routing vehicles in a bidirectional AGV network. Bidirectional AGV networks have intrigued practitioners and researchers because of their performance advantage over conventional unidirectional flow paths. When traffic congestion is not a factor, vehicles operating on a bidirectional network will always take the shortest paths. However, the nature of bidirectional systems introduces the problem of segment contention, since a vehicle moving in a segment essentially blocks another vehicle from entering the same segment from the opposite direction. This paper provides as algorithm for resolving this problem. The chapter by J.M.A. Tanchoco and Christine Co discusses the operational control of multi-load AGVs. The use of multi-load AGVs gives opportunities for reducing the transfer batch size, which may significantly decrease work-in-progress inventory. However, the resulting vehicle management problem becomes severe. Some practical algorithms are needed to efficiently dispatch these multi-load carriers. The last chapter in this section, by Cyrus Chang and James Lin, introduces the load routing problem in tandem AGV systems. The tandem system is based on assigning pickup and delivery stations to single-vehicle closed loops. Additional transit area is provided as an interface between loops. The operational issue addressed in this paper is the determination of which loops to pass when a load is moved from a pickup station in one loop to a delivery station in another loop which is not adjacent to the first loop.

The last section addresses *tooling requirements and transport equipment.* Much of the reported work on flexible manufacturing systems has focused on the routing of workpieces among computer-controlled machining centers. Often, the assignment of parts to machines is dominated by the availability of the tools that are already in the tool magazine. Lawrence Leung and Suresh Khator address the problem of tool automation in a flexible manufacturing system. The last chapter in this book is by Colin Besant on the subject of guidance and navigation techniques for guided and autonomous vehicles. Discussion on factory automation naturally leads to automation in material transport system. The use of autonomous vehicles within a factory environment continues to challenge researchers in developing both the hardware and software necessary for their effective use. The chapter by Colin Besant discusses several of the technologies needed in the development of both guided and autonomous vehicles.

It is my hope that this book will provide further impetus to more in-depth studies in material flow systems in manufacturing. It is also my hope that the readers of this book who come from industry will find the subjects covered to be of value to current and future planning of their factory flow system. In many cases, the studies conducted which resulted in these papers were triggered by problems observed in industrial settings, and some were conveyed by company engineers involved in material flow planning.

Finally, I wish to thank all the participants in this book project for their contributions.

J.M.A. Tanchoco
Lafayette, Indiana
November 1993

PART ONE

Design Integration and Justification

CHAPTER 1

An integrated framework for the design of material flow systems

Bernhard F. Rembold and J.M.A. Tanchoco

1.1 INTRODUCTION

Material flow is a significant factor in the design of manufacturing systems. The designer of a material flow system is faced not only with the specification of individual system components but also with the overall objective of the manufacturing system. The association between components and the interaction of the material flow system with the manufacturing system are the basis by which its performance is judged. A material flow system design may be optimal in itself, but if the design cannot be integrated into the overall manufacturing system, it may have a negative impact on the manufacturing system performance. Therefore, the designer is expected to analyse the role of each component as a part of the total system and consider its influence on the overall system performance.

Since the design of material flow systems involves achieving a comprehensive set of design goals, the designer is required to keep track of large amounts of information. A key issue that makes reaching these goals difficult is the cognitive limits to the human mind. The average human mind is capable of processing only a limited amount of information, far less than is required to come up with a good material flow system design in a timely and efficient manner. For this reason, a computer-based design framework is necessary to assist the designer.

Aside from the large amounts of data and the interdependent design goals, the designer is confronted with a large selection of design tools. Many researchers have performed studies on solving partial problems in material flow systems design such as layout optimization, equipment selection and facilities planning. Implementation of such research work exists in the form of computer programs written in a variety of programming languages. Each of these applications can be expected to have unique input data requirements.

The material in this chapter is obtained from Rembold, B. F. and Tanchoco, J. M. A. (1994a, 1994b, 1994c).

The output reports will all have differing data formats so that information generated by an application cannot be directly input into another. The designer faces the problem of choosing among a set of design tools, each of which has its own mathematical concepts, data requirements and user interface. The difficulty associated with selecting design tools as well as sequencing these tools based on data precedence can be reduced through computer-based decision support.

Once a material flow system model has been developed, the designer must deal with the evaluation of the model with respect to the expected performance. If the performance specifications are not met, the designer needs to diagnose the problem, find an action to improve model performance and select a tool with which to implement this action.

Design frameworks provide an environment for the management of design complexity. Within such a framework, the knowledge and experience gained in research and practice concerning specific topics of material flow system design can be integrated. The following list summarizes the characteristics that a framework for developing material flow systems should display:

- The ability to provide the designer with a comprehensive set of tools with which to design material flow systems.
- Design applications can reside on the workstation and on hosts throughout a computer network. The designer should have the possibility to run software on special purpose hosts in order to take advantage of high-speed computing or graphical capabilities. Also, design applications developed at one institution can be made available to the public through remote execution techniques. In this way, results of research work developed throughout the research and application community can be shared, and a much larger variety of tools becomes accessible to the designer.
- A common database accessible to all design tools.
- The ability to integrate a changing selection of design tools. The design workstation manager should be able to add new tools, remove obsolete tools and upgrade existing tools with minimal effort.
- The ability to identify the data requirements of design applications and sequence these according to data precedence.
- The ability to take the task of running applications out of the designer's hands.
- The ability to assist the designer in the analysis of the model and to make recommendations for improvements.

1.2 LITERATURE REVIEW

1.2.1 The role of material flow in manufacturing

The role of the material flow system in manufacturing can be likened to that of the cardiovascular system in living organisms. Its primary function is to provide the components of the manufacturing system with raw materials, tools and consumables. The material handling system works across departmental boundaries to connect individual parts of the manufacturing system to form a whole. The effectiveness of the material handling system directly affects the performance of the manufacturing system within which it is embedded.

Despite its integrating character, the material flow system is not a value adding factor in manufacturing. The material flow system increases work-in-process (WIP), production time and production costs. Between 13% and 30% of production costs can be attributed to material handling (Baumgarten, 1989; Kuprat, 1990). The long-term goal for material flow should therefore be its elimination. Since this goal can never be achieved, steps should be taken to minimize its need.

This need can be reduced in the design of parts to be produced in the manufacturing system. Any steps which can be taken to reduce the number of setups and types of processes used in producing a part will reduce the amount of required material handling. Ideally, the part should be made at one machine. Hollingum (1988) refers to this as 'one-shot manufacturing'. Using versatile machines which are capable of performing many different types of operations will also reduce the amount of material handling. The material flow system itself should be designed and operated such that the amount of time parts spend waiting and in transit is minimized. Tompkins and White (1984) formulate this definition for material handling which sets realistic goals:

> Material handling uses the right method to provide the right amount of the right material at the right place, at the right time, in the right sequence, in the right position, in the right condition, and at the right cost.

1.2.2 Strategies for the design of material flow systems

Implementing capital projects such as manufacturing and material flow systems is expensive. Computer controlled machining centers can cost anywhere upward from $150 000 without any form of automated material handling. A complete manufacturing system including several workcenters, a material flow system, tooling and fixturing and a computer control center can cost several million dollars. In addition, it has been found that most of the growth, for example in Europe, is attributed to small and medium sized businesses (Schönheit and Wiegershaus, 1990). Insufficient planning on the side of management and engineers in implementing a flexible manufacturing system can easily ruin a small company. Two-thirds of the problems occurring in the use of flexible manufacturing can be attributed to deficient planning.

As a guiding philosophy for designing material flow systems, Apple (1972) refers to Nadler in describing the 'ideal systems approach'. Throughout the design process, the designer should aim for the theoretical ideal system. This system may never be achieved, and represents the perfect system in which material is moved in zero time and at zero cost. Based on the ideal system, the designer conceptualizes the 'ultimate ideal system'. This system will be both technically and financially feasible some time in the future. Research and development invested in this area should be directed towards the creation of the ultimate ideal system. The third step is the design of a system which is technically and financially feasible in the current setting. This system should be the basis for the development of the ultimate ideal system.

It is very unlikely that a material flow system model will be successfully designed in one pass. Therefore an iterative approach should be taken. An example of such an iterative design procedure is shown in Fig. 1.1 (Schönheit

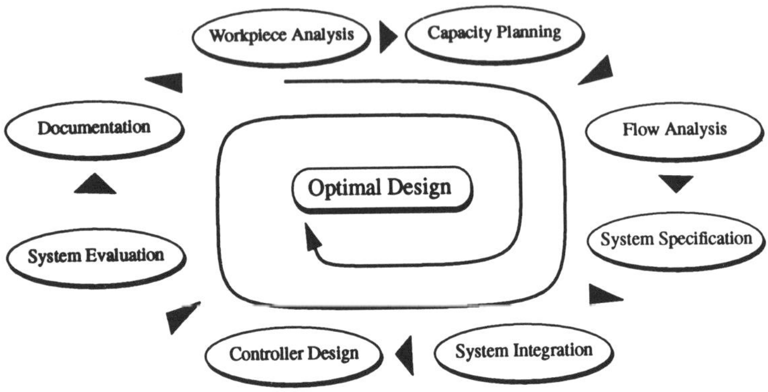

Fig. 1.1 Iterative strategy for the design of flexible manufacturing systems (Schönheit and Wiegerhaus, 1990).

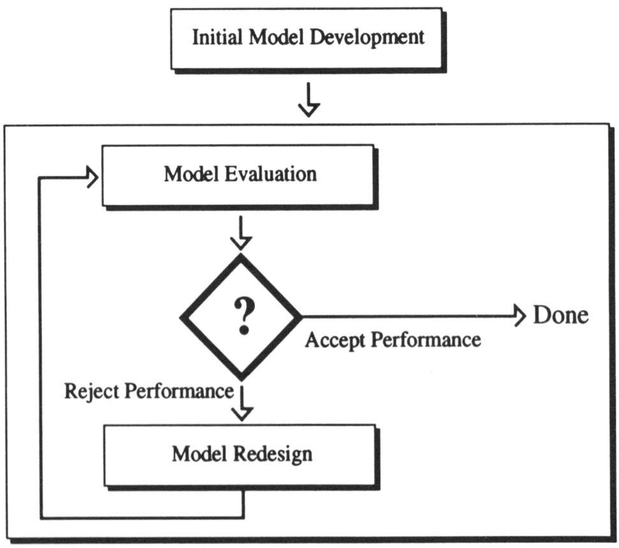

Fig. 1.2 Two-stage methodology for designing complex systems.

and Wiegershaus, 1990). The best results can be achieved by following through this planning cycle several times. Practical applications of this strategy have shown that repeating the cycle a second and third time increases the time and effort put into planning by only 10–20% while significantly improving the system design. It should also be noted that the further the project advances through the execution phase, the more costly these iterations become. Therefore, the least expensive changes are made in the model building phase.

A two stage methodology for this design task is proposed by Dixon *et al.* (1984), which can be applied to this problem domain as shown in Fig. 1.2. The goal within the model development phase is to build an initial, yet complete material flow system model. In the second stage, the designer makes an initial performance evaluation of the design object within the context of the system into which it is integrated. The designer then reconfigures the model based on the evaluation results and reanalyses the model performance. This cycle is repeated until the design meets the required performance standards.

1.2.3 Issues in designing material flow systems

The design of material flow systems involves the specification of six relevant components (Maxwell and Muckstadt, 1982; Müller, 1983), as shown in Fig. 1.3. The conveyance devices are the primary medium for the flow of material throughout the manufacturing system. The layout of the conveyance devices describes the paths within the facility along which material is moved. Load transfer points are positions within the layout at which material is loaded onto or unloaded from the conveyance devices. The load transfer operations are performed by the auxiliary conveyance devices. Buffers can be set up throughout the material flow system for the temporary storage of loads. Finally, the control and dispatching system coordinates the activities within the material flow system.

Müller lists specific spatial considerations and timing conditions which should be included in the design of material flow systems. In analysing the spatial requirements, the focus of the designer should first be directed towards

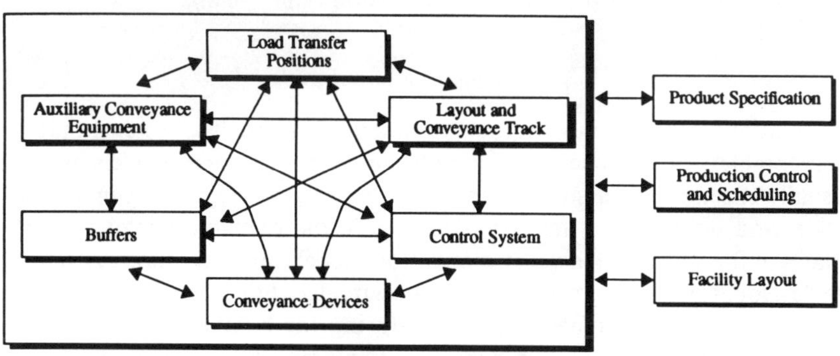

Fig. 1.3 Components of a material handling system.

the space available for the material flow systems. The space will be primarily constrained by the physical dimensions of the facility. The speed of conveyance and the frequency with which loads can be picked up from load transfer stations are also important contributing factors.

1.2.4 Tools for material flow system design and analysis

This section describes a selection of the many tools used in material flow systems design. They are grouped according to the type of underlying models used to describe the material handling and manufacturing systems.

(a) Analytical tools

Static analytical models were developed by Maxwell and Muckstadt (1982) for estimating the required number of carriers in an AGV system, for determining the effects of vehicle blocking and estimating the storage capacity requirements at load transfer stations. Müller (1983) and Egbelu (1987) propose carrier fleet size estimation techniques with varying degrees of sophistication. These formulae are based upon expected material flow quantities and distances. Though these models have been designed with AGV systems in mind, they can be readily be applied to other asynchronous discrete flow material handling systems.

Mathematical programming is an approach taken in developing and improving material flow system layouts. Gaskins and Tanchoco (1987) and Kaspi and Tanchoco (1990) have developed integer programming models which find the best flow directions within a material flow system layout in terms of minimizing the total flow distance. A mixed integer and linear programming model has also been proposed by Leung *et al.* (1987) that is used to estimate the number of vehicles in an AGV system.

Design tools based on queueing networks are frequently used for capacity planning in manufacturing systems. CAN-Q (Solberg, 1981), MVAQ (Suri and Hildebrandt, 1984) and QNA (Segal and Whitt, 1989) can be used to give initial estimates of the number of carriers required in the system (Tanchoco *et al.*, 1987) and perform load transfer station buffer sizing.

(b) Simulation-based tools

Simulation models are useful tools to describe and analyse systems that are too complex for analytical modeling techniques. Analytical models are also usually based on the assumption that the material flow system is in a steady state, and do not give an understanding of the dynamic behavior of the material flow system. However, simulation models are more time consuming to develop, validate and run than analytical models. For this reason, analytical models are an important means for reducing the number of viable design alternatives before simulation runs are attempted. The following list shows some of the issues in material flow system design that can be addressed using simulation:

- verification of material flow system function and performance;
- analysis of material flow system behavior under various operating conditions;
- analysis of the effect of equipment failure on material flow system performance;
- dimensioning of conveyance device capacities;
- analysis of material flow layouts; and
- evaluation of routing and dispatching procedures.

Since the material flow system is an important part of the manufacturing system, the simulation of these systems must be embedded within a manufacturing system simulation. Simulation software that can be used for material flow system simulation can be grouped into two categories: general purpose simulation languages with material handling extensions (e.g. SLAM II and SIMAN IV), and models specifically designed for use in a manufacturing environment.

Law and Haider (1989) and Law and Kelton (1991) performed surveys of commercially available simulation software packages. In the analysis of available material handling modules they find that only AutoMod II provides the flexibility to model a full range of conveyance devices. This system provides elements describing conveyors, AGVs, robots, cranes, AS/RS and transporters (e.g. forklifts). Other simulation packages such as ProModel, SIMFACTORY II.5, WITNESS and XCELL+ are more restrictive in that they do not offer this variety. Other simulation packages specifically developed for manufacturing systems include MAST (CMS Research, 1990), MODELMASTER (Miller, 1987) and MUSIK (Warnecke *et al.*, 1986).

Simulation packages designed for the general simulation of manufacturing systems often lack the detail needed for the accurate analysis of material flow systems. The simulation software AGVSim (Egbelu and Tanchoco, 1982a), AGVSim2 (Gaskins and Tanchoco, 1989) and SattControl (Andersson, 1985) are specially created for analysing specific types of material flow systems, e.g. AGV systems.

(c) Data input and output tools

One of the most time consuming tasks in using tools for designing material flow systems is the creation of input data files and the compilation and interpretation of output data. Most of the commercially available simulation packages have elaborate graphical input capabilities and are able to present output data in the form of graphs, tables and animation. A great number of front-end interfaces have also been developed at research institutions to simplify the use of design tools. Schroer (1989) describes a simulation assistant which offers the user a set of predefined GPSS simulation macros and automatic code generation for the simulation of manufacturing systems.

Electronic spreadsheets (Lotus Development, 1990; Borland International, 1989) can be programmed to provide an easy to use interface for less user-friendly applications. Wysk *et al.* (1987) present a spreadsheet-based front-end for rough cut AGV system analysis using CANQ (Solberg, 1981). Many spreadsheets provide graphics capabilities which can be used to display information such as machine utilization as a function of input parameters. Lesch (1990) presents a rough cut capacity analysis program displaying relevant capacity information in the form of bar charts. This type of analysis could readily be implemented using spreadsheets. Even though spreadsheets can be useful in both data input and output analysis, they are limited in their mathematical and graphical capabilities.

Mathematical analysis programs such as Mathematica (Wolfram, 1988), MathCAD (Anderson, 1989) and MatLab (MathWorks, 1990) are interactive software for scientific and engineering numerical calculation. Applications can be preprepared which use output data from simulation or analytical models. This information can be compiled and displayed in the form of graphs, contour plots and three-dimensional graphics. These programs are not as easy to use as spreadsheets, but offer greater output analysis capabilities.

Graphical design programs such as AutoCAD (Autodesk, 1985) are useful for describing material flow paths. McGinnis (1989) centers an AGV system design workstation around this application. Rembold and Tanchoco (1991) have developed a graphical editor with which the designer can model a material flow network against a converted AutoCAD drawing of a facility layout.

1.2.5 Integration of design tools into a single workstation

The activities surrounding the design of material flow systems not only involve the conception of the end product. A great deal of time is spent on tasks that only indirectly affect the design. These tasks include gathering and entering data, learning to use and using design tools and interpreting design data. Much of this work is tedious and diverts the attention of the designer from the primary problem: coming up with a good material flow system design. The integration of design into a single workstation is one means of helping the designer focus on the global task without getting lost in details

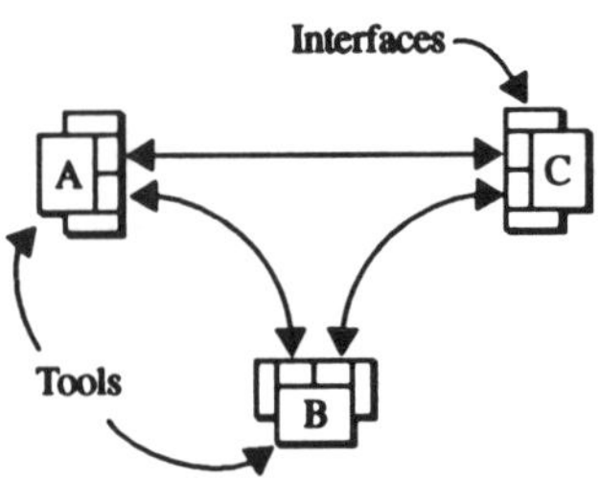

Fig. 1.4 The $(n-1)!$-interface model.

(Carver, 1989). The development of a framework within which such a workstation can be implemented is a key goal of this study.

Using such a workstation, the designer can ideally develop the complete material flow system model. All applications created to solve partial problems are made available through the workstation. The designer should not be expected to deal with the creation and analysis of data files, and need not know how to run an application. A decision support system aids the designer in identifying problems, and lists available tools that can help in solving those problems.

The information system within the design workstation is an important factor. The information system is the means for attaining design tool integration. Three models for the connecting design tools are shown in Figs. 1.4–1.6. In the first model (Fig. 1.4) each application communicates with all other applications through a special set of interfaces. This method is impractical since a total of $(n-1)!$ interfaces must be developed to link n applications.

The second model (Fig. 1.5) assumes that a common language or protocol has been established to which every application has an interface. This reduces the number of interfaces and simplifies the introduction of new tools into the workstation. McGinnis (1989) uses a common data protocol in the integration of heterogeneous software components. Solberg and Heim (1989) propose the use of a blackboard concept for communication between software control modules within a manufacturing system. Each entity in the system accesses

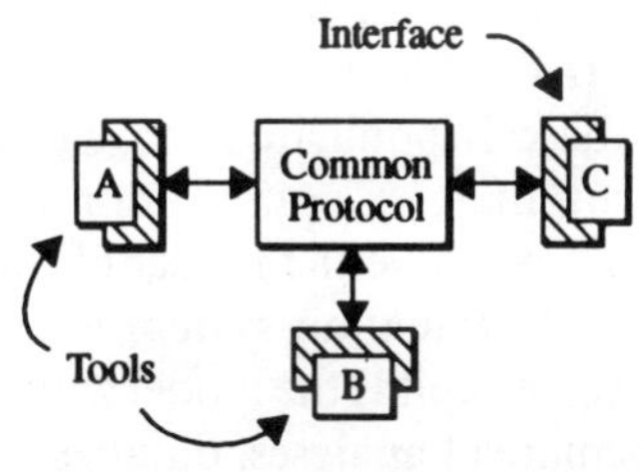

Fig. 1.5 The n-interface model.

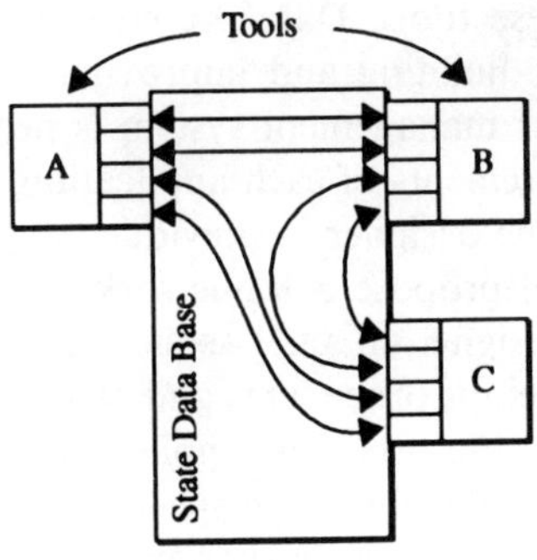

Fig. 1.6 The element connection model.

the information through a strict communications protocol. Naylor and Volz (1988) suggest that all software to be integrated into a large system should be written in a common distributed language and in an associated software environment. The problem with this approach is that many tools have been already implemented in a variety of languages. Translating these applications into a common language would require a considerable amount of time and effort.

In the third design tool integration model, Heim (1990) proposes a framework in which elements corresponding to each other in various models are linked together, as shown in Fig. 1.6. Any time an element changes in one model, it is automatically updated in all others using the same element. A state database is maintained which contains references to these links.

Daniell and Director (1989) integrate VLSI component design applications by modeling them as objects. Each tool is encapsulated by a layered front-end. This front-end knows how to use the tool and what its I/O requirements are. The combination of the tool with the front-end is referred to as a 'CAD tool knowledge object'. The common medium between all objects is a blackboard. The user selects a design tool by posting task requirements to the blackboard. All those objects capable of performing the task essentially bid for the job. The designer selects the tool best suited for the task. Haabma (1988) and Gottheil *et al.* (1988) describe general operating environments for CAD-based design tools. Both support the integration of design tools by providing a standard user interface and a unified database and operating system. McGinnis (1989) outlines an engineering workstation (EWS) that is centered around an ASCII flat file database specifically built for modeling AGV systems. This database is populated through a CAD package and a dialog window. Rough cut analysis, optimization and simulation packages can be used to analyse and improve the model of the AGV system. A more general model building and evaluation system is outlined by Pritsker and Associates (1986). The designer builds the model within a database. A number of applications for mathematical analyses, database queries, simulation and data presentation are available to the designer.

When running design applications, it is necessary to maintain the data precedence between these tools. Data precedence is closely related to the precedence of tasks for building and improving the material flow system model. A design process management system is necessary that is not only aware of the data requirements of each application, but also keeps track of the data generated by the designer in previous design tool runs.

Frydman *et al.* (1989) propose a framework within which two types of specialists assist the designer in task sequencing and tool selection: the algorithm specialist develops procedures and algorithms that aid the user in the VLSI component design task. The application builder, who needs only to know how to use what the algorithm specialist has developed, establishes a rule base describing under which circumstances an application is used. Based on the description of the problem by the designer, an inference engine is used to find the proper sequence of design tasks. The disadvantage to this separation of design tool sequencing and application is that the designer cannot make up tool sequences based on his or her own preference and experience. Moser *et al.* (1988) present a knowledge-based tool box management system for sequencing design tasks in the development of VLSI circuits. The tool box management system provides a framework for describing data dependencies of existing tool input and output data. A rule base is used to analyse these dependencies and propose the use of design tools to reach the design goal. The decision support in this application is based on the precedence of data.

An integrated design workstation should be able to run applications that distribute computational loads among a group of networked computers. Many design tasks such as generating simulation replications involve making several independent runs of an application on a series of data sets (Krishnamurthy, 1989). By distributing the computational load, the time the designer waits for computational results can be reduced. This allows the designer either to complete the design in a shorter time or explore a larger number of design alternatives. Brentano (1984) and Cardinal (1985) discuss a different approach to the distribution of design tools. The design applications are resident on computer systems best suited to running the specific application.

1.3 A FRAMEWORK FOR MATERIAL FLOW SYSTEM DESIGN

A design framework for building material flow system models has been developed by the authors to take key steps towards the management of design complexity. Within this framework the concept of object-oriented analysis and design is used as a tool for modeling material flow systems and a design workstation.

Talavage and Hannam (1988) and Sharp and Liu (1990) propose a sequential approach to designing flexible manufacturing systems, of which the design of material flow systems is an integral part. The designer begins by building a rough cut model of the system and gradually adds details by advancing to more sophisticated modeling techniques, including queueing analysis and simulation. In this method for building manufacturing and material flow system models, the designer spends much time transforming the rough cut model into, for example, a queueing network model, and then into a simulation model. In the transformation process, additional information is fed into the new, more complex model. The nature of this information is determined not by the designer or the design problem, but by the model upon which the software used in each stage is based.

This methodology stands in contrast to the approach taken here for designing discrete material flow systems, as shown in Fig. 1.7. Instead of being faced with a number of models predetermined by the design software, the designer deals only with one central, generalized material flow system model. Design tools are treated as black boxes. Only their data requirements, function and the nature of the underlying model are known. These specialized tools

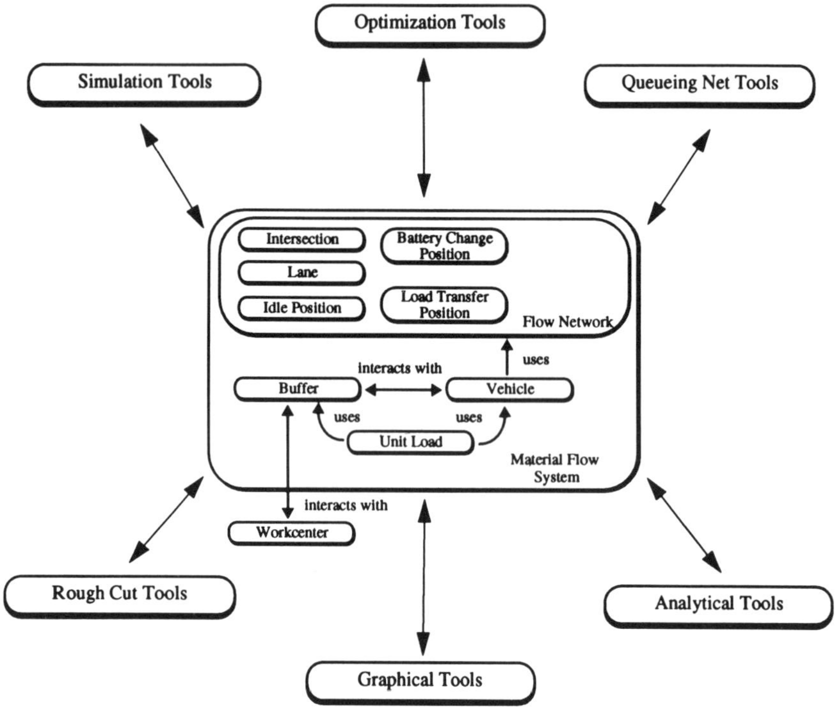

Fig. 1.7 Cooperative model building.

are applied to the model to instantiate individual model components and evaluate and improve the performance of the complete model based on specific measures. A larger selection of design tools can be applied to the model, thus opening up more options to the designer. The required conversion between the central goal model and the model upon which the tool is based is now hidden from the designer. This responsibility is transferred from the designer to the developer of the design tool.

1.3.1 Approach

Figure 1.8 shows the configuration of the design workstation in which the proposed framework is implemented. This workstation addresses the requirements outlined in the introductory section. From a single workstation, the designer is able to access an integrated set of design tools and take advantage of the computational power of a group of networked computers.

The philosophy behind this workstation is twofold: as much of the material flow system design process should be completed from this workstation; also, the structure of the design workstation should be flexible enough to accommodate a gradually evolving set of design tools. This allows the workstation to function as a testbed for new design applications.

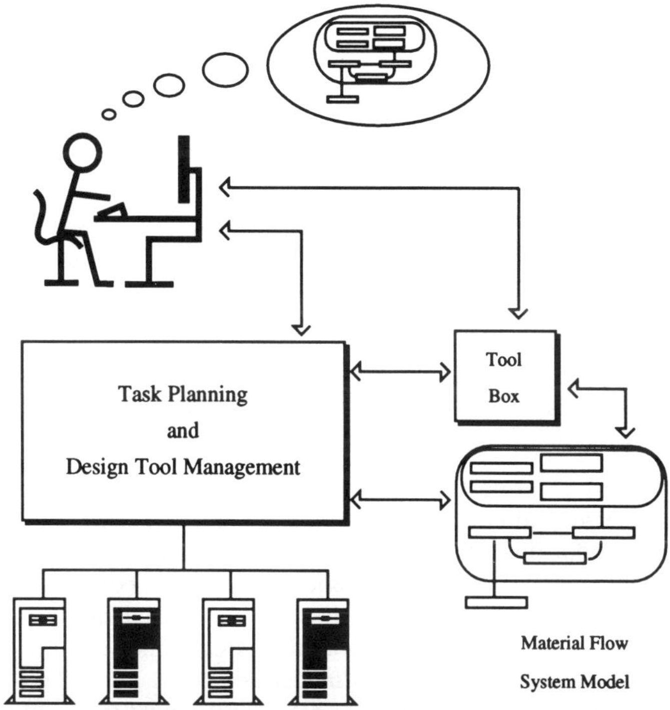

Fig. 1.8 Configuration of the material flow system design workstation.

The common interface between all design tools is the operating system and a flat file ASCII database that is closely linked to the material handling system object model. The only conditions which need to be met by a design tool is that it runs on one of the networked computers and that all data input and output takes place through the flat file database. This form of data representation was selected to ensure data compatibility among the computers used to run applications.

The task planning and design tool management module shown in Fig. 1.8 helps the designer identify those steps involved in building the material handling system model. These tasks include setting the design goals, establishing the criteria by which the model is evaluated, selecting design tools for building and analysing the object model, and specifying the components of the object model.

Throughout the design process, the designer should not be expected to deal directly with the design tools and their data requirements. This work is taken over by the task planning and design tool management module. This module is responsible for preparing all the data files necessary for running a design tool and extracting information from the resulting output reports. Frequently, multiple runs of a design tool are necessary, for example to evaluate the performance of the model under varying conditions. The design

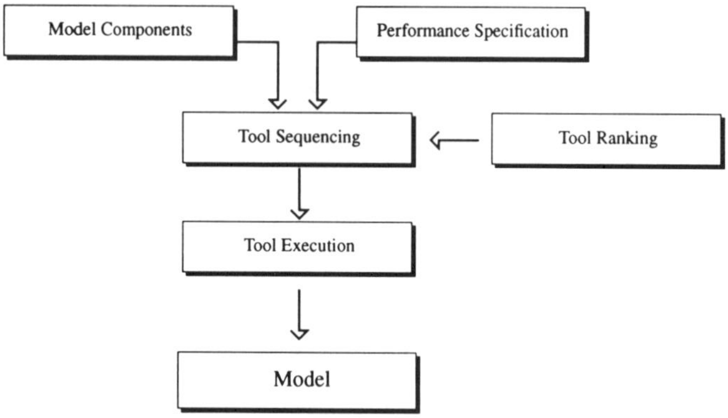

Fig. 1.9 Stages in material flow system model building.

tool resource management module can distribute the computational load among the networked computers to decrease the designer waiting time.

The design tool box contains applications for data entry, optimization, model analysis, simulation and material flow system control. Each tool addresses a partial problem within the design of material handling systems.

The process of designing the material flow system model is divided into two phases: model development and model improvement. The goal within the model development stage shown in Fig. 1.9 is to build an initial material flow system model and make a performance evaluation based on user-defined specifications. First the designer determines which model components to create and by what performance measures the model is to be evaluated. This information is then mapped onto a set of parameters that need to be generated and then entered into the tool sequencing module. The resulting tool sequence is then executed by the designer. In the model improvement phase of this 'generate and test' design procedure (Dixon *et al.*, 1984), the attention of the designer is directed towards redesigning the material flow system model until it meets the user-defined performance specifications.

Figure 1.10 outlines the tasks involved in evaluating and redesigning the material flow system model. The designer begins with a complete model of the material flow system and a list of performance specifications the model is expected to meet. This model could be an initial description of a new material flow system as well as an existing facility in need of redesign. The designer then selects a design goal that the material flow system should achieve. After evaluating the model with respect to that goal, the designer must decide whether or not that goal was met. If the goal was achieved, the designer proceeds to the next design goal. Should the contrary be the case, the designer performs a diagnosis on the material flow system model to determine the root causes of the problem. These causes are mapped onto a set of model improvement actions. Next, the designer must rank the model improvement actions according to their potential contribution to the set

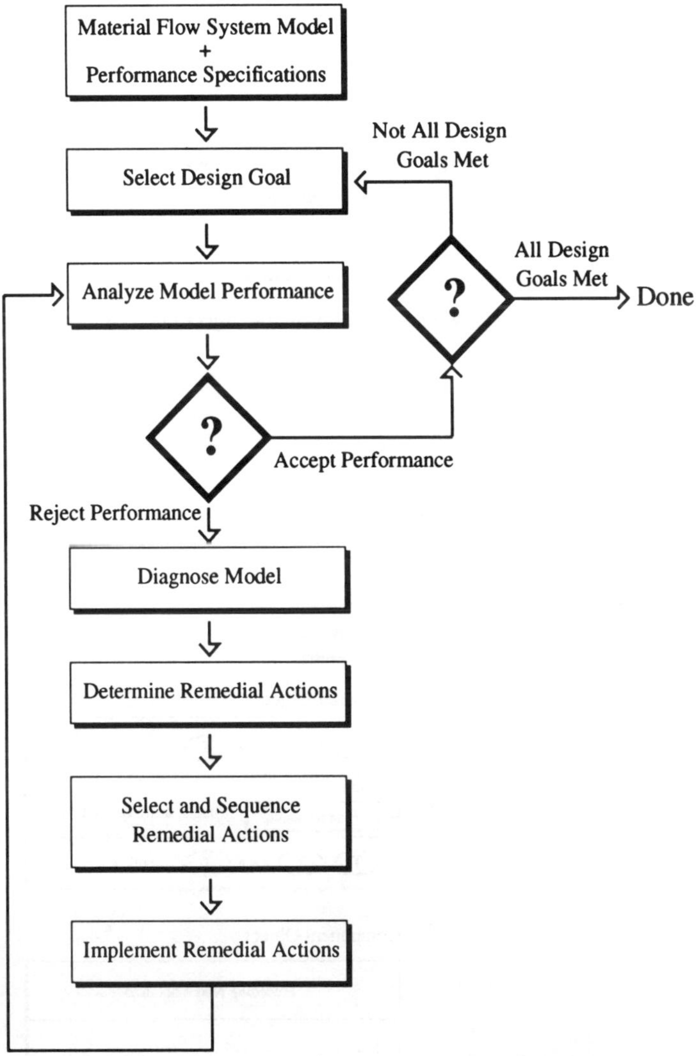

Fig. 1.10 Procedure for model evaluation and redesign.

of design goals. The most promising model improvement action is then implemented. The designer then moves back to the model evaluation stage of this procedure. The evaluation-redesign procedure concludes when all the performance specifications (i.e. design goals) are met.

1.3.2 The underlying object model

Within this design framework, design tools and data are modeled as objects. The object class hierarchy is shown in Fig. 1.11 and the characteristics of

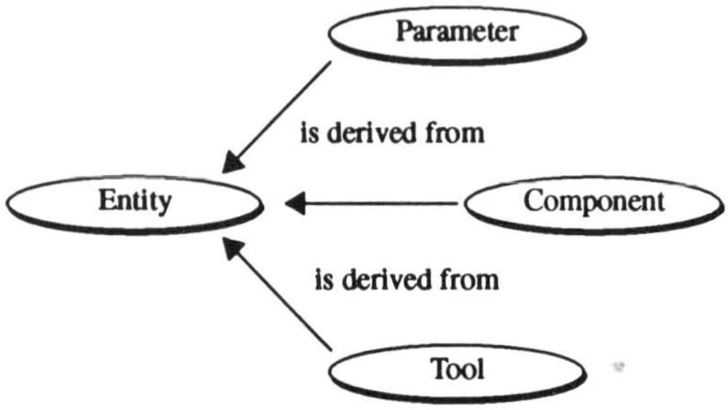

Fig. 1.11 Tool, parameter and component object class hierarchy.

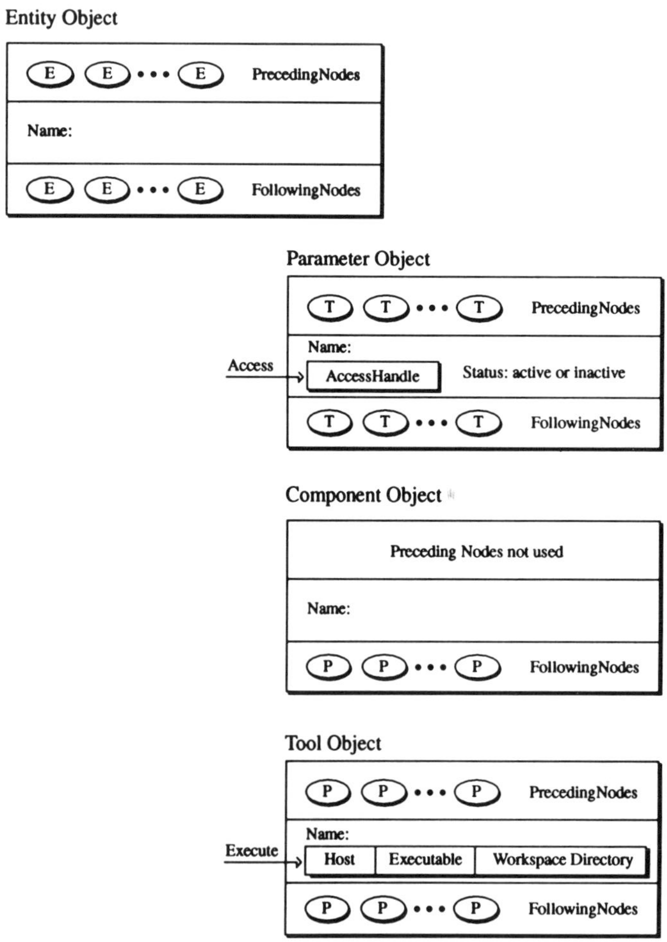

Fig. 1.12 Entity, tool, parameter and component objects.

these object classes are illustrated in Fig. 1.12. The entity object class is the basic class in this object model. Its properties include a name, a list of preceding entities and a list of following entities, as shown in Fig. 1.12. Entity objects are used to model components in a directed graph. The reference lists of preceding and following entities represent the directed arcs in a graph. Three object classes are derived from the entity class: the tool, parameter and component object classes.

The parameter object represents a logical unit of data. This data may describe model components, expected model performance, actual model performance and information specific to a design application. A parameter object, as shown in Fig. 1.12, does not actually contain the information it represents. Instead, it shows the designer where the information can be accessed. The parameter object maintains a list of design tools with which

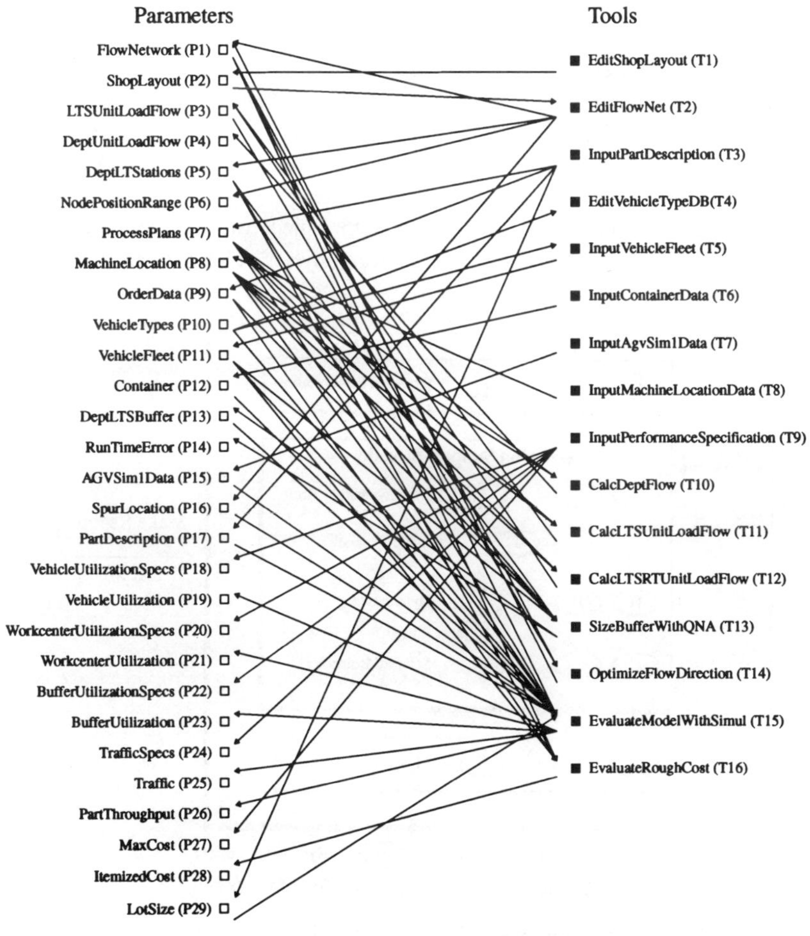

Fig. 1.13 Example of a tool–parameter, graph.

the information it represents can be generated. It also keeps a list of tools that use this data for input. Parameters can be either active or inactive. If a parameter is inactive, the design data represented by that parameter has either not been created or is not current. As soon as this information is made current, the corresponding parameter object is activated.

Tool objects, as shown in Fig. 1.12, represent design software applications that create and process information in the form of parameter objects. The tool object contains information outlining the computer hosts on which this application can be launched, the location of the executable code and in which directory the data processed by the tool is located. As this tool is derived from the entity object, it maintains two lists. The first list contains the parameter objects representing the data required to run the tool. The second list of parameter objects states which data is generated by the tool.

The tool–parameter graph combines tool objects and parameter objects to form a bipartite directed graph, as can be seen in Fig. 1.13. A directed arc pointing from a tool object to a parameter object indicates that the parameter object can be activated using the tool represented by the tool object. Parameter objects not preceded by design tools contain generic, design-independent information, e.g. a list of available load carrier types. Tool objects without preceding parameters represent editors or other means for obtaining information from outside the scope of the design system.

Components of the material flow system object model are specified by design data represented through parameter objects. The mapping of model

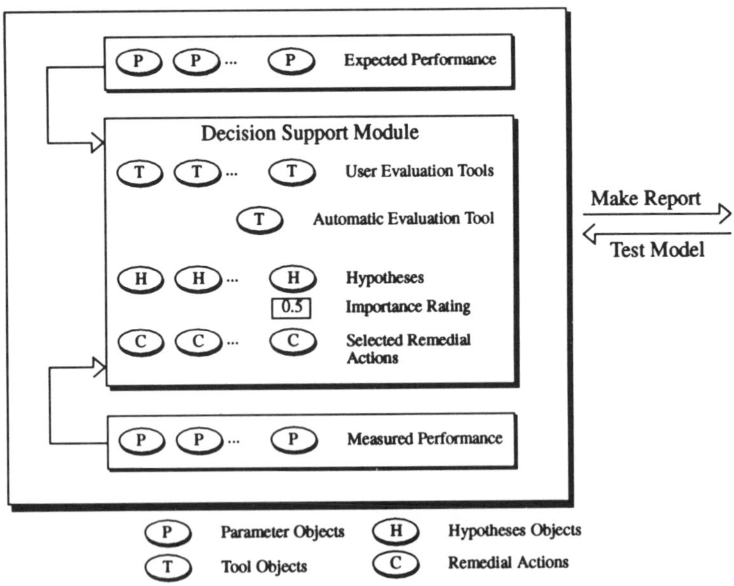

Figure 1.14 The specification object. P = parameter objects; T = tool objects; H = hypothesis objects; C = remedial actions.

components onto parameter objects takes place through component objects. The component object is derived from the entity object, as shown in Fig. 1.11. Figure 1.12 shows the internal structure of the component object. The list of parameters that make up the model component is stored in the 'FollowingNode' list inherited from the entity object.

The performance specifications by which the material flow system model is judged are described within the framework using specification objects, as shown in Fig. 1.14. This object has lists for storing expected and measured performance, as well as an automatic evaluation tool and a list of tools for performance evaluation by the user. In addition to these features, the specification object maintains a list of hypotheses describing postulated causes for a specific performance measure not being met. A fourth list within the specification object is used to store model improvement actions retrieved from proven hypotheses.

1.3.3 Evaluating and ranking design tools

Within every stage of the material flow system design process, the designer may need to choose from among design tools that can perform a task at hand. Ideally, the designer will be familiar with the capabilities and limitations of all the tools within the design system. Based on this knowledge, the designer should be able to select the most appropriate tool. In reality, designers will at best be able to recall this information only for the most frequently used tools. By rating design tools based on key criteria, the designer can be given a feel for the capabilities of a tool and how well the tool will perform under given circumstances.

(a) Rating design tools

Design tools may be rated by intrinsic attributes. Rating a tool by its characteristics has the advantage that it gives the designer an understanding of the strengths and weakness of a tool when applied to a specific situation. Each characteristic is assigned a rating scheme. Fishburn (1967) presents a number of different methods for rating the characteristics of a decision situation. The direct rating method was selected for this purpose for its simplicity and clarity to the user of the design system. The score given under such a scheme can be a numerical value that describes the quality of the tool characteristic under evaluation. In many cases, the characteristics are intangibles to which no numerical value can be assigned. In this case, scoring can be done on a relative scale by comparing it to other tools or to a benchmark. Figure 1.15 shows two examples of this rating scheme for measuring the utility of a design tool with respect to tool attributes. Using the first scheme, the tool characteristics are rated on a scale of one to ten. The second scheme is commonly encountered in the evaluation of consumer goods. A symbolic rating value is assigned to the characteristic instead of a numeric value.

Table 1.1 Criterion matrix

	Context	*T*1	*T*2	*T*3	*T*4	*T*5	*T*6	*T*7	*T*8	*T*9	*T*10	*T*11	*T*12	*T*13	*T*14	*T*15	*T*16
Model sophistication	G	–	–	–	–	–	–	–	–	L	L	L	L	L	H	VH	VL
Processing speed	G	VL	VL	VL	VL	VL	VL	VL	VL	VL	VH	VH	VH	VH	VL	VL	VL
Level of user interaction	G	VH	VH	M	M	M	M	M	M	M	VL	VL	VL	VL	VL	VL	VL
Increase number of vehicles	I	–	–	–	–	VH	–	–	–	–	–	–	–	–	VL	–	–
Increase production capacity	I	–	–	–	–	–	–	–	VH	–	–	–	–	–	–	–	–
Increase buffer capacity	I	–	–	–	–	–	–	–	–	–	–	–	–	VH	–	–	–
Add bypasses	I	–	M	–	–	–	–	–	–	–	–	–	–	–	–	–	–
Improve flow network	I	–	L	–	–	–	–	–	–	–	–	–	–	–	VH	–	–
Add spurs to load transfer stations	I	–	VH	–	–	–	–	–	–	–	–	–	–	–	–	–	–
Remove lanes	I	–	VH	–	–	–	–	–	–	–	–	–	–	–	–	–	–
Reduce number of vehicles	I	–	–	–	–	VH	–	–	–	–	–	–	–	–	–	–	–

G General ranking criteria; I Model improvement criteria; – Not applicable; VL Very low; L Low; M Medium; H High; VH Very high.

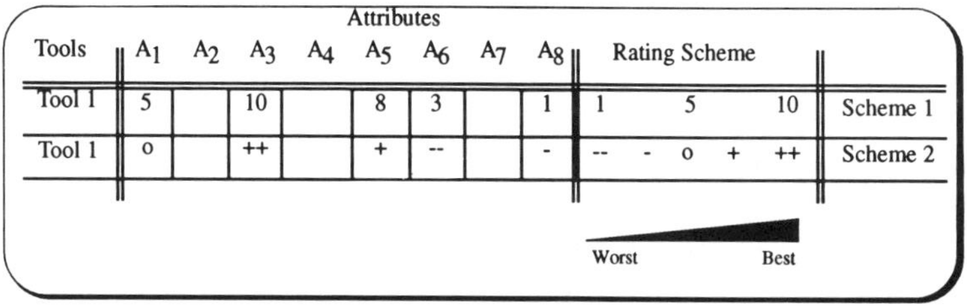

Fig. 1.15 Examples of rating schemes.

When selecting design tools for building, evaluating and improving material flow system models, the designer is primarily concerned with finding the set of design tools that produce the required data. The following tool characteristics may be used to evaluate and choose among candidate design tools:

1. The sophistication of the model upon which the tool is based: this rating implies that the accuracy or quality of the output produced by a design tool stands in direct relation to the complexity of the underlying model. The rating for a tool can take on any of the following five values: rough cut = VERY LOW, analytical = LOW, queueing-network-based = MEDIUM, optimization-based = HIGH, simulation-based = VERY HIGH. This judgement may be misleading as the quality of the output is not only dependent upon the model on which a tool is based, but is also determined by the quality of the input data.
2. The speed of the design tool: an absolute value can usually not be specified, as processing time is dependent upon the size of the design problem. The tool processing speed should be rated in general terms: VERY LOW, LOW, MEDIUM and HIGH and VERY HIGH.
3. The amount of user interaction required to operate this tool: this characteristic should be rated in general terms, e.g. VERY LOW, LOW, MEDIUM and HIGH and VERY HIGH.

The criteria matrix is a means for representing the ratings of design tools with respect to their characteristics and their potential contribution to model improvement recommendations. Each tool occupies a column in the criteria matrix. Tool performance criteria make up the rows of this matrix. The entries in a particular column of the criteria matrix represent evaluation of the tool with respect to each characteristic. The example shown in Table 1.1 illustrates how general model evaluation criteria as well as model improvement recommendations are included in this matrix. A separate column is introduced into the matrix to allow the specification of the context in which the criteria is to be used. The general criteria are applied in sequencing design tools and as a decision aid when the designer is required to choose among tools (Rembold and Tanchoco, 1994b). Model improvement criteria are used to select tools for implementing a model improvement action.

(b) Ranking design tools

If the designer is expected to decide among tools and is only interested in one aspect of design tool performance, the candidate tools can be easily ranked by the rating value of the key attribute. Multi-attribute comparison may be used when the designer is concerned with the performance of the tool with respect to more than one characteristic. A common approach to making multi-attribute decisions is to develop a linear utility function that aggregates the rating values of key attributes (MacCrimmon, 1973). This approach allows the designer to establish the priority of one tool characteristic over another. Mitta (1991) applies hierarchical additive weighting to the evaluation and comparison of expert systems.

These utility functions have the following form:

$$U_i = \sum_{j=1}^{n} w_j a_{ij}$$

where n = number of attributes; m = number of options; U_i = utility of option i; w_j = importance rating of attribute j; a_{ij} = rating of option i with respect to attribute j.

If applied to the selection of design tools, the user must determine the importance weights for each tool characteristic in the given context. The design tools are ranked by the value calculated from the utility function. In a different approach outlined by MacCrimmon (1973), lexicographical sorting of the decision options is performed based on the importance ranking of the characteristics. The designer must select the important tool characteristics and rank them according to preference without assigning any numerical weights. In using this sorting procedure, the tools are first sorted by their primary sort key, i.e. the most important characteristic. Ties are broken by sorting according to the less important tool characteristics.

1.3.4 Design tool sequencing

In this section two approaches for sequencing design tools are presented. The designer has a list of parameters that are given and a second list of parameters that must be generated, as well as a set of tools that must be included in the tool sequence. The objective is to find a sequence of design tools that uses given data to create specified output data, all while considering tool selection criteria chosen by the designer. The first technique involves building a zero-one integer programming model; the second applies a search through the tool–parameter graph.

The sequence of design tools applied to specify and build a material flow system model is governed by available data and the data requirements of the design tools that are used by the designer. Figure 1.16 shows how the selection of design tools affects the design tool sequence. To specify the carrier fleet, the designer must activate the CarrierFleet parameter object. This

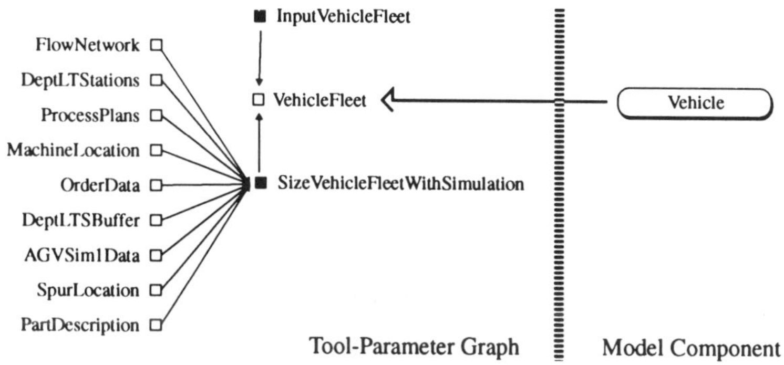

Fig. 1.16 Alternate means for instantiating model components.

parameter can be activated directly by the InputCarrierFleet tool. If the designer chooses to apply the simulation-based tool, all the parameters preceding this tool must be activated.

Due to the effect of design tool use on the sequence of model component instantiation, using PERT charts (Eppinger *et al.*, 1992), SADT/IDEF0 (Eppinger *et al.*, 1992; Mackulak, 1984) or precedence matrices (Eppinger *et al.*, 1992; Middendorf, 1989) is not appropriate for task sequencing in this environment. These techniques are prescriptive, in that the task sequence is preset prior to the design session. These models cannot describe alternate means for achieving a design goal. Also, the prescriptions must be updated any time improvements are made to the design procedures or new design tools are introduced to the design system.

(a) Zero–one integer program-based design tool sequencing

The objective of the zero–one integer program formulation shown in Fig. 1.17 is to minimize the overall cost of applying tools to instantiate model components. The variables in the integer program formulation are defined as follows:

p = number of parameters in the tool–parameter graph;
p_g = number of given parameters;
p_r = number of required (goal) parameters;
p_n = number of all remaining parameters;
$p = p_g + p_r + p_n$;
P_i = zero–one variable representing parameters:
$P_i = 1$ if tool is used, otherwise $P_i = 0$:
representing given parameters $i = 1 \cdots p_g$,
representing required (goal) parameters $i = p_g + 1 \cdots p - p_n - 1$,
representing all other parameters $i = p - p_n \cdots p$;
l_i = number of tools preceding parameter P_i;
t = number of tools in the tool–parameter graph;
T_i = zero–one variable representing tool i in the tool–parameter graph:

$$\text{Minimize } \sum_{i=1}^{t} c_i T_i$$

Subject to the constraints:

1. Variables representing selected tools and parameters that are either required or given must be set

$$P_i = 1; \quad \forall i = 1 \cdots p_g$$
$$P_j = 1; \quad \forall j = p_g + 1 \cdots p - p_n - 1$$

$T_r = 1$

2. Input requirements of all selected tools must be met

$$n_i T_i \leqslant \sum_{k=1}^{n_i} P_k; \quad \forall i = 1 \cdots t \qquad P_k = \text{parameter preceding tool } T_i$$

3. Selected tools activate all following parameters

$$m_i T_i \leqslant \sum_{j=1}^{m_i} P_j; \quad \forall i = 1 \cdots t \qquad P_j = \text{parameter following tool } T_i$$

4. Parameters that are not given need at least one preceding tool

$$P_i \leqslant \sum_{j=1}^{l_i} T_j; \quad \forall i = p - p_n \cdots p \qquad T_j = \text{Tool preceding parameter } P_i$$

Fig. 1.17 Zero–one integer program for design tool selection and sequencing.

$T_i = 1$ if tool is used, otherwise $T_i = 0$;
T_r = zero–one variable representing a tool that must be included in the sequence;
n_i = number of parameters preceding tool T_i;
m_i = number of parameters following tool T_i;
c_i = cost associated with using tool T_i.

If the cost c_i is set equal to one for all tools, the integer program produces the smallest possible list of design tools with which to build the model. Setting a high cost for design tools that are either difficult to use or require large amounts of computational power will discourage the use of that specific tool. Within the integer program model, the variables representing the given parameters and the parameters to be created by the tools sequence are set equal to one, as shown in the first constraint in Fig. 1.17. The second constraint makes sure that, if a tool is selected, all the input requirements of that tool are met. The third constraint indicates that, if a tool is selected, all its output parameters are activated. Except for the given parameters, all parameters that are activated must have at least one preceding tool that has been executed before the parameter can be used. This is reflected in the final constraint.

The integer program does not generate the sequence in which the selected design tools are applied. The following algorithm is used to create that sequence:

1. Start with a copy of the tool–parameter graph. All parameters are inactive.

Activate the parameters that were already available when the integer program model was built. Remove all the design tools from the tool–parameter graph that are not in the list of selected tools, *L*, provided by the integer program. The tools sequence list, *S*, is initially empty.
2. Find the set of tools, *T*, within the tool–parameter graph that have no preceding parameters or have all the input requirements met, i.e. all input parameters are active. Append the set of tools *T* to the end of sequence list *S*. Activate all the output parameters of the tools within the set *T*.
3. Repeat step 2 until all specified output parameter are active.

(b) Graph-based tool sequencing

The second tool sequencing approach is based on a modification of pre-order tree traversal (Sedgewick, 1992) through the tool–parameter graph. The goal of this search procedure is to create an ordered set of trees with goal parameters as roots and any one of a specific set of terminal nodes as leaves. Terminal nodes can be any of the following nodes within the tool–parameter graph: given parameters, tools without input parameters and parameters that are members of existing trees and parameters following tools already members of existing trees. The sequence of design tools is determined from the ordered set of trees.

The tools sequencing procedure is presented in the following example. Given a tool–parameter graph, as shown in Fig. 1.18, with P_1, P_2 and P_3 as goal parameters, the root of the first tree is chosen to be parameter P_1. This parameter can be activated by either tool T_1 or tool T_2. The user must now

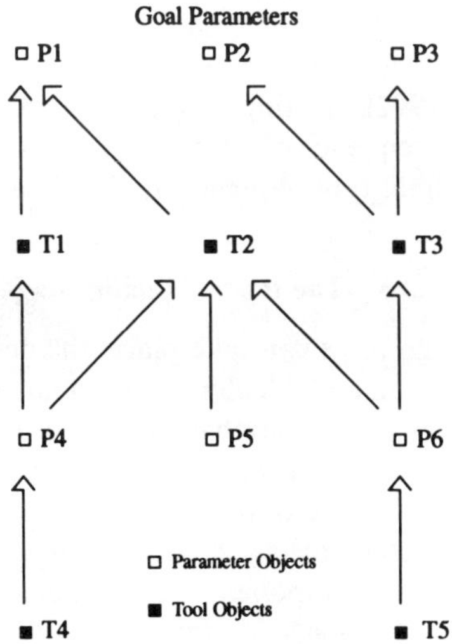

Fig. 1.18 Example tool–parameter graph for graph-based tool sequencing. □ = parameter objects; ■ = tool objects.

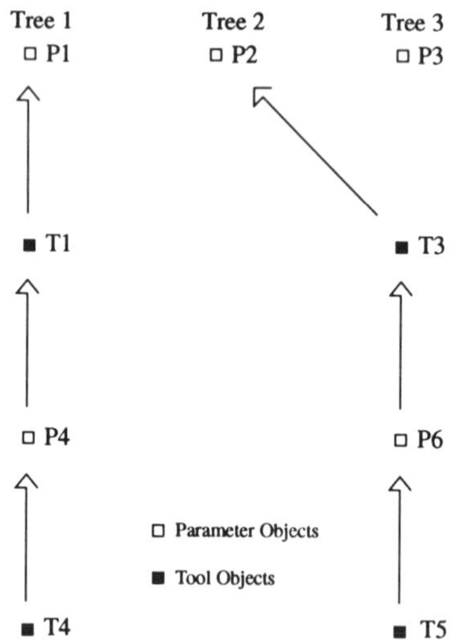

Fig. 1.19 Trees created in graph-based tool sequencing. □ = parameter objects; ■ = tool objects.

choose between these tools. Assuming T_1 is selected, parameter P_4 must be activated as it is a prerequisite to this tool. This parameter can only be created by tool T_4. The first tree for activating parameter P_1, as shown in Fig. 1.19, has this node as its root, and contains tools T_1 and T_4 and parameter P_4.

The next tree has P_2 as its root. Tool T_3 is the only tool capable of activating this parameter. To run tool T_3, input parameter P_5 must be activated using tool T_5. The tree growing from parameter P_3 terminates at that node as tool T_3 is already a component of the second tree, and activates this parameter. The final sequence of design tools is determined by performing a post-order traversal (Sedgewick, 1992), starting at the leaves of the first tree down to the root and noting the sequence of tools. The same is done for all subsequent trees, resulting in the final tool sequence T_4, T_1, T_5 and T_3.

1.3.5 The model building stage

Before sequencing design tools can take place, the designer must determine which parameters are given and which must be created by the tool sequence. Within the model building stage, emphasis in tool sequencing is put on being able to build an initial version of the material flow model quickly and efficiently. The model is described by a group of component objects, each of which contains a list of parameter objects. These parameter objects represent the data that outline the model component specifications and their relationships to other components. The goals for this stage of the design process are established by extracting the parameter objects from the model component objects. An example is shown in Fig. 1.20. The data requirements for evaluating

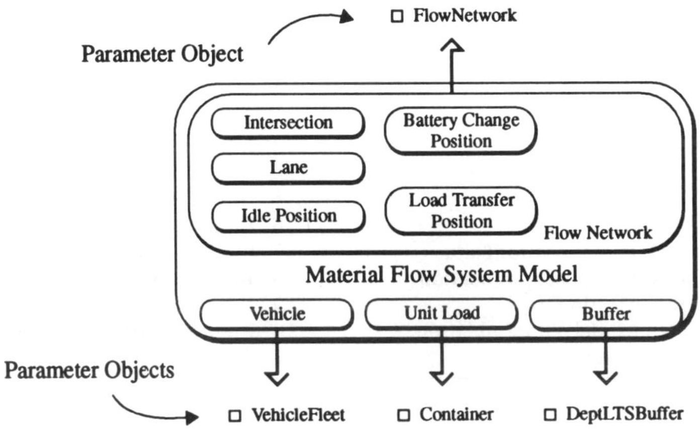

Fig. 1.20 Mapping material flow system model components to parameter objects.

model performance and comparing it to a reference are obtained through specification objects. These have been selected by the designer to describe the desired model performance characteristics. The parameter objects contained in the expected and measured performance lists represent the goal parameters for these performance characteristics. Figure 1.21 shows an example of cost and part throughput specifications being mapped into goal parameter objects.

The next step in the model building stage involves finding a sequence of design tools that uses the available data to generate all model component and performance data. Both the zero–one integer program and graph-based sequencing procedure can be applied. The following list in Table 1.2 shows the sequence of tools that are generated using both methods. The initial model is built by executing all these tools in the proper sequence.

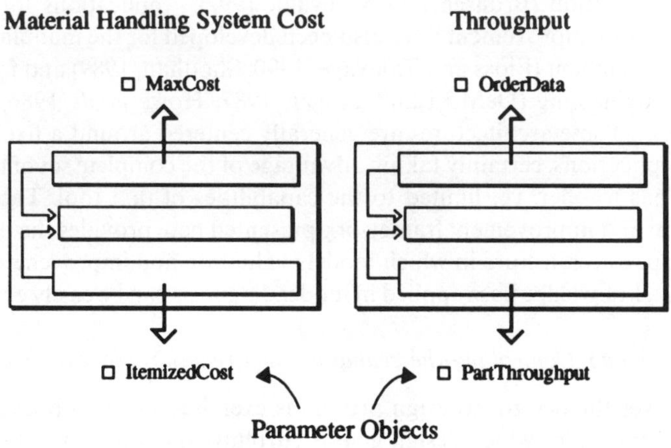

Fig. 1.21 Mapping performance specifications to parameter objects.

Table 1.2 Tool execution

Zero–one integer program-based sequencing	*Graph-based sequencing*
1. EditShopLayout	1. EditShopLayout
2. EditFlowNet	2. EditFlowNet
3. InputPartDescription	3. InputPartDescription
4. InputVehicleFleet	4. InputVehicleFleet
5. InputContainerData	5. InputContainerData
6. InputAGVSim1Data	6. InputMachineLocationData
7. InputMachineLocationData	7. SizeBufferWithQNA
8. InputPerformanceSpecification	8. InputAGVSim1Data
9. SizeBufferWithQNA	9. EvaluateModelWithSimulation
10. EvaluateModelWithSimulation	10. InputPerformanceSpecification
11. EvaluateRoughCost	11. EvaluateRoughCost

1.3.6 The model evaluation and improvement stage

This section addresses the situation in which the designer has built an initial model of a material flow system and specified a set of performance specifications the model is expected to meet. Assuming that the model given does not exhibit the expected performance, the designer is faced with diagnosing the problems, finding suitable model improvement actions and implementing these actions. Conflicting design goals, interrelationships between model components, the large amounts of data associated with the model and a large number of design tools available to the designer make this a very complex task.

Due to the complex nature of the evaluation and redesign problem, the designer can benefit substantially from having the computer assist in fault diagnosis and suggesting improvement strategies. Current applications for material flow system design offer substantial help to the designer simply by integrating graphical, numerical and simulation-based design applications into a workstation (Brunsen and Maiwald, 1992). Applications for model evaluation and improvement have also been developed for the manufacturing design environment (Floss and Talavage, 1990; Shodhan, 1989) and for other fields of engineering (DeMori and Prager, 1987, Howe *et al.*, 1986; Dixon *et al.*, 1984). These architectures are generally centered around a fixed set of design applications, certainly taking advantage of the complete set of features the tool has to offer, yet limited to the capabilities of that tool. The model evaluation and improvement framework presented here provides the designer with an open architecture in which model evaluation and improvement tools as well as a knowledge base applied in model diagnosis can be easily extended.

(a) Controlling the evaluation and redesign process

Control over the design–redesign process is exercised at the top level when the designer decides which design goal is currently the most important with respect to the performance of the material flow system model. The critical factors in this task are the interrelationships between design goals.

Mostow (1985) identifies four types of design goals, classified by how they affect each other: independent, cooperative, competing and interfering. Independent design goals can be dealt with in any order. Unfortunately, purely independent and cooperative design goals of this nature are rarely encountered in the design of material flow systems. Dealing with competing goals involves making trade-offs. For example, material flow throughput can often be increased by adding unit load carriers to the system. As a consequence, this action will drive up the cost of the material flow system. The designer must decide if the achieved increase in material flow capacity is justifiable by the increase in cost. Formal methods for making trade-offs are outlined by MacCrimmon (1973) and MacCrimmon and Siu (1974).

In most situations, the designer will be faced with a set of interfering design goals. The relationships between interfering goals can be cooperative, competing or independent, depending upon the state of the design process and actions taken to meet the individual goals. As an example, in the case when a material flow system displays a low level of traffic congestion, increasing the number of unit load carriers can increase throughput, but it will also increase the cost of the system. If a high level of congestion is present, decreasing the number of carriers may increase throughput and also the cost, thus making the two goals cooperative. These goals may be viewed as independent if the designer manages to improve throughput by optimizing the directionality of the material flow network (Kaspi and Tanchoco, 1990) without adding lanes and additional cost.

Mostow (1985) proposes a set of strategies for working with interfering design goals. In a sequential approach, the designer brings the model to meet one design goal before advancing to another. The key problem with this approach is the effect that commitments made while achieving one design goal have on the other goals. Therefore, the designer must iterate between individual goals to meet the complete set of goals. By choosing the right design goal to address, this process can be shortened and improved. Strategies for goal selection include:

1. Choosing the goal requiring the least amount of commitments (Mostow, 1985).
2. Choosing the goal constraining the design the most (Mostow, 1985).
3. Working on the goal having the highest priority first (Howe *et al.*, 1986).

There is no best strategy for coordinating these design activities. Which approach works best depends largely on the design objective and also the personal style of the designer. For this reason, the design system should be flexible enough to allow the designer to apply these strategies and provide adequate decision support.

(b) Analysing model performance

Four distinct tasks can be identified in model performance evaluation: gathering information, compiling data, interpreting data and comparing these data with reference values. Information gathering involves applying measurement tools to obtain raw performance data. This information must be compiled and brought into a form that allows a meaningful comparison with performance expectations.

Objective, numerical performance values are easiest to obtain, process and compare. These can be measured using analytical, statistical or simulation-based approximations. Examples of this type of data include machine utilization, vehicle utilization, average queue lengths at load transfer stations and unit load waiting times. The reference values are typically expressed as a maximum or minimum allowable performance, e.g. maximum machine or vehicle utilization. Performance evaluation based on these measures can be automated by having computer programs make comparisons with reference data.

Some material flow system model performance data will be subjective in nature. These are quantified through a rating scheme if these data are to be processed by the computer. By applying direct rating (Fishburn, 1967), the performance of the model is rated, for instance, on a scale of one to ten. Maximum and minimum threshold values are also set on this scale.

Even though measuring and rating material flow system performance is in itself a complex task, defining reference data to which this measured data is compared can be more difficult. Some information such as the maximum allowable equipment utilization is easily obtained. On the other hand, how is the designer to assess an acceptable level of lane traffic or carrier interferences at nodes within the material flow network? If either the measured or the reference data concerning an aspect of material flow system model performance cannot be meaningfully quantified, the design system can at least contribute by providing the designer with pertinent information to make a decision as to whether a performance specification is met. Numerical information for decision-making can be presented within a spreadsheet. The graphical capabilities of the spreadsheet application allow the designer to present the information as charts and graphs. Illustrative examples of how such information can be displayed are shown by Lesch (1990) and Spur *et al.* (1983). Graphical information such as material flow network congestion can be displayed using CAD-based applications.

(c) Model fault diagnosis

Diagnosing a material flow system model not meeting performance standards is a critical task within the evaluation–redesign process. Due to the complex nature of the material flow system model, the most appropriate techniques are found in the domain of fault diagnosis of technical and industrial systems. The primary function of a fault diagnosis system is to draw inferences from available data and additional measurements, and attempt to pinpoint the source of the problems (Tzafestas, 1987).

Knowledge-based diagnostic systems have proved beneficial in identifying problems within complex systems (Milberg *et al.*, 1992; Hofmann *et al.*, 1986). They provide capabilities for describing not only numerical and procedural data, but also non-quantifiable information such as rules, heuristics and causal models of system behavior. This information is typically extracted from a domain expert. By applying a knowledge-based approach to search for problems within a model, unlikely problem sources can be systematically eliminated, thus improving the speed and efficiency of the diagnosis. Additional benefits include the ability to give reasoning for the diagnosis. The reasoning feature aids in the validation of the diagnosis, and provides the designer with an opportunity to learn from the domain experts.

The hypothesis formulation and hypothesis testing approach models the manner in which humans perform diagnostic tasks. This method is applied by Hofmann *et al.* (1986) and Fink *et al.* (1985). First, a hypothesis on the cause of a problem is postulated. Evidence is then collected in the form of process measurements to either prove or disprove the hypothesis. This is done by looking for symptoms that can cause the postulated problem to occur. Data related to symptoms are measured and compared to references in the same way as model performance data. Appropriate rules help guide the diagnostic process and reduce the time required to determine the validity of a hypothesis. These rules combine the presence or absence of certain symptoms using logical operators. Modularity is an important benefit to using the hypothesis formulation and hypothesis testing approach. Modules for diagnosing specific problems can be developed and applied independent of each other, and are easily integrated within the design framework.

The diagnostic system proposed for the design framework is based on the hypothesis formulation and testing concept. Symptom objects are the basic building blocks from which rules for proving or disproving hypotheses are constructed, as illustrated in Fig. 1.22. The two lists in this object are used

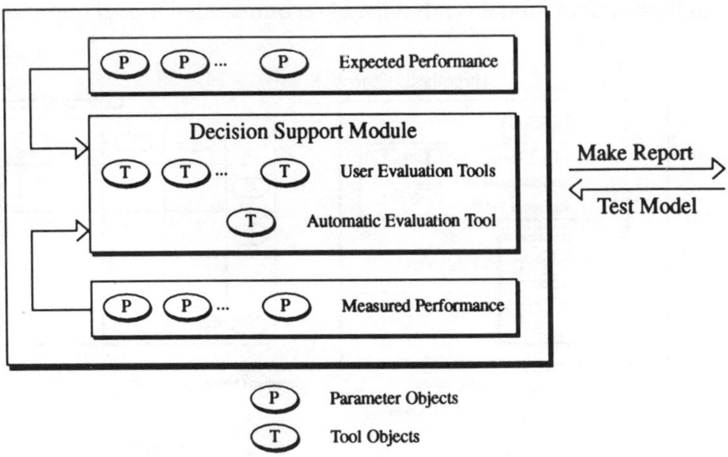

Fig. 1.22 The symptom object. P = parameter objects; T = tool objects.

to store parameter objects representing reference data and measured model performance data. In addition to these lists, the symptom object contains a reference to a tool that can perform an automatic evaluation of the model to determine whether a symptom is present. This tool updates a special file represented by a decision parameter object. The file contains the result of the test, either TRUE or FALSE, as well as a summary of the test for the designer to read. A third list is maintained by the symptom object that contains tool objects. The tools represented by these objects are intended to help the designer decide upon the presence of a particular symptom. This feature becomes necessary, especially when an automatic test cannot be performed.

The hypothesis object class is used to model reasons for why the material flow system model does not meet a performance specification. It describes how the postulated cause can be either proved or disproved. A rule is used to perform this proof. The syntax of these rules allows the use of the logical operators AND, OR and NOT, as well as parentheses to control the evaluation of the logical expression. The operands of the rule are references to symptom objects. The values of these operands are determined by testing the symptom through either the automatic tool or a decision of the user. Once all the symptoms of a rule have been established, the hypothesis can be tested by evaluating the logical expression. If the hypothesis is proven to be true, a list of possible model improvement actions can be retrieved from the hypothesis object. This list contains references to rows within the criteria matrix representing model improvement actions. The structure of the hypothesis object and its relationship to the criteria matrix is shown in Fig. 1.23.

(d) Rating and ranking model improvement actions

The hypothesis formulation and testing approach applied to diagnosing the material flow system model yields a list of recommended model improvement

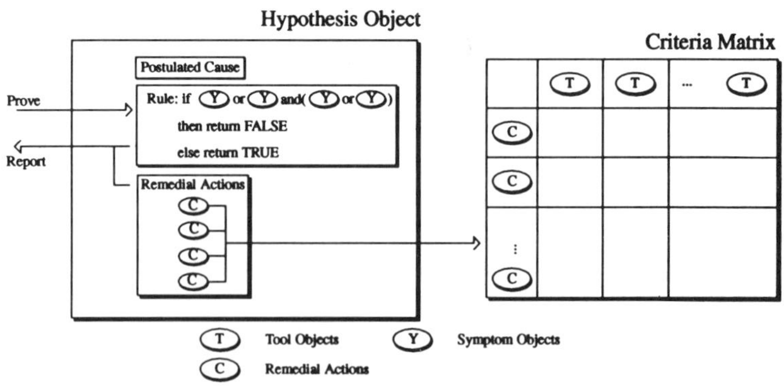

Fig. 1.23 The hypothesis object and criteria matrix. T = tool objects; Y = symptom objects; C = remedial actions.

actions that is presented to the designer. When implemented, these actions should bring the material flow system model a step closer to meeting expected performance specifications. The user of the design system must evaluate these recommendations in terms of their potential contributions to the improvement of the object under analysis. The model improvement action promising to have the greatest effect in moving the design towards compliance with all performance specifications should be chosen and implemented. This hill climbing approach is applied in a number of design systems (Howe *et al.*, 1986; Shodhan, 1989; Milberg *et al.*, 1992).

Model improvement actions can be assigned to one of three classes: quantifiable and incrementable, nonincremental and conceptual. Quantifiable and incrementable actions are the easiest to deal with. This type of model improvement action proposes changing a system control variable by a specific amount. The required level of change is either predetermined or predicted on the basis of the current design. Predictions can be made using a mathematical model of the design object (e.g. Howe *et al.*, 1986) or a simulation experiment. Examples of quantifiable model improvement actions include increasing or decreasing vehicle fleet size, adding machining capacity to the facility or increasing buffer storage capacity. The second type of model improvement recommendation is nonincremental, but the implementation options are enumerable. Changing a specific property of the material flow system model such as control and dispatching strategies falls into this category of recommendations. It is typically not possible to predict which state will provide the best performance without launching an experiment testing the model for each possibility. Conceptual model improvement actions present the greatest difficulty from the viewpoint of automatically determining the effect of the action. Redesign recommendations of this type involve applying the experience, intuition and creativity of the designer. Reconfiguring the material flow network to relieve congestion is an example of such a task. The effects of these activities on the design are nearly impossible to predict, since they can affect model performance for better or for worse.

The relationship between design goals adds another dimension to the problem of selecting appropriate model improvement actions. A matrix can be formed showing the effect that the implementation of each model improvement action may have on performance specifications. The quantities entered into this dependency table are the bases for rating and ranking model improvement tasks. Given this matrix, lexicographical ordering or other multi-attribute decision-making techniques can be applied to determine the most appropriate model improvement action for a set of prioritized design goals. This table is initially set up by a domain expert (Howe *et al.*, 1986; Shodhan, 1989). In the framework developed by Howe *et al.* (1986), this matrix continuously evolves as the design progresses. Every time a step was taken to improve the model, the effect of this action was recorded and entered into the dependency table.

Unfortunately, the effects of a significant number of the model improvement actions cannot be quantified in this manner. Therefore, a hybrid approach to designer decision support in selecting model improvement actions must be taken. Wherever possible, the designer should be given the option to have the design system measure the effects of a model improvement action on the design goals. In the cases in which this cannot be done, the design system provides decision support by presenting the designer with a dependency table showing the relationships between design goals based on past experience of the evaluation of a domain expert. The recommendations given through this pre-prepared matrix will naturally not apply equally to every possible design situation. It can be expected that as the user gains experience with the design problem and the design system, the designer will be able to rely on personal judgement instead of this information.

Within the proposed design framework, the row and column entries of the dependency table represent model improvement actions and model performance specifications. The elements of the table are modeled using dependency objects, as shown in Fig. 1.24. Their location within the dependency table are given by a reference to a model improvement action, as listed in the criterion matrix and a performance specification object. A symbolic rating is presented to the designer to assist in evaluating the effect of the model improvement action. The rating scale ranges from a strong negative influence (− −) over no influence (0) to a strong positive influence (+ +). The symbolic rating is generated from a numerical rating value and a set of three cutoff points. This allows for the comparison of model improvement actions for which the measured potential impact on the design have different units of

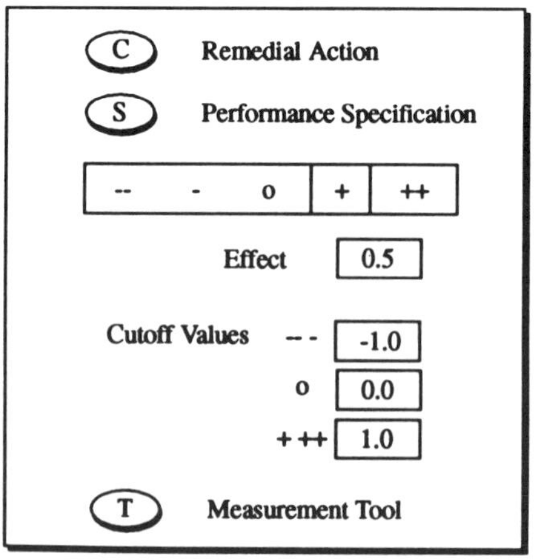

Fig. 1.24 The dependency object. T = tool objects; S = performance specification; C = remedial actions.

measure. The following if clause illustrates this connection:

```
IF NumericalRating < MinusCutoff THEN SymbolicRating = '− −'
ELSE IF NumericalRating < ZeroCutoff THEN SymbolicRating = '−'
ELSE IF NumericalRating = ZeroCutoff THEN SymbolicRating = '0'
ELSE IF NumericalRating < PlusCutoff THEN SymbolicRating = '+'
ELSE SymbolicRating = '+ +'
```

Numerical rating and cutoff values can be determined by launching a design tool specifically built for measuring the effect of the particular model improvement action on a performance specification. A reference to this tool is contained in the dependency object. If such a tool does not exist, a default setting of cutoff values is used. In such a case, the designer or a domain expert can enter a numerical rating value.

With this dependency table it is possible to rank the model improvement actions by their symbolic rating values. Two ordering schemes are made available to the user: lexicographical ordering and ranking by utility. Using the lexicographical ordering scheme, the model improvement actions are first ranked according to what is perceived as the most important performance criteria. Any ties are broken using the secondary performance specifications. This ranking method is applicable when the performance specifications complement each other, or when one performance measure is far more important than the others.

If a trade-off must be made between performance measures, the model improvement actions should be ranked by their utility to the designer. Each performance specification is assigned an importance rating value, as shown in the specification object in Fig. 1.12. The symbolic rating values stored in the dependency objects are transformed into numerical, i.e. '− −' to −2, '−' to −1, '0' to 0, '+' to 1 and '+ +' to 2. The utility of each model improvement action is calculated by summing the weighted rating values of a row. The model improvement actions are then ranked based on this calculated utility.

(e) Implementing model improvement actions

Two issues in the implementation of model improvement actions must be addressed: selecting design tools to carry out the model improvement action, and ensuring that the data set representing the material flow system model and its performance remains consistent and up-to-date.

Matching model improvement tasks to design tools is a matter of applying the criteria matrix. Each design tool occupies a column in this matrix while all possible model improvement actions make up the rows. The entries in the matrix represent the ratings of tools with respect to the contributions the tool can make in implementing the model improvement action. Using the matrix, the designer can choose appropriate design tools by scanning the row corresponding to the proposed model improvement action and selecting the tool with the highest rating.

After selecting a design tool to perform a model improvement action, the designer needs to find a sequence of design tools that includes the model improvement tool, updates the material flow system model and the data describing model performance, and carries the effect of the model improvement action throughout the complete set of design data through the data dependencies. The next step in the model improvement action implementation process is to create a sequence of design tools that includes the selected model improvement tool, tools generating required input data for that tool, and the tools evaluating the effect of the change to the model. The tool zero–one integer program-based sequencing procedure is applied. At the beginning of the model evaluation and improvement process, the following information is required:

- a tool-parameter graph in which all parameters are initially deactivated;
- a list of all tools that have already been applied to build, evaluate and improve the model;
- a list of model components describing the design object; and
- a list of performance specifications that the model is expected to meet.

An example of these data is shown in Fig. 1.25. To determine which parameters have been specified in the model building stage, the output parameters referenced by the tools in the tool sequence are activated, as shown in Fig. 1.26. The output parameters of the selected model improvement tool are then marked on the tool–parameter graph to record the effects of the application of this tool. In the example shown in Fig. 1.27. parameter P_4 is marked as it is the output of selected model improvement tool T_6. Next,

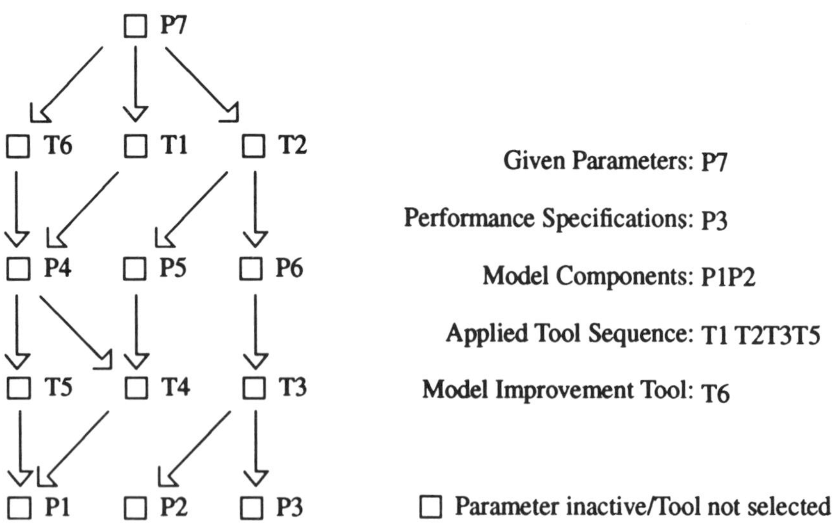

Fig. 1.25 Initial tool–parameter graph. □ = parameter inactive/tool not selected.

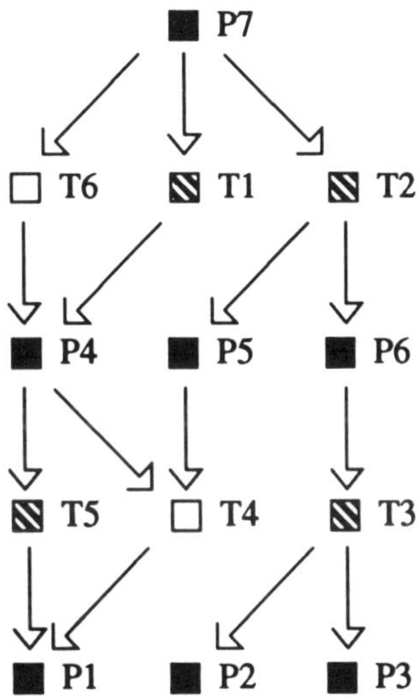

Fig. 1.26 Tool–parameter graph after parameter activation. □ = parameter inactive/tool not selected; ■ = parameter active; □ = tool selected.

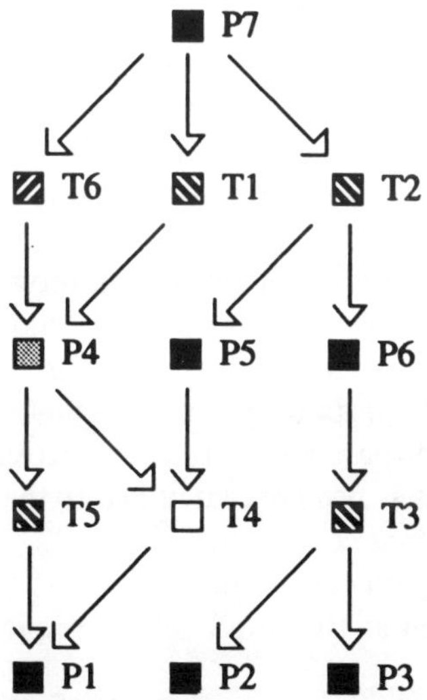

Fig. 1.27 Marking output parameters of selected model improvement tools. □ = parameter inactive/tool not selected; ■ = parameter active; ▦ = parameter marked; ⧅ = tool selected; ▨ = model improvement tool.

the output parameters of all tools having marked input parameters are marked. This step is repeated until the effect of applying the model improvement tool is carried through the complete tool–parameter graph. In the example, the input requirements of tool T_5 are affected. The output parameter of this tool is therefore also marked, as shown in Fig. 1.28.

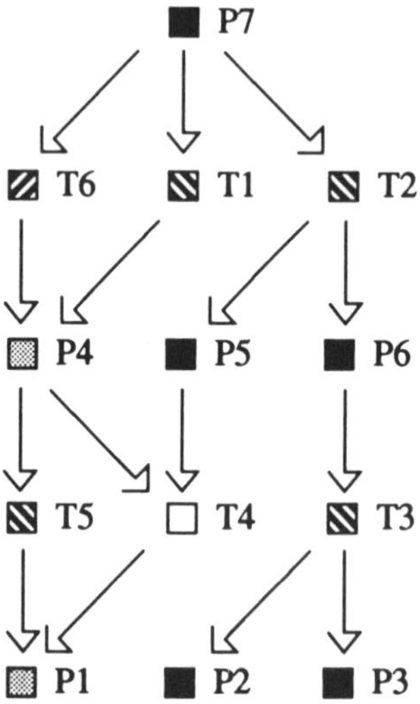

Fig. 1.28 Marking output parameters of other affected tools. □ = parameter inactive/tool not selected; ■ = parameter active; ▩ = parameter marked; ▧ = tool selected; ▨ = model improvement tool.

At this point, the zero–one integer program can be formulated. In this case, the cost c_i is set equal to one for all tools, and the integer program produces the smallest possible list of design tools to update the model. Within the integer program, the variables representing the given parameters and the parameters to be created by the tools sequence as well as the selected model improvement tool T_r are set equal to one, as shown in the first constraint in Fig. 1.17. The second constraint makes sure that, if a tool is selected, all the input requirements of that tool are met. The third constraint indicates that, if a tool is selected, all its output parameters are activated. Except for the given parameters, all parameters that are activated must have at least one preceding tool that has been executed before the parameter can be used. This is reflected in the final constraint.

Figure 1.29 shows the integer program formulation for the example given in Figs. 1.25–1.28, as well as the results of the sequencing procedure.

1.3.7 Implementation of the design framework

Figure 1.30 shows the user interface of the tool sequencing and selection module. The goal parameters are displayed in the top view. These are typically obtained through the component and performance specification module shown in Fig. 1.31. The user can also edit the goal parameter list through a separate window. This window is activated by clicking the 'Set...' button. The entries in the list of given parameters can be edited in the same fashion. The parameters without preceding design tools in the tool–parameter graph are automatically entered here.

```
MIN
1.00 T1 + 1.00 T2 1.00 T3 + 1.00 T4 + 1.00 T5 + 1.00 T6
S.T.
\Parameters that are given are active
P5 = 1, P6 = 1,P7 = 1
\Goal Parameters must be active
P1 = 1, P2 = 1, P3 = 1
\Remedial Tools must be selected
T6=1
\Input requirements of all tools must be met
T1 - P7 <=0
T2 - P7 <=0
T3 - P6 <=0
2T3 - P4 - P5 <=0
T5 - P4 <=0
T6 - P7 <=0
\All output parameters of a selected tool are activated
T6 - P4 <=0
T1 - P4 <=0
2T2 - P5 - P6 <=0
2T3 - P2 - P3 <=0
T4 - P1 <=0
T5 - P1 <=0
\Parameters not given need at least one tool
P4 - T1 - T6 <=0
P1 - T4 - T5 <=0
BOUNDS
0<=Pi<=1∀i=1...7
0<Tj<=1∀j=1...6
END

SOLUTION
T1=0,T2=0,T3=0,T4=0,T5=1,T6=1
P1=1,P2=1,P3=1.P4=1,P5=1,P6=1,P7=1,

SOLUTION TOOL SEQUENCE: T6, T5
```

Fig. 1.29 Example tool sequencing integer program formulation.

The tool sequencing module allows the user to choose between automatic and interactive tool selection. When the automatic tools sequencing mode is selected, a window appears as shown in Fig. 1.32, prompting the user to choose a tool selection criterion. The objective function of the integer program is constructed based on the selected criterion.

The first of these criteria weights all tools equally. In this case, the integer program simply minimizes the number of design tools. The second criterion allows the user to create an objective function based on multiple tool attributes.

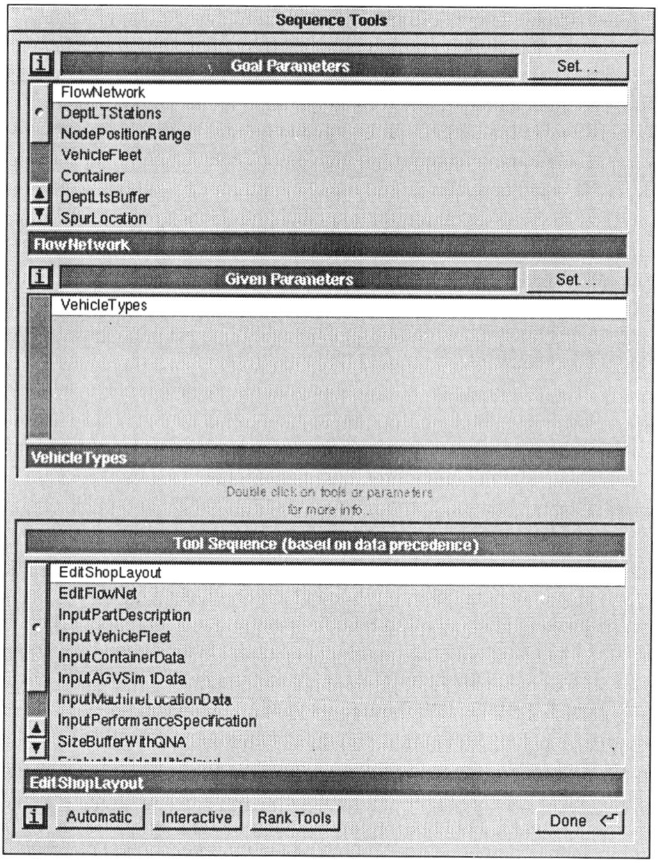

Fig. 1.30 Tool sequencing module.

If this option is chosen, the tool ranking window appears. The user then chooses and ranks the displayed criteria. From this ranking, the weights of each tool in the objective function are created. All the other selection criteria for automatic tool sequencing are derived from the general characteristic entries in the criteria matrix.

When the user clicks the 'Done' button, the integer program is formulated and solved. The selected tools are sequenced using the data precedence set in the tool–parameter graph. The resulting tool sequence is then presented in the bottom view of the tool sequencing module, and is made available to other modules within the design system.

The interactive tool sequencing method requires the user to choose among tools, if more than one are available to activate a parameter. In this case, a window opens, displaying the choices. The user can rank these tools using the ranking module. The resulting tool sequence is displayed in the lower view of the tool sequencing module.

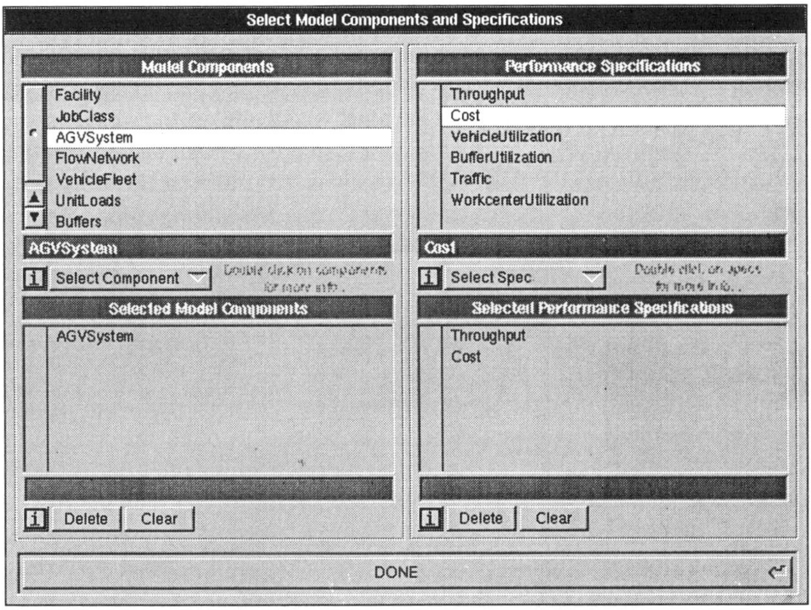

Fig. 1.31 Model component and performance specification module.

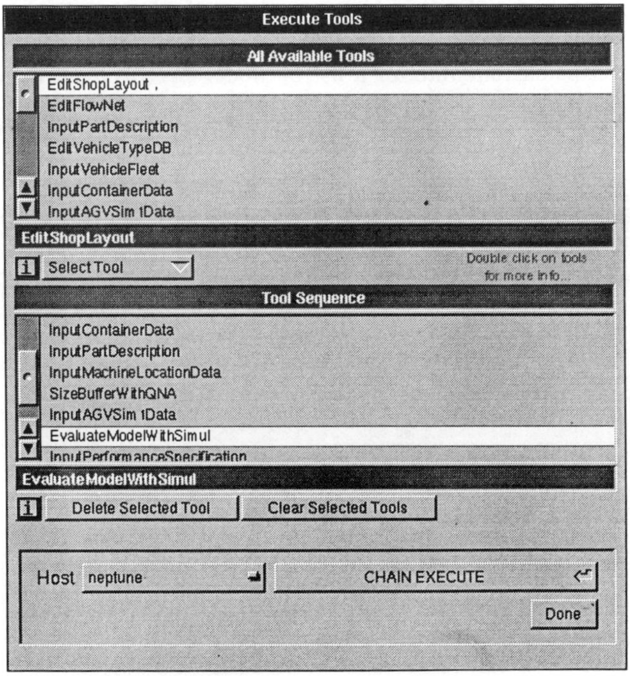

Fig. 1.32 Tool execution module.

The tool sequencing module shown in Fig. 1.30 lets the designer run the tools from the design workstation, regardless of whether the application resides on the workstation or on a remote host. This module has two views: the top view lists all the design tools currently available on the workstation. The second view displays the tool sequence created by the sequencing module. This sequence can be modified by the designer by transferring tools from the top view into the second view. Tool are executed highlighting an entry in the bottom view with the mouse, selecting the host on which to run the application and clicking the 'Chain Execute' button. If the tool is located on a remote host, the module copies all the data required by the application into a pre-specified workspace directory, runs the application and copies the output data back from the remote host.

In the implementation of the concepts presented in this chapter, the designer controls the model evaluation and redesign process through the main user interface, as shown in Fig. 1.33. When the model evaluation and improvement application is launched, a list of performance specifications that the model is to meet is read. This list is displayed in the top left view of the user interface. The specifications are initially untested, as indicated in the label preceding the specification names. The list entries can be ordered by importance by 'control dragging' them into proper position.

Performance specifications are tested by selecting an entry in the specification list and clicking the 'Evaluate Model on Selected Parameter' button. The selected specification is passed into the performance evaluation and diagnosis module. The analysis of model performance can result in either an acceptance of the current state or a list of model improvement actions that

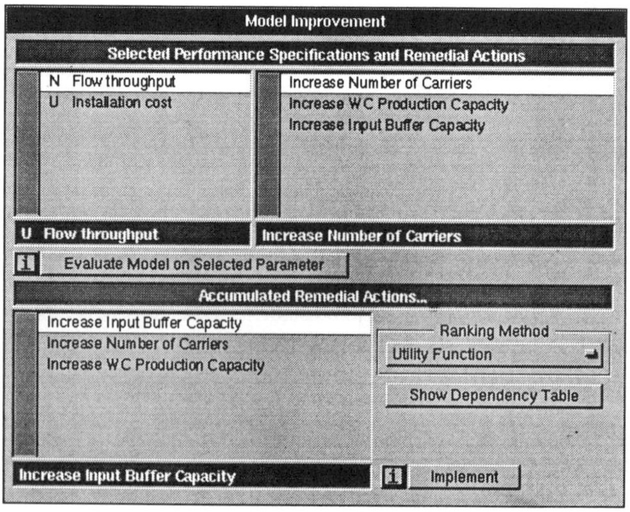

Fig. 1.33 Main control window.

can potentially improve the model. When the performance is accepted, the label preceding the specification name in the model improvement interface is set to 'A'. In the second case, the label is set to 'N', and the list of model improvement actions associated with that specification is displayed in the top right view.

Figure 1.34 shows the user interface of the performance testing and diagnosis module. This user interface should be viewed as a worksheet that helps the designer decide whether the model performance has been met, and to investigate the failings of the model.

The decision to accept the model performance is logged by the user through the 'Performance Test Result' button on the 'Test Performance' window. The 'View Performance Data' and 'Perform Automatic Test' buttons start up applications that allow the user to view raw performance data or perform a

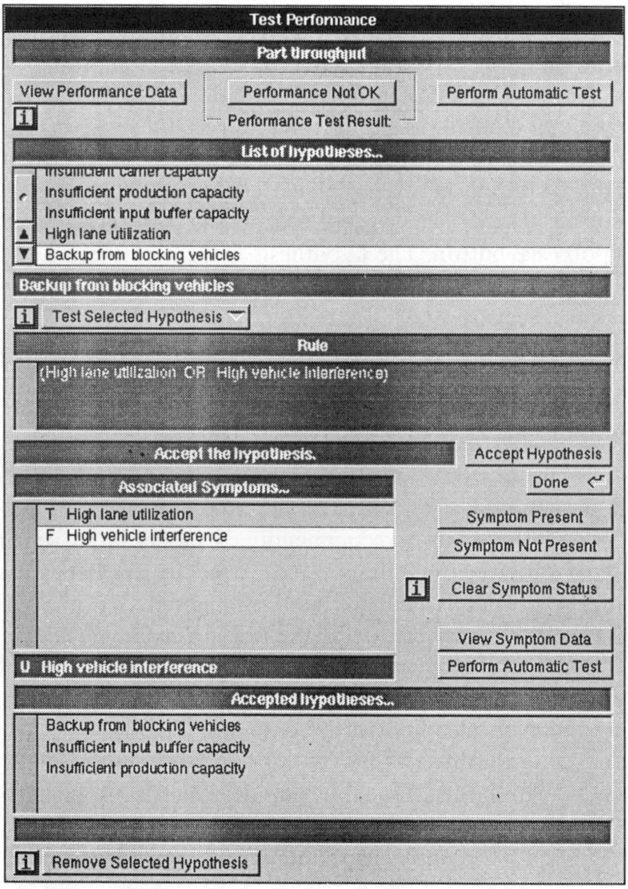

Fig. 1.34 Performance evaluation and model diagnosis module.

test based on prespecified reference data. References to these applications are stored within the specification object associated with the performance characteristic under investigation. After completing the automatic test, the title of the 'Performance Test Result' button is set to reflect the test result.

The list of hypotheses projecting causes for the performance specification not being met are displayed in the view below the performance testing button cluster. The designer can test one of these hypotheses by selecting an entry in the hypotheses view and clicking the 'Test Selected Hypothesis' button. The rule for verifying the hypothesis is displayed in the rule view. The field under this view shows the evaluation of the rule. The associated symptom view presents the list of symptoms that need to be specified before the rule can be successfully tested.

The designer tests these symptoms in the same manner as the performance specifications. Tools for displaying symptom data and performing an automatic test are launched by clicking the 'View Symptom Data' and 'Perform Automatic Test' buttons. After completing an automatic test, the status of the highlighted symptom is set to reflect the test result. The user can also set this status using the 'Symptom Present' and the 'Symptom Not Present' buttons after analysing the appropriate raw data. The status of all tested symptoms remains static to avoid the need for retesting every time a different performance specification is tested.

After all the symptoms in this list have been set, the hypothesis is evaluated and the result displayed. The designer accepts the hypothesis by clicking the 'Accept Hypothesis' button. The hypothesis is then transferred to the list at the bottom of the user interface. The designer completes the performance evaluation worksheet by clicking the 'Done' button. The model evaluation module extracts the model improvement actions from the objects representing the selected hypotheses. The resulting list is copied into the performance specification object for which the model diagnosis was made, and displayed in the main user interface shown in Fig. 1.33.

A cumulative list of model improvement actions is displayed in the bottom of the main user interface. This list can be ranked using three methods: the number of verified hypotheses recommending a model improvement action; lexicographical ordering by rating values listed in the dependency table; and ordering by a utility calculated as the weighted sum of rating values. The ranking method is selected with the pop-up menu to the right of the cumulative model improvement action view. As a result of the ranking, the model improvement action at the top of the list is the one most likely to advance the design to meeting the design goals. The designer also has the option of analysing the effect of model improvement actions on each of the design goals through the decision table module, which is displayed by clicking the 'Show Dependency Table' button.

Figure 1.35 shows the implementation of the dependency table module. The model improvement actions available for choice are listed in the rows.

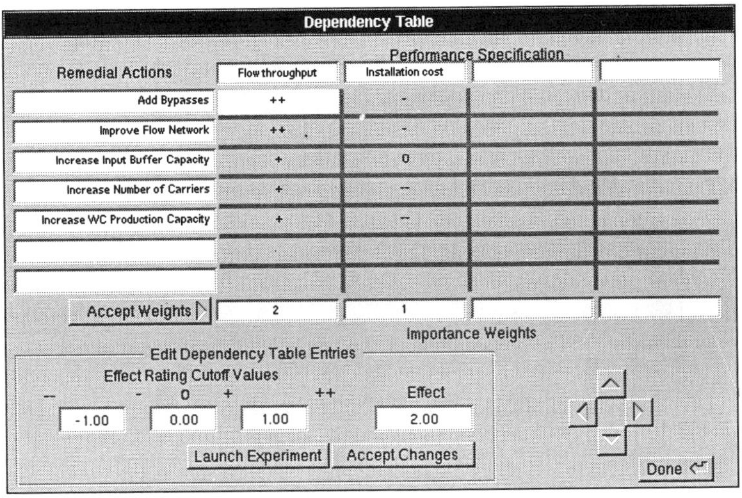

Fig. 1.35 Dependency table module.

The performance measures are shown in the column labels at the top of the matrix. The importance rating values of each performance measure are displayed, and can be edited in the fields below the dependency matrix. The symbolic rating of each dependency object is displayed within the table entries. The numerical rating of the effect as well as the cutoff values are edited by first selecting an entry in the dependency table. The current values are displayed in the fields beneath the table. The designer can edit any of the values in these fields. The tool for performing an analysis of the effect is launched by clicking the 'Launch Experiment' button. This button is disabled if no such tool is available. At the end of the experiment, the effect and cutoff fields are updated and a more detailed report is presented through a separate text display window. The designer can now accept these changes by clicking the 'Accept Changes' button. The symbolic rating value is recalculated and updated within the dependency table.

A model improvement action is implemented by selecting an entry in the cumulative model improvement action list, as shown in Fig. 1.33, and clicking the 'Implement' button in the main user interface. This action launches the model improvement action implementation module. As previously outlined, the major factors in implementing model improvement actions are the selection of an appropriate design tool and maintaining data consistency after the selected tool has been applied. The user interface of the module addressing these issues is shown in Fig. 1.36. The tool selection process that takes place at the top of the user interface makes use of the criteria matrix, as shown in Table 1.2.

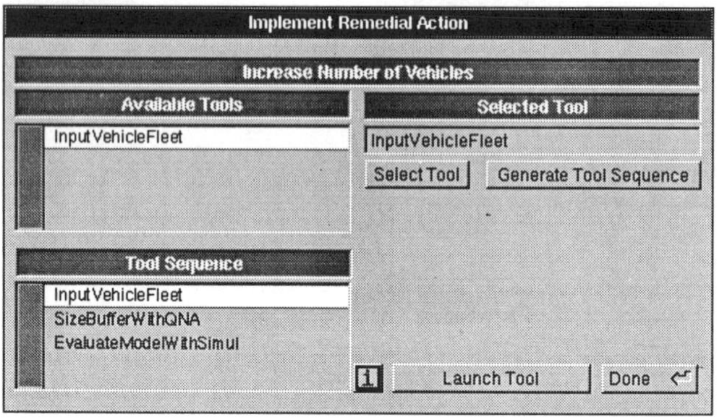

Fig. 1.36 Remedial action implementation module.

A model improvement action selected by the user is passed into the implementation module when it is launched. This action is displayed in the top field of the model improvement action user interface. Each model improvement action has its own entry in the criteria matrix in which the potential contributions of the available design tools in implementing the action are noted. Applicable tools for the model improvement actions are extracted from the matrix, sorted by the rating value in the corresponding row and displayed in order in the view underneath the model improvement action name display field. The user can select a tool from this list by clicking the 'Select Tool' button.

Once a model improvement tool has been selected, the user clicks the 'Generate Tool Sequence' button to find and sequence all the tools needed to implement the model improvement action, and reevaluate the performance of the material flow system model. This is done using the integer program-based method discussed previously. The resulting tool sequence is displayed in the bottom view of the implementation module. The tool sequence is presented to the designer in the bottom view of the user interface in Fig. 1.36. The tools in this sequence are executed by selecting an entry in the tool sequence view and clicking the 'Launch Application' button. A small panel appears where the user is requested to select the computer on which the application is to be executed. If the design tool resides on the workstation, it is immediately executed. When a remote host was selected, all the data required by the design tool are copied onto that host. The tool is then run remotely. The resulting data sets are copied back onto the workstation once the design tool has terminated. After executing all tools in the sequence, the designer has not only implemented the proposed model improvement action, but has also made a performance measurement of the material flow system model. The designer can now return to evaluating the performance and making further fault diagnoses.

REFERENCES

Anderson, R.B. (1989) *The Student Edition of MathCAD*, Addison Wesley, Reading, MA.

Andersson, M. (1985) *AGV system simulation – A planning tool for AGV route layout.* Proceedings of the 3rd International Conference on AGV Systems, Stockholm, Sweden, pp. 291–6.

Apple, J.M. (1972) *Material Handling System Design*, The Ronald Press Co. New York, NY.

Autodesk Inc. (1985) *AutoCAD User's Manual.*

Baumgarten, H. (1989) *Trends in Logistics.* Proceedings of the 7th International Conference on AGV Systems, 13–14 June, Berlin, Germany, 3–10.

Borland International (1989) *Quattro Pro User's Guide.*

Brentano, L. (1984) Distributed CAD/CAM: Myth and Reality. *IEEE Computer Graphics and Applications*, **4**(8), 18–22.

Brunsen, H. and Maiwald, R. (1992) INPAS – Integriertes Packereiplanungssystem. Short Report, Fraunhofer-Institut für Materialfluß und Logistik, Joseph-von-Fraunhofer-Str., 2–4, 4600 Dortmund 50, Germany.

Cardinal, D.J. (1985) File Server Offers Transparent Design Tools. *Computer Design*, June, 147–54.

Carver, B. (1989) Frameworks: the Design Environment of the 1990's. *Electronic Products*, September, 19–28.

CMS Research (1990) *MAST Simulation Environment – User's Manual Version 3.0.*

Daniell, J. and Director, S.W. (1989) *An Object Oriented Approach to CAD Tool Control Within a Design Framework.* Proceedings of the 26th ACM/IEEE Design Automation Conference, Las Vegas, NV, Paper 14.1, pp. 197–202.

DeMori R. and Prager, R (1987) An expert system for verification and tuning of simulation models, in *Artificial Intelligence in Engineering: Tools and Techniques*, (eds D. Sriram and R.A. Adey), Computational Mechanics, Southampton Boston, pp. 160–75.

Dixon, J.R., Simmons, M.K. and Cohen, P.R. (1984) *An Architecture for Applications of Artificial Intelligence to Design.* Proceedings of the 21st IEEE Design Automation Conference, Albuquerque, NM, pp. 634–40.

Egbelu, P.J. and Tanchoco, J.M.A. (1982a) AGVSim User's Manual. *Technical Report No. 8204*, Department of Industrial Engineering and Operations Research, Virginia Polytechnic Institute and State University, Blacksburg, VA, USA.

Egbelu, P.J. (1987) The use of non-simulation approaches in estimating vehicle requirements in automated guided vehicle based transport systems. *Material Flow*, **4**, 17–32.

Eppinger, S.D., Whitney, D.E. and Gebala, D.A. (1982) *Organizing the Tasks in Complex Design Projects: Development of Tool to Represent Design Procedures.* Proceedings NSF Design and Manufacturing Systems Conference, Atlanta, GA, January 8–10.

Fink, P.K., Lusth, J.C. and Duran, J.W. (1985) A general expert system design for diagnostic problem solving. *IEEE Transactions on Pattern Analysis and Machine Intelligence*, **7**(5), 353–560.

Fishburn, P.C. (1967) Methods of estimating additive utilities. *Management Science*, **13**(7), 435–53.

Floss, P. and Talavage, J. (1990) A knowledge-based design assistant for intelligent manufacturing systems. *Journal of Manufacturing Systems*, **9**(2), 87–102.

Frydman, C., Giambasi, N., Gatumel, M. *et al.* (1989) *DeBuMa: Description, Building and Management of Applications.* Proceedings of the 26th ACM/IEEE Design Automation Conference, Las Vegas, NV, pp. 203–208.

Gaskins, R.J. and Tanchoco, J.M.A. (1987) Flow path design for AGV systems. *International Journal of Production Research*, **25**(6), 667–76.

Gaskins, R.J. and Tanchoco, J.M.A. (1989) AGUSIM2 – a development tool for AGVS controller design. *International Journal of Production Research*, **27**(6), 915–26.

Gottheil, K., Kachel, G., Kathoefer, T. *et al.* (1988) The CADLab Workstation CWS – An open system for tool integration, in *Tool Integration and Design Environments*, (ed F.J. Rammig), North Holland, Netherlands.

Haabma, J. (1988) The NMP-CAD-Based System – An open and integrating framework for CAD tools, in *Tool Integration and Design Environments*, (ed F.J. Rammig), North Holland, Netherlands.

Heim, J.A. (1990) Integration of distributed heterogenous models for design of manufacturing systems. PhD Thesis, School of Industrial Engineering, West Lafayette, IN, USA.

Hofmann, H., Caviedes, J., Bourne, J. *et al.* (1986) Building expert systems for repair domains. *Expert Systems*, **3**(1), 4–11.

Hollingum, J. (1988) Renishaw probes new manufacturing methods. *FMS Magazine*, **6**(2), 75–7.

Howe, A., Cohen, P., Dixon, J. *et al.* (1986) Dominic: A domain-independent program for mechanical engineering design, in *Applications of Artificial Intelligence in Engineering Problems*, (eds D. Sriram and R.A. Adey), Springer-Verlag: Berlin, Germany, pp. 289–99.

Kaspi, M. and Tanchoco, J.M.A. (1990) Optimal flow path design of unidirectional AGV systems. *International Journal of Production Research*, **28**(6), 1023–30.

Krishnamurthy, E.V. (1989) *Parallel Processing – Principles and Practice*, Addison Wesley, Reading, MA.

Kuprat, T. (1990) Fördertechnik im Fertigungsbetrieb. *VDI-Z*, **133**(6), 98–106.

Law, A.M. and Haider, S.W. (1989) Selecting simulation software for manufacturing applications: Practical guidelines and software survey. *Industrial Engineering*, May, 33–46.

Law, A.M. and Kelton, W.D. (1991) *Simulation Modeling and Analysis*, McGraw Hill, New York.

Leung, L.C., Khator, S.K. and Kimbler, D.L. (1987) Assignment of AGVs with different vehicle types. *Material Flow*, **4**(1&2), 65–72.

Lesch, U. (1990) FFS-Planer. *VDI-Z*, **132**(1), 26–31.

Lotus Development Corporation (1990) *1-2-3 For Sun: User's Reference*.

MacCrimmon, K.R. (1973) An overview of multiple objective decision making, in *Multiple Criteria Decision Making*, (eds J. Cochrane and M. Zeleny), University of South Carolina Press, Columbia, SC, pp. 18–44.

MacCrimmon, K.R. and Siu, J.K. (1974) Making trade-offs. *Decision Sciences*, **5**, 680–703.

Mackulak, G.T. (1984) High level planning and control: An IDEFo analysis for airframe manufacture. *Journal of Manufacturing Systems*, **3**(2), 121–33.

MathWorks (1990) *Pro-MatLab User's Guide*.

Maxwell, W.L. and Muckstadt, J.A. (1982) Design of automated guided vehicle systems. *IIE Transactions*, **14**(2), 114–24.

McGinnis, L.F. (1989) Computer-aided facility design revisited: A prototype design workstation for AGV systems, in *Progress in Material Handling and Logistics*, (eds J.A. White and I.W. Pence), Springer-Verlag, Berlin, pp. 67–93.

Middendorf, W.H. (1989) *Design of Devices and Systems*, 2nd edn, Marcel Dekker, New York.

Milberg, J., Burger, C. and Zetlmayer, H. (1992) Flexible Regelung der Produktion mit Entscheidungsunterstützenden Systemen. *VDI-Z*, **134**(5), 140–5.

Miller, R.K. (1987) *Automated Guided Vehicle Systems*, SME Publications, Dearborn, MI.

Mitta, D.A. (1991) A methodology for quantifying expert system usability. *Human Factors*, **33**(2), 233–45.

Moser, E., Rust, H. and Golze, U. (1988) Knowledge-based tool box management system, in *Tool Integration and Design Environments*, (ed F.J. Rammig), North Holland, Netherlands, pp. 211–21.

Mostow, J. (1985) Towards better models of the design process. *AI Magazine*, Spring, 44–57.

Müller, T. (1983) *Automated Guided Vehicles*, IFS Publications, Bedford.

Naylor, A.W. and Volz, R.A. (1988) Integration and flexibility of software for integrated manufacturing systems, in *Design and Analysis of Integrated Manufacturing Systems*, (ed W.D. Compton), National Academy Press, Washington, DC.

Pritsker, A.A.B. (1986) *Introduction to Simulation and SLAMII, 3rd ed*, Holsted Press (Wiley), New York.

Rembold, B.F. and Tanchoco, J.M.A. (1992) *The Next Step in AGVS Design Workstations.* Proceedings of the First IIE Industrial Engineering Research Conference, Chicago, IL, pp. 113–6.

Rembold, B.F. and Tanchoco, J.M.A. (1994a) A modular framework for the design of material flow systems. *International Journal of Production Research*, **32**(1), pp. 1–21.

Rembold, B.F. and Tanchoco, J.M.A. (1994b) Selecting and sequencing design tools in developing material flow systems models. *International Journal of Production Research*, **32**(2), pp. 243–62.

Rembold, B.F. and Tanchoco, J.M.A. (1994c) Material flow system model evaluation and improvement. *International Journal of Production Research*, forthcoming.

Schönheit, M. and Wiegershaus U. (1990) Flexible Fertigungssysteme für Mittelständische Unternehmen. *VDI-Z*, **132**(9), 58–65.

Schroer, B.J. (1989) A simulation assitant for modelling manufacturing systems. *Simulation*, November, 201–6.

Sedgewick, R. (1992) *Algorithms in C++*, Addison Wesley, Reading, MA.

Segal, M. and Whitt, W. (1989) A queueing network analyzer for manufacturing, in *Teletraffic Science for New Cost-Effective Systems, Networks and Services*, (ed M. Bonati), North Holland, Netherlands, pp. 1146–52.

Sharp, G.P. and Liu, F.-H.F. (1990) An analytical method for configuring fixed-path, closed loop material handling systems. *International Journal of Production Research*, **28**(4), 757–83.

Shodhan, R.H. (1989) COMAND: A computer consultant for design, operation and control of flexible manufacturing systems. PhD thesis, Purdue University, School Industrial Engineering, West Lafayette, IN, USA.

Solberg, J.I. (1981) Capacity planning with a stochastic workflow model. *AIIE Transactions*, **13**(2), 116–22.

Solberg, J.J. and Heim, J.A. (1989) Managing information complexity in material flow systems, in *Advanced Information Technologies for Industrial Material Flow, Systems*, (eds S.Y. Nof and C.L. Moodie) NATO ASI Series, Vol. F53, Springer-Verlag, Berlin.

Spur, G., Hirn, W. and Seliger, G. (1983) *The Role of Simulation in Design of Manufacturing Systems.* Proceedings of the 5th International IFIP/IFAC Conference on Programming Research and Operations Logistics in Advanced Manufacturing Technology, Leningrad, USSR, pp. 349–73.

Suri, R. and Hildebrandt, R.R. (1984) Modelling FMS's using mean value analysis. *Journal of Manufacturing Systems*, **3**(1), 27–38.

Talavage, J. and Hannam, R.G. (1988) *Flexible Manufacturing Systems in Practice*, Marcel Dekker, New York.

Tanchoco, J.M.A., Egbelu, P.J. and Taghaboni, F. (1987) Determination of the total number of vehicles in an AGV-based material transport system. *Material Flow*, **4**, 33–51.

Tompkins, J.A. and White, J.A. (1984) *Facilities Planning*, John Wiley, New York.

Tzafestas, S.G. (1987) A look at the knowledge-based approach to system fault diagnosis and supervisory control, in *System Fault Diagnosis, Reliability, and Related Knowledge-Based Approaches*, 2, (ed S.G. Tzafestas), D. Reidel, Boston, MA, pp. 3–15.

Warnecke, H.J., Steinhilper, R. and Zeh, K.-P. (1986) Simulation as an integral part of FMS planning, in *Simulation Applications in Manufacturing*, (ed R.D. Hurrion), IFS Publications, Berlin, Germany, pp. 131–47.

Wolfram, S. (1988) *Mathematica – A System for Doing Mathematics by Computer*, Addison-Wesley, Reading, MA.

Wysk, R.A., Egbelu, P.J., Zhou, C. *et al.* (1987) Use of spread sheet analysis for evaluating AGV systems. *Material Flow*, **4**, 53–64.

FURTHER READING

Ashayeri, J., Gelders, L.F. and Van Looy, P.M. (1985) Micro Computer Simulation in Design of Automated Guided Vehicle Systems. *Material Flow*, **2**, 37–48.

Bakkalbasi, O. and McGinnis, L.F. (1988) *ABC's of Preliminary In-House Planning of AGVS Applications.* Proceedings of the AGVS 88 Forum and Exhibit, Cincinnati, OH, Session 8, 13–48.

Bastide, R. and Sibertin-Blanc, C. (1991) Modeling a Flexible Manufacturing System by Means of Cooperative Objects, in *Computer Applications in Production and Engineering: Integration Aspects*, (eds G. Doumeingts, J. Browne and J. Tomljanovich), North Holland: Elsevier, Netherlands, pp. 593–601.

Booch, G. (1991) *Object Oriented Design With Applications*, Benjamin/Cummings, Redwood City, CA.

Bozer, Y.A. and Srinivasan, M.M. (1991) Tandem AGV Systems: A Partitioning Algorithm and Performance Comparison With Conventional AGV Systems. *Unpublished Paper*, Department of Industrial and Operations Engineering, The University of Michigan, Ann Arbor, MI, USA.

Coad, P. and Yourdon, E. (1991) *Object Oriented Analysis*, 2nd edn, Yourdon Press, Englewood Cliffs, NJ.

Duffau, B. and Bardin, C. (1985) *Evaluating AGVS Circuits by Simulation.* Proceedings of the 3rd International Conference on AGV Systems, Stockholm, Sweden, pp. 229–45.

Edwards, W. (1977) The use of multiattribute utility measurement for social decision making, in *Conflicting Objectives in Decision*, (eds D.E. Bell, R.L. Keeney and H. Raiffa), John Wiley, Chichester, pp. 274–76.

Egbelu, P.J. and Tanchoco, J.M.A. (1982b), Operational Consideration for the Design of AGV-Based Material Handling Systems. *Technical Report No. 8201*, Department of Industrial Engineering and Operations Research, Virginia Polytechnic Institute and State University, Blacksburg, VA, USA.

Egbelu, P.J. and Tanchoco, J.M.A. (1984) Characterization of AGV dispatching rules. *International Journal of Production Research*, **22**(3), 359–74.

Fishburn, P.C. (1965) Independence in utility theory with whole product sets. *Operations Research*, **13**(1), 28–45.

Fisher, E.L. (1986) An AI-based methodology for factory design. *AI Magazine*, Fall, 72–85.

Glassey, C.R. and Adiga, S. (1990) Berkeley library of objects for control of manufacturing (BLOCS/M), in *Applications of Object Oriented Programming*, (eds L.J. Pinson and R.S. Wiener), Addison-Wesley, Reading, MA, pp. 1–27.

van Haarpen, N.T. (1987) *Integrated Systems for Computer Aided Design*. Proceedings of International Conference on Engineering Design, Boston, MA, pp. 426–33.

Jones, A.H. and Burge, S.E. (1989) An expert system design using cause-effect representations and simulation for fault diagnosis, in *Knowledge-Based Systems Diagnosis and Control*, (ed S.G. Tzafestas), Plenum Press, New York, pp. 71–80.

Keeney, R.L. and Raiffa, H. (1976) *Decisions With Multiple Objectives: Preferences and Value Tradeoffs*, John Wiley, New York.

Kiesewetter, S.A., Dörken, T.P., Melchert, M. *et al.* (1991) Flexible Fertigung im Jahresübersicht. *VDI-Z*, **133**(8), 58–75.

King, C.U. and Fisher, E.L. (1986) *Object Oriented Shop Floor Design, Simulation and Evaluation*. Proceedings of the Fall Industrial Engineering Conference, Dallas, TX, pp. 131–7.

Koff, G.A. and Boldrin, B. (1985) Automated guided vehicles, in *Materials Handling Handbook*, (ed R. Kulweic), John Wiley, New York.

Mahadevan, B. and Narendran, T.T. (1990) Design of an automated guided vehicle based material handling system for a flexible manufacturing system. *International Journal of Production Research*, **28**(9), 1611–22.

Martinez, J., Alla, H. and Silva, M. (1986) Petri nets for the specifications of FMSs, in *Modelling and Design of Flexible Manufacturing Systems*, (ed A. Kusiak), North Holland, Netherlands.

Matson, J.O., Swaminathan, S.R. and Mellichamp, J.M. (1990) *Knowledge-Based Material Handling Equipment Selection*. Proceedings of the International Industrial Engineering Conference, San Francisco, CA, pp. 212–7.

Rembold, B.F. and Tanchoco, J.M.A. (1991) A Graphical Editor for the Design of Material Flow Systems, *Technical Report*, School of Industrial Engineering, Purdue University, West Lafayette, IN, USA.

Ritz, G.J. (1990) *Total Engineering Project Management*, McGraw Hill, New York.

Schönheit, M., Wiegershaus, U. and Kiesewetter, S. (1990) Fachgebiete im Jarhresübersicht: Flexible Fertigung. *VDI-Z*, **132**(10), 92–109.

Shewchuk, J.P. and Chang, T.C. (1991) *An Approach to Object-Oriented Discrete-Event Simulation of Manufacturing Systems*. Proceedings of the 1991 Winter Simulation Conference, Phoenix, AZ, pp. 302–11.

Shubin, H. and Ulrich, J.W. (1982) *IDT: An Intelligent Diagnostic Tool*. Proceedings of the National Conference on Artificial Intelligence, Pittsburgh, PA, pp. 290–5.

Taghaboni, F. and Tanchoco, J.M.A. (1988) A LISP-based controller for free-ranging automated guided vehicle systems. *International Journal of Production Research*, **26**(2), 173–88.

Tanchoco, J.M.A. and Agee, M.H. (1981) Plan unit loads to interact with all components of warehouse systems. *Industrial Engineering*, June, 36–48.

Tanchoco, J.M.A. and Sinriech, D. (1992) OSL – Optimal Single Loop guide paths for AGVS. *International Journal of Production Research*, **30**(3), 655–81.

Wright, J.A. (1989) *Systems Thinking: A Guide to Managing in a Changing Environment*, Society of Manufacturing Engineers, Dearborn, MI.

CHAPTER 2

Design justification of material handling systems

J.S. Noble and J.M.A. Tanchoco

2.1 INTRODUCTION

2.1.1 Material handling system design problems

Material handling systems perform a key integrating function in today's dynamic and integrated manufacturing environment. Failure to fully integrate material handling design within the larger manufacturing system design problem results in designs that are unable to compete successfully in today's markets. However, typical approaches for material handling system (MHS) design have tended to isolate the analysis with respect to specific objectives and material handling technologies.

This chapter focuses on the material handling component of the manufacturing system design problem from both a performance and economic perspective. Figure 2.1 illustrates the range of interactions, both internal and external, that affect MHS design. Internal interactions are those such as system control, equipment selection and specification and unit load size. External interactions are those outside the MHS with which it must interact, such as the product specification, manufacturing processes, layout and production control/scheduling. This chapter specifically addresses the system alternative (selection) and system size (specification) problems, while maintaining an integrated perspective.

The system integrating nature of a MHS design requires that a design methodology maintain the perspective that the MHS supports the larger manufacturing system design problem. The material handling design problem predominately interacts with the facility layout and system control/scheduling problems. The facility layout affects the physical aspects of material flow, while production control/scheduling affects the logical aspects of material flow. The MHS design problem as a whole requires that the logical and physical aspects of material flow be combined by means of material handling equipment, and that the design be justified from both a performance and an economic perspective.

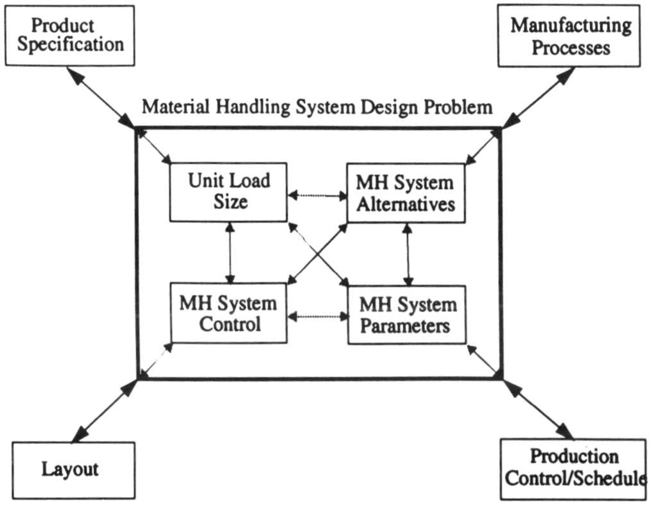

Fig. 2.1 Material handling system design.

The economic justification of technology in integrated systems is a major concern in MHS design. Much has been written about the failure of traditional justification procedures to capture the relevant information necessary to justify investments in an integrated manufacturing system. Even with the justification methodologies that have been proposed to quantify the subjective benefits that integrated technology provides, the underlying justification issues that result from integration remain.

2.1.2 Motivating issues for research in economic MHS design

Material handling systems are not a directly productive component of the manufacturing system; rather, they serve as a support function to the value added activities (where value added is defined as activities that transform the product). Tompkins and White (1984) have estimated that between 20–50% of the operating cost within manufacturing can be attributed to material handling type functions, depending upon the type of product and processes. Given the magnitude of the cost impact of material handling, it seems clear that great effort should be taken in the design of material handling systems to minimize costs. Therefore, in this research the goal of designing a MHS is to minimize its cost impact and to ensure that it support the objectives of the overall manufacturing system.

Manufacturing systems are highly complex and integrated by nature. Since MHSs are an integrating component of a manufacturing system, all complexity that is inherent in the manufacturing problem is translated to the material handling problem. The integrated and complex nature of a MHS results in a tremendous amount of design detail that grows quickly as the problem expands. Associated with the MHS design problem growth is the complexity of the economic justification process. All design effort is of no value if the resulting design is economically infeasible. Therefore, the breadth of the MHS design problem must include economic justification of the design; however, much of the recent research relegates the economics of the design to a secondary issue that is evaluative in nature. This research addresses this through the development of a design methodology that considers the economic justification of MHSs concurrently with the design of the MHS.

Past research has taken the MHS design problem, focused it on different material handling alternatives (i.e. conveyor, fork truck and automated guided vehicle systems), and solved these problems independently with a variety of analytical approaches, thereby reducing the scope of the systems approach. The approaches that have been proposed to select equipment for an entire system have been analytic in nature, and have not captured the interactions and dynamics of the material handling system. Rather, such approaches have made overly simplifying assumptions concerning system behavior. Therefore, this research proposes a MHS selection and specification methodology that goes beyond highly abstract modeling approaches and considers system dynamics.

2.1.3 Past work on the economic design of MHSs

Different design approaches that incorporate economic considerations have been developed for both specific material handling components and for a complete MHS. The following briefly notes the design approaches that both explicitly and implicitly incorporate economic criteria within the MHS design process. A more extensive review of economic design in manufacturing can be found in Noble and Tanchoco (1993a).

An economic approach to the belt conveyor specification problem was developed by Tanchoco *et al.* (1979). They formulated the problem to minimize total capital cost such that the operating constraints are satisfied. Scott *et al.* (1983) developed a combined performance and cost model design methodology for unit load conveyors based on a cost estimate used for the design solution and operating condition development. Ziai and Sule (1989) developed an equipment selection methodology for selecting between conveyors and forklift trucks which used a mixed integer program to obtain an initial solution, and a cost-driven improvement heuristic to improve equipment utilization.

The complete material handling selection problem was first addressed using an economic approach by Webster and Reed (1971). The problem of minimizing capital, operating and unit load changing cost with respect to material flow requirements and transporter capacity was modeled as an integer program, and solved using a heuristic solution procedure. Hassan *et al.* (1985) reformulated Webster and Reed's model, and then solved the model using a construction algorithm.

Multi-attribute or knowledge-based methodologies include economic issues in an implicit manner. Shelton and Jones (1987) used 23 attributes (including cost) in an additive form of a multi-attribute value function for the selection of automated guided vehicles. Liang *et al.* (1989) developed a multi-attribute approach to the specification of material handling equipment that utilizes the analytic hierarchy process combined with a database. A knowledge based multi-attribute design approach was proposed by Gabbert and Brown (1987) for specifying the types of material handling equipment to use in a given manufacturing scenario. An expert system for the selection of material handling equipment was developed by Fisher *et al.* (1988). Economic factors related to the selection of material handling equipment were implicitly embedded in the selection rules in the knowledge base.

A variety of different cost models have been developed to analyse MHSs at different levels. An overall MHS cost framework was discussed by Apple (1972). Müller (1981) explored cost relationships for a variety of material handling equipment. Dahlstrom (1981) explored the applicability of automated guided vehicles (AGVs), forklift trucks and fixed roller conveyors based on general cost models for each type of equipment. General cost models for AGVs have been developed by Daum (1987), and for conveyors by Shultz (1978) and Wiltse (1979).

Each material handling design approach mentioned above considers economic factors. However, none explore the ability to economically guide design decisions, nor do they provide an explicitly integrated economic/performance basis for MHS design and justification. Therefore, this chapter develops an approach whereby these abilities would be provided to the MHS designer.

2.2 ECONOMIC DESIGN APPROACH

The goal of economic MHS design is the integration of economics in the generation of robust material handling systems. A design procedure for economic MHS generation is developed based upon an integrated economic design approach termed **design justification**. The following describes design justification, followed by a discussion of the design analysis tools required for material handling systems.

2.2.1 Development of the design justification approach

Design justification refers to an economic design procedure where the economic ramifications of design decisions are considered concurrently with design development (Fig. 2.2). The goal of design justification is to guide the designer to a design that is justifiable from both a performance and an economic perspective. The underlying conceptual approach of design justification requires justification at each design decision to ensure a functionally and economically viable design.

During the development of the design justification concept, it was noted

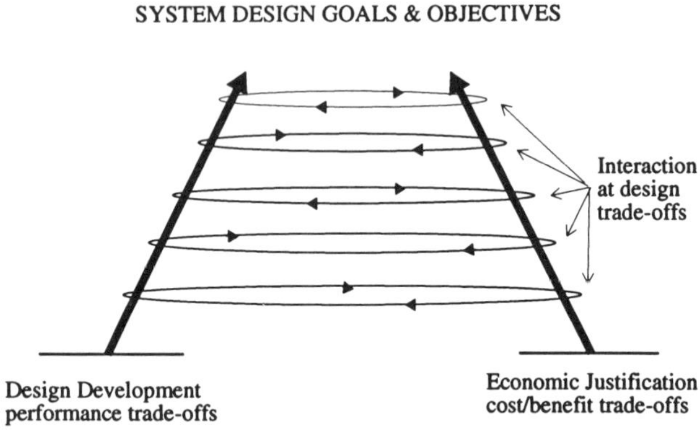

Fig. 2.2 Concurrent design justification interaction (Noble and Tanchoco, 1990).

that there are three different approaches to engineering design: analytical, procedural and experimental (Boyle, 1989). Design justification utilizes a procedural design approach. The procedural approach provides an advantageous framework for dealing with the uncertainty and complexity in system design. The uncertainty associated with system design is dealt with by refining the objective as the design proceeds. Feedback information from design iteration serves to reduce the level of uncertainty in the design, and to direct the solution process to meet design objectives. Complexity is dealt with by using design knowledge in a structural manner (i.e. proper application of design tools according to design phase) and through assigning tasks where they are most appropriate (i.e. computationally intensive procedures are dealt with automatically, while decision-making and the handling of intuitive informaton is dealt with by the designer).

The concurrent design approach in design justification is in contrast to the traditional sequential design and justification approach (Fig. 2.3). Both approaches handle the development of design specifications and constraints in a similar fashion. However, there are three major differences between design justification and traditional design and justification. First, in design justification the evaluation of design performance and economics occurs concurrently, whereas in traditional design and justification they are performed serially. A concurrent consideration of performance and economic issues provides trade-off information that is more readily usable within the design process. Second, information generated within the evaluation process in design justification is used in a design modification strategy. The modification strategy assists the designer in knowing where to alter the design to reach design objectives. In design justification, economic issues are considered as a key decision variable in all phases of design, and are used to determine how the design should proceed, whereas in traditional design and justification economic issues are limited to the analysis and evaluation phases. Third, design justification utilizes a knowledge guided design synthesis that is a

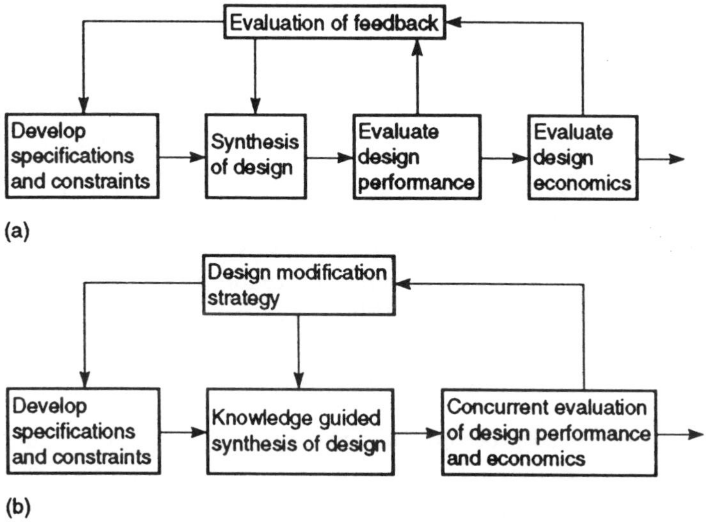

Fig. 2.3 (a) Traditional design and justification process vs. (b) design justification (Noble and Tanchoco, 1993).

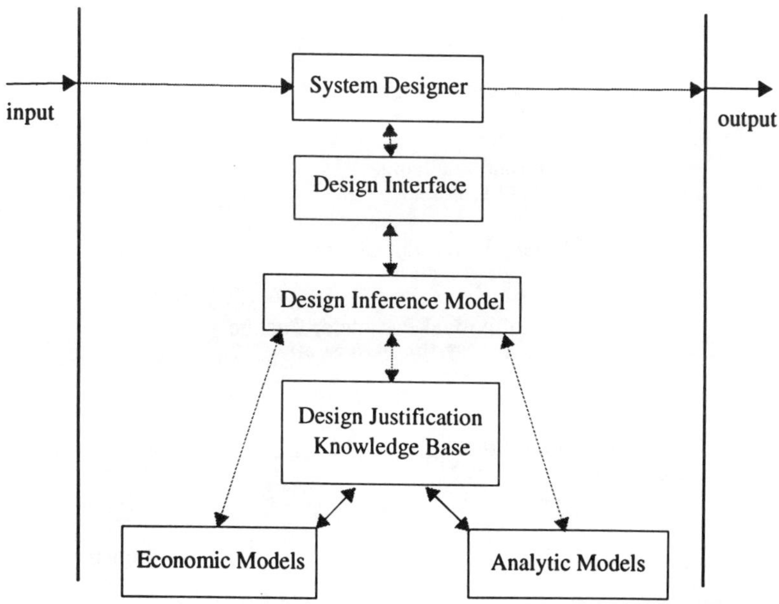

Fig. 2.4 Design justification conceptual framework.

combination of top-down and bottom-up design approaches. Traditional designs are synthesized using either a top-down approach (defined as algorithmic methods, with simplifying assumptions) or a bottom-up approach (defined as unguided experimental search).

There are four characteristic elements of design justification:

1. Integrated engineering and economic design of the system so that both occur simultaneously (Fig. 2.4).
2. Goal directed design so that the goals/objectives of the design are achieved through a branching and bounding procedure that guides the designer throughout the design process (Fig. 2.3).
3. Decision-maker centered design to provide an information processing ability and presentation mode that enables a designer to make good decisions.
4. Knowledge guided/procedural design to provide a design procedure where critical points in the redesign or modification of the current design have a modification strategy (Fig. 2.5) which will guide the design efficiently.

2.2.2 Development of the MHS design justification framework

There are three major phases within the material handling design justification procedure that is based upon the framework. Figure 2.6 illustrates the overall

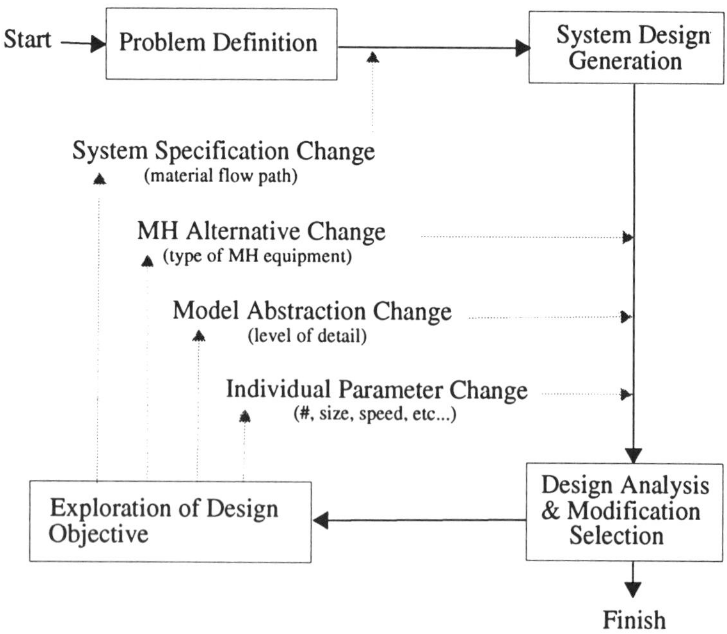

Fig. 2.5 Material handling design justification modification strategy (Noble and Tanchoco, 1993b).

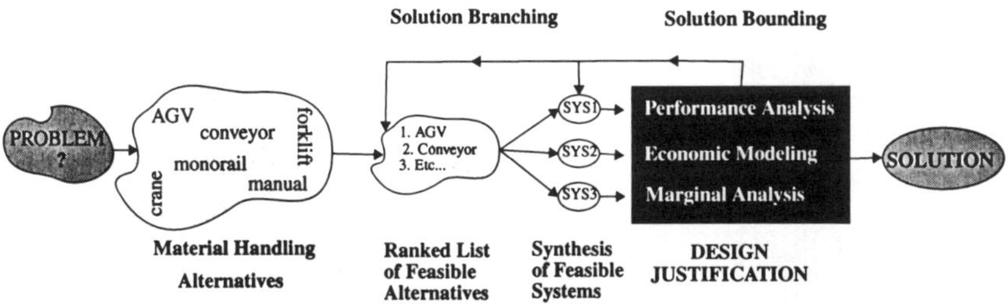

Fig. 2.6 Material handling design justification procedural approach (Noble and Tanchoco, 1993b).

procedural approach embedded in the design justification framework. The following describes each phase in the procedure.

(a) Manufacturing system data

The first phase of the design justification procedure is the importation and preparation of manufacturing system data. The MHS design justification procedure requires an instance of the: product description (size, weight, quantity); process plan (operation types, sequence and times; number of machines at each operation); physical layout (departmental assignments, spatial relationships); and unit load description (size, weight, quantity). A data structure for product, process and unit load was developed so that a consistent representation exists for the various functions within the design procedure. The data for the physical layout is developed using CAD software. CAD data files are imported into a graphical interface for the analysis of a MHS developed by Rembold and Tanchoco (1991) (Fig. 2.7).

(b) MHS alternative generation

The second phase of the design justification procedure is a two step procedure to develop MHS alternatives. As shown in Fig. 2.6, this is a solution branching phase where the MHS alternatives are generated. The first step reduces the number of equipment options for each material flow so that the second step can configure a complete system efficiently.

The first step of the alternative generation process ranks equipment alternatives for each material flowpath in the system using a knowledge-based procedure. A material handling equipment knowledge-base was developed that is a variation of MATHES (Fisher *et al.*, 1988). The knowledge-base uses four parameters as the basis for ranking equipment: path type, path distance, flow rate and unit load size. The ranking procedure utilizes a subset of the EMYCIN certainty factor calculus to obtain a measure of how well a given equipment type matches the material flow. The results of the equipment ranking are passed to the equipment selection/system generator step of the solution branching stage. The second step of the alternative generation

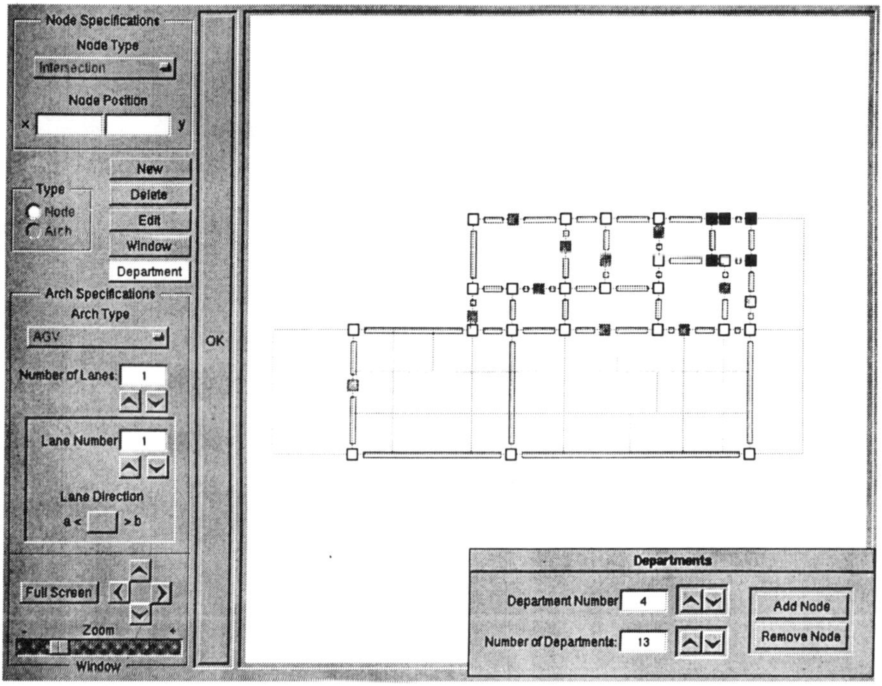

Fig. 2.7 Material flow path layout on network editor (Noble and Tanchoco, 1993b).

process utilizes an integer programming formulation of the equipment selection problem that is solved using a modification of the construction algorithm developed by Hassen *et al.* (1985).

Several issues must be addressed concerning the use of the generation procedure. First, if the rules in the rule base are appropriate, then obtaining the highest level of certainty would be the goal. However, based on the variety of material flows that can occur in a system, experimentation with different certainty factor cutoffs is required. Second, the formulation of the model requires an aggregate fixed cost for capital expenditures and an aggregate variable cost for operating expenses. The result is that the cost values associated with a generated system do not always accurately reflect the actual system cost. Rather, it tends to be a good indicator of the relative comparison between different alternatives. Third, since the results of the formulation are based primarily on capacity requirements, system dynamics are not captured. This effect is found when the model overspecifies the number of each equipment type, since it does not material flow routing.

(c) MHS analysis and modification

The objective of design justification is to enable the designer to proceed through the design process so that each design decision is justifiable based on the design objectives. Therefore, the crux of the material handling design justification framework is the analysis and modification phase of the design procedure. The MHS designer's primary role is to provide both economic and performance-based solution bounds.

The analysis and modification stage comprises four functions: system classification, system performance modeling, system cost and flexibility modeling and combined performance and economic analysis. The following describes each of these four functions individually, and discusses how they are implemented within the material handling design justification framework.

MHS classification

The design and analysis of a MHS is unique based on the equipment composition of the system. A MHS classification scheme was developed based upon each equipment's share of the total system material flow. The classification scheme developed only considers horizontal material handling equipment of four general types: manual, truck, conveyor and automated guided vehicle.

MHS performance modeling

The MHS classification scheme provides the basis from which to model material handling systems at different levels of model abstraction (Fig. 2.8). This allows for systems to be modeled at the appropriate level to obtain useful information on system performance, and reduces the complexity and time associated with model development. The designer is quickly guided to the abstraction of the system model that reflect the analysis detail required. Two functions performed in the design procedure are related to the abstraction of the system model. First, models are generated for the initial abstraction of the MHS based on the initial alternative system selection and specification by the designer. Second, models are generated to provide lower levels of abstraction of the MHS to allow the designer to automatically go to the next appropriate level of abstraction if it is merited, or to allow the designer to change the MHS specification and have the model regenerated to represent the altered system.

Three levels of system abstraction were implemented: deterministic capacity

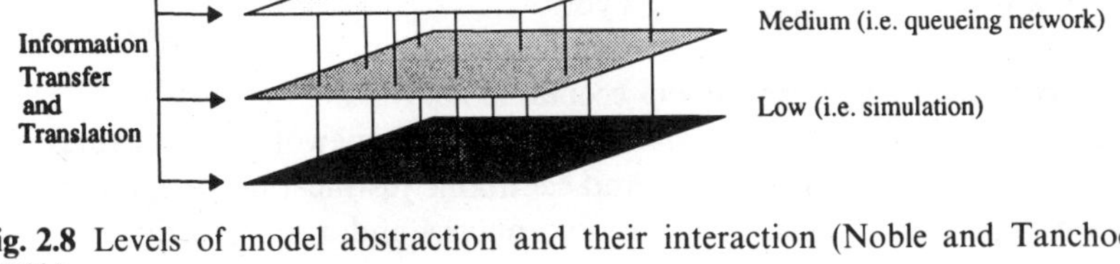

Fig. 2.8 Levels of model abstraction and their interaction (Noble and Tanchoco, 1993b).

analysis, queueing analysis and simulation. In each level, an automatic model generator was developed. The deterministic capacity analysis calculates both operation capacity and material handling capacity. The closed queueing network program CAN-Q (Solberg, 1977) was used as the basis of a queueing network generator. The SLAM simulation language with the material handling extension was used as the basis of a simulation model generator (Pritsker, 1986). The generated simulation model is a mixed network-discrete event model with statistical significance testing.

MHS cost and flexibility

Two other measures of MHS performance are used within the material handling design justification framework. The first is a MHS flexibility metric which was developed to measure the ability of a system to adapt to change. The second is an extensive cost modeling structure for each type of material handling equipment which was developed to measure the economic impact of design decisions.

Because the material handling function is a primary integrator within a manufacturing system, the ease with which a facility can be adapted is critically linked to MHS design. MHS flexibility has been defined as the ability to move different part types efficiently for proper positioning and processing through the manufacturing facility it serves (Sethi and Sethi, 1990). The development of different metrics to measure MHS flexibility has been limited, therefore a metric was developed to measure the combined ability of a material handling system to reconfigure, so as to handle new material flows and additional material flow. The metric of MHS flexibility is defined as:

$$F_{\mathrm{MHS}} = \sum_{i=1}^{N} x_i u_i \beta_i \tag{2.1}$$

where x_i is the number of each type of equipment, u_i is a measure of capacity based upon unit load, equipment speed and flow path distance, and β_i is a measure of the relative rerouting cost based upon the inverse of the approximate installation cost.

Detailed cost models were developed so that an accurate assessment of the economic performance of the system could be obtained. The goal of modeling the cost of the different types of equipment is to capture not just capital costs, but also installation, operating and maintenance costs. It was found in the modeling process that there are three major categories of models based upon the commonality shared between model structure: automated guided vehicles, trucks/manual and conveyors.

MHS combined performance and economic analysis

The mechanism within the design justification framework that enables it to meet the combined performance and economic justification design objective is a combination of system sensitivity analysis and an interactive marginal analysis directed modification strategy.

There are two main areas of sensitivity analysis supported in the material handling design justification framework: capacity and design objective. Capacity sensitivity is explored through the development of a unit cost curve which provides the designer with a considerable amount of information on cost performance (Fig. 2.9). The unit cost curves provide a visual perception of the system cost flexibility with respect to material flow rate and the cost of changing system capacity requirements. Sensitivity analysis on the design objective entails analysing the system with respect to minimum job makespan or maximum MHS flexibility. These two types of analysis are facilitated through a variety of decision graphics that are provided to the designer.

Determining where to alter a system design so that the net effect will satisfy the design objectives is a key component of the MHS design justification framework. The implementation of a multi-level modification strategy builds upon, and utilizes, all of the components discussed (Fig. 2.5). The first level modification addresses the alteration of individual parameters of a system and instantiates the analysis of a MHS. The second level modification pertains to the alteration of the analysis tool used to set solution bounds. This alteration requires returning to the first level to restart the analysis procedure. The third level modification explores altering the equipment selection. This alteration requires returning to the first level to restart the analysis procedure once the equipment changes are determined. The fourth level modification of a system design re-specifies the material flow paths within the system. This alteration first returning to the third level to obtain a new system design,

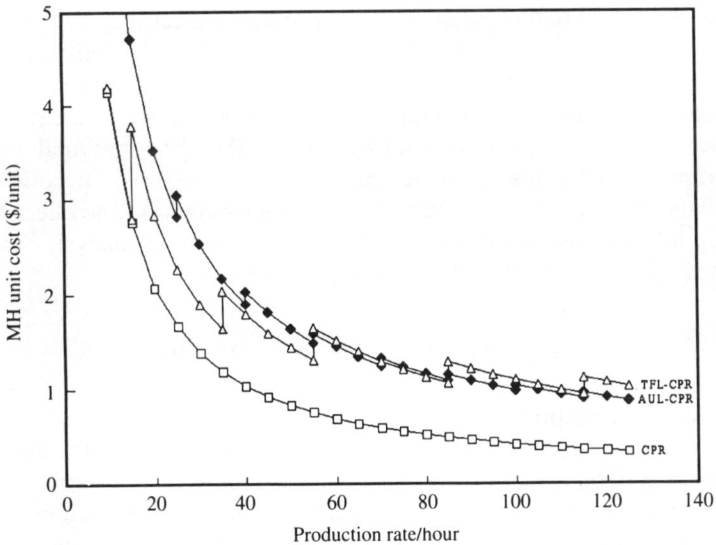

Fig. 2.9 Detailed unit cost relationships for CPR (□), AUL-CPR (◆) and TFL-CPR (△) systems.

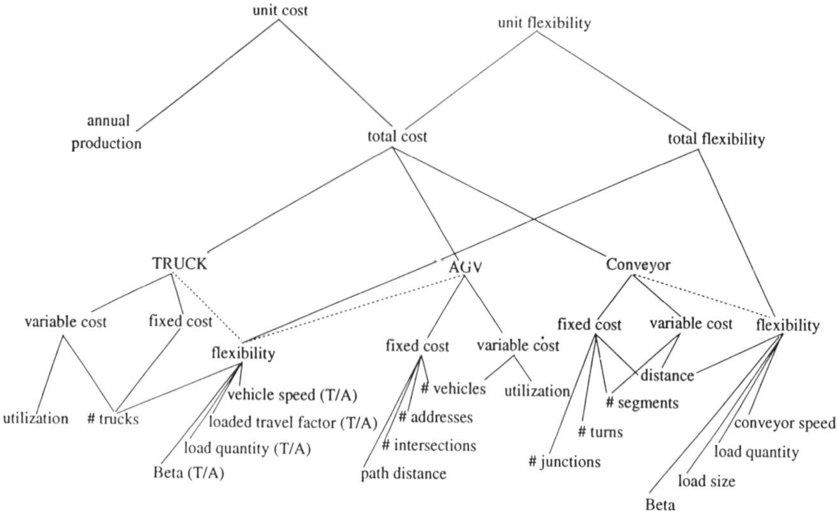

Fig. 2.10 Marginal analysis parameter relationships.

which is then followed by returning to the first level to restart the analysis procedure.

At the first, third and fourth modification levels there are performance and economic trade-offs that must be considered. The mechanism within the framework for this analysis is a marginal analysis network that shows the marginal or relative incremental effect of any absolute or incremental percentage alteration to the system on the performance measures of throughput, total MHS cost, material handling unit cost, MHS flexibility and unit flexibility cost. Figure 2.10 illustrates the MHS marginal analysis network developed. The marginal analysis used in the modification strategy utilizes the incremental calculus developed by Eilon (1984). The marginal analysis was implemented using a spreadsheet to provide for ease in conducting sensitivity analysis and presenting graphical information. The effect of exploring different marginal parameter changes on the overall analysis measures are used by the designer to direct the design process.

2.3 ECONOMIC DESIGN OF MHSs

The design justification framework for economic design of material handling systems is illustrated in the following example. To highlight the economic design aspects of the example, the generation of system alternatives was limited to combinations of three types of material handling equipment: unit load automated guided vehicles (AUL), powered roller conveyors (CPR) and/or forklift trucks (TFL). Figure 2.11 shows the results of generating different alternatives and the dominance relationships between alternatives.

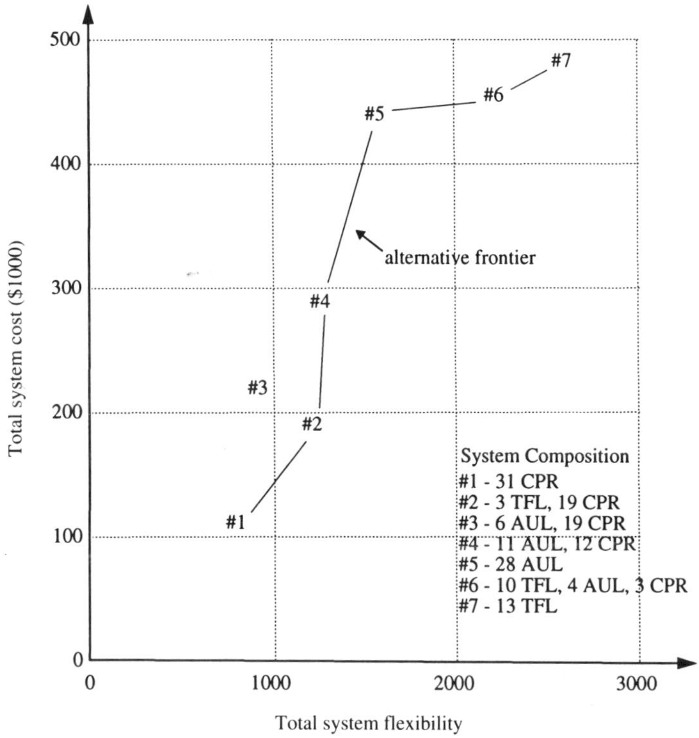

Fig. 2.11 Total cost vs. total flexibility of generated MHS alternative.

The economic design procedure continues by requiring that the designer determine the objectives of the design process. The nature of procedure requires that the designer re-evaluates the design objectives and conducts an analysis on the effect of changing the design objectives. The justification of each design provides a quantitative analysis of the reasoning supporting each design decision, which can then be used as the basis of an overall justification argument. Due to the overall design objectives, the procedure starts bounding the solution using system alternative #2.

The analysis of a MHS was conducted utilizing the three-level design analysis framework. Based on the system composition, a deterministic capacity analysis conducted, followed by a stochastic simulation. Using the simulation results, a detailed unit cost curve is developed and the marginal analysis within system alternative #2 is initiated.

The marginal analysis is conducted by incrementally adding or deleting one unit of forklift from alternative #2. From this, the relative incremental effect on the performance measures are noted. The interpretation of the results of the marginal analysis reveals several interesting characteristics of the alternative (Fig. 2.12). First, the number of forklifts, total system cost and

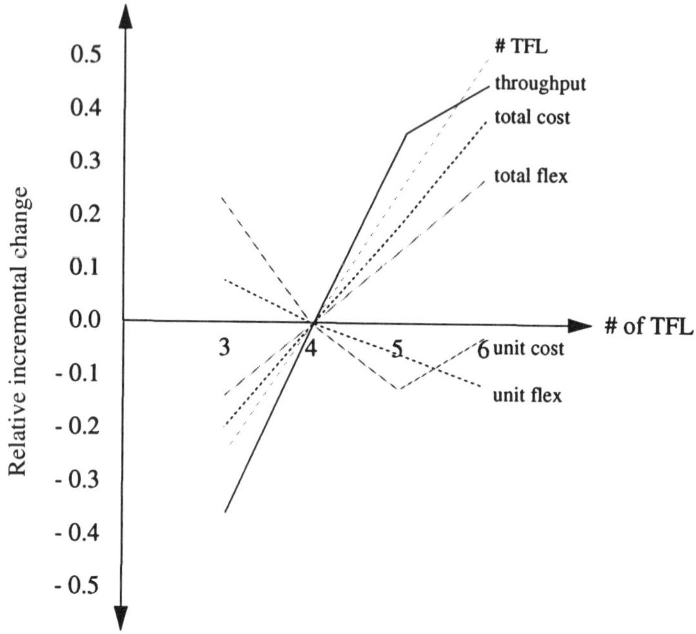

Fig. 2.12 Relative incremental changes for alternative #2 performance measures.

total system flexibility all follow a linear relationship. Hence, there is no increasing or decreasing marginal benefit associated with these changes. Second, the relative incremental unit cost values display a decreasing, followed by a negative, marginal benefit from increasing the number of forklifts. Third, unit flexibility exhibits a decreasing marginal benefit behavior. The designer concludes the following with respect to changing the number of forklifts in the system:

1. The system displays acceptable economic and performance characteristics in the 55 to 115 throughput/hour range.
2. A higher marginal flexibility to cost ratio is desirable.
3. The system throughput capacity expands easily within the 55 to 115 range.

A second area of analysis within alternative #2 concerns the benefit that could be derived from changing conveyor speed. Simulations are conducted comparing the throughput of alternative #2 operating at both 60 fpm and 120 fpm. A 90% Welch confidence interval was constructed on the difference of the throughput values, revealing that conveyor speed changes do not result in statistically better throughput values for this system, implying a negative marginal benefit due to changing conveyor speed.

The marginal analysis has so far served to inform the designer of the performance ranges of MHS alternative #2 and the type of changes which result in positive marginal benefit. Next, the marginal analysis explores equipment changes that would better achieve the design objective.

The marginal analysis between alternatives begins with complete simulation and statistical significance test results for system alternatives #1, #3, #5, and #7. The marginal analysis now focuses on the effect of changing from alternative #2 to another system alternative (Fig. 2.13). The ratio of the relative incremental total flexibility to the relative incremental total cost (TF/TC) is used as a primary measure of overall change. The TF/TC ratio

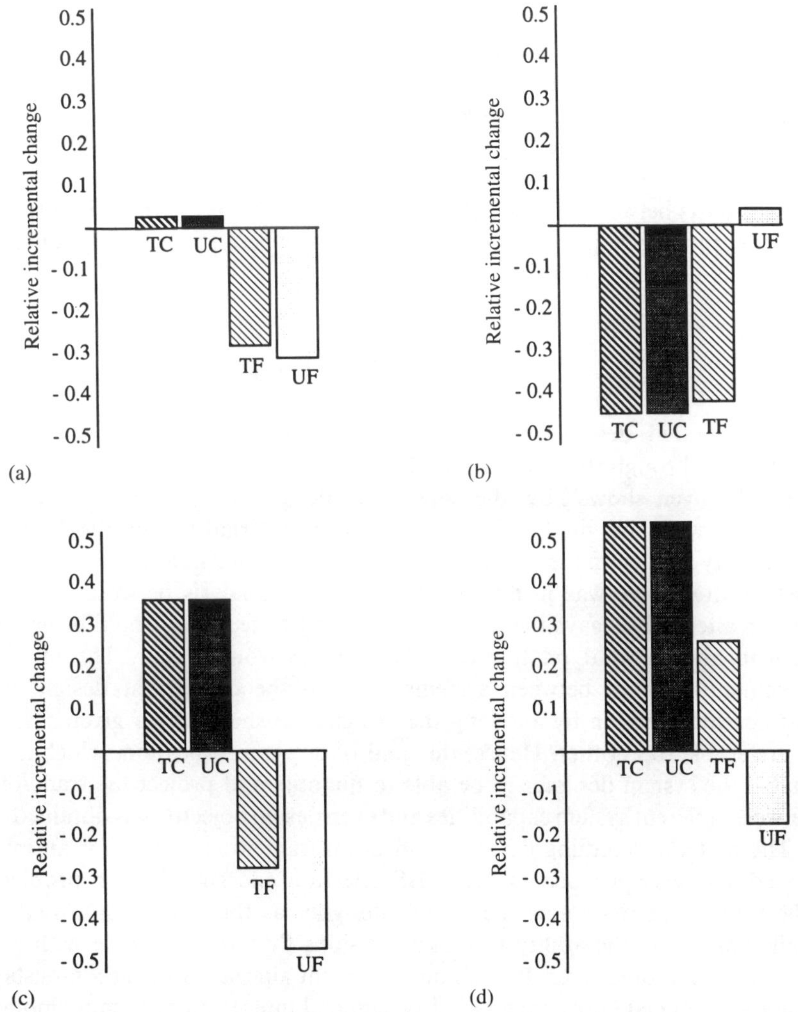

Fig. 2.13 Marginal analysis results of changing system alternative from alternative #2. (a) Alt #3 (TF/TC = – 8.99); (b) Alt #1 (TF/TC = 0.92); (c) Alt #5 (TF/TC = – 0.74); (d) Alt #7 (TF/TC = 0.49).

for changing to alternative #3 implies that there is a highly negative marginal benefit with respect to alternative #2 (i.e. alternative #3 has a higher total cost and a lower system flexibility). The same result is found for alternative #5. Alternatives #1 #7 both have a positive TF/TC ratio, yet for different reasons. In alternative #1 there is a decrease in both total cost and total flexibility, with the decrease occurring in approximately the same proportions. For alternative #7 there is an increase in both total cost and total flexibility, with the cost increasing at a greater rate than the flexibility.

The designer has now explored some of the marginal effects of changes both within the system and between systems. The decision remains as to which alternative best satisfies the design objectives. The initial system, alternative #2, provides a compromise solution based upon the objective. It might be valuable for the analysis to continue further by selecting either alternative #1 or alternative #7 for more detailed analysis. A further marginal analysis could be pursued by altering the facility layout, repeating the MHS alternative analysis for the new layout, and then comparing the marginal benefit/costs between layout/MHS alternatives. Whichever analysis path the designer pursues, a quantified decision path of the performance and economic trade-offs will exist to justify the final design.

2.4 SUMMARY

The integrated design and economic justification of a MHS has been illustrated through the application of the design justification procedure. The example given shows how the concurrent design and justification occurs. Specifically, the marginal analysis within a given alternative gave the designer the ability to determine whether changing the performance ranges of the system alternative was justifiable. The marginal analysis between different system alternatives gave the designer the ability to see what would be gained or lost, and at what relative rate the changes would occur. The relative incremental values between systems perform the concurrent design and justification function by allowing the designer to show why a given system is preferable to another. Hence, the goal of design justification, which is to enable the system designer to be able to quantify and project the trade-offs between different system capabilities and even design objectives, is obtained.

The material handling design justification framework allows for an integrated systems approach to the MHS selection and specification problem. The framework maintains the system designer as the key decision maker, while enhancing the ability to make decisions that are justifiable with respect to design objectives. Branching on system alternatives is accomplished using a combined knowledge-based system and integer programming model. Bounding system design solutions occurs using automatically generated models within a marginal analysis directed modification strategy. The combined effect of the branching and bounding scheme enables the designer to develop a MHS that provides the material handling capabilities that are justifiable from both a performance and economic perspective.

REFERENCES

Apple, J.M. (1972) *Material Handling Systems Design*, Ronald Press, New York.

Boyle, J.-M. (1989) Interactive engineering systems design: A study for artificial intelligence applications. *Artificial Intelligence in Engineering*, **4**(2), 58–69.

Dahlstrom, K.J. (1981) *Where to Use AGV Systems, Manual Fork Lifts, Traditional Fixed Roller Conveyor Systems Respectively*, Proceedings Ist International Conference on Automated Guided Vehicle Systems, (ed R.-H. Hollier), Stratford-upon-Avon, UK.

Daum, M. (1987) *Operating Costs of AGV Systems*, Proceedings 5th International Conference on Automated Guided Vehicle Systems (ed T. Takahashi), Tokyo, Japan, pp. 311–28.

Eilon, S. (1984) *The Art of Reckoning – Analysis of Performance Criteria*, Academic Press, London.

Fisher, E.L., Farber, J.B. and Kay M.G. (1988) MATHES: An expert system for material handling equipment selection. *Engineering Cost and Production Economics*, **14**(4), 297–310.

Gabbert, P.S. and Brown D.E. (1987) *A Knowledge-Based Approach to Materials Handling System Design in Manufacturing Facilities.* Proceedings 1987 International Industrial Engineering Conference, Washington, DC, pp. 445–51.

Hassan, M.M.D., Hogg, G.L. and Smith, D.R. (1985) A construction algorithm for the selection and assignment of materials handling equipment. *International Journal of Production Research*, **23**(2), 381–92.

Liang, M., Dutta, S.P. and Abdou, G. (1989) *A New Approach to Material Handling Equipment Selection in a Manufacturing Environment.* Proceedings 1989 International Industrial Engineering Conference, Toronto, Canada, pp. 225–30.

Müller, T. (1981) *Comparison of Operating Costs Between Different Transport Systems.* Proceedings Ist International Conference on Automated Guide Vehicle Systems, (ed R.-H. Hollier), Stratford-upon-Avon, UK, pp. 145–55.

Noble, J.S. and Tanchoco, J.M.A. (1990) Concurrent design and economic justification in developing a product. *International Journal of Production Research*, **28**(7), 1225–38.

Noble, J.S. and Tanchoco, J.M.A. (1993a) A framework for material handling system design justification. *International Journal of Production Research*, **31**(1), 81–106.

Noble, J.S. and Tanchoco, J.M.A. (1993b) Design justification of manufacturing systems – a review. *International Journal of Flexible Manufacturing Systems*, **5**(1), 5–25.

Pritsker, A.A.B. (1986) *Introduction to Simulation and SLAM II*, 3rd edn, Halsted Press, John Wiley, New York.

Rembold, B. and Tanchoco, J.M.A (1991) Network Editor for Material Handling System Design. *Technical Report*, School of Industrial Engineering, Purdue University.

Schultz, G.A. (1978) Estimating total cost of a package-conveyor system. *Plant Engineering*, **32**(8), 149–52.

Scott, H.A., Tanchoco, J.M.A. and Agee, M.H. (1983) Unit load (belt) conveyor operating design and cost. *International Journal of Production Research*, **21**(1), 95–107.

Sethi, A.K. and Sethi, S.P. (1990) Flexibility in manufacturing: A survey. *International Journal of Flexible Manufacturing Systems*, **2**(4), 289–328.

Shelton, D. and Jones M.S. (1987) A selection method for automated guided vehicles. *Material Flow*, **4**(1), 97–107.

Solberg, J.J. (1977) *A Mathematical Model of Computerized Manufacturing Systems.* Proceedings 4th International Conference on Production Research, Tokyo, Japan, pp. 22–30.

Tanchoco, J.M.A., R.P. Davis, Wysk, R.A. *et al.* (1979) *Component Selection and Operating Specifications for Belt Conveyor Systems.* Proceedings 1979 Spring Industrial Engineering Conference. San Francisco, CA, pp. 112–9.

Tompkins, J.A. and White, J.A. (1984) *Facilities Planning*, John Wiley, New York.

Webster, D.B. and Reed, R. Jr. (1971) A material handling system selection model. *AIIE Transactions*, **3**(1), 13–21.

Wiltse, J. (1979) Eight simple steps to an accurate forecast of conveyor system costs. *Modern Materials Handling*, **34**(2), 60–3.

Ziai, M.R. and Sule, D.R. (1989) Computerized materials handing and facility layout design. *Computers and Industrial Engineering*, **17**(1–4), 55–60.

PART TWO

Cell Design and Material Handling Considerations

CHAPTER 3

Cell design strategies for efficient material handling

C.L. Moodie, J. Drolet, Y.-C. Ho and G.M.H. Warren

3.1 INTRODUCTION TO CELLULAR MANUFACTURING

3.1.1 Product versus process layout

Not too long ago, production environments were laid out according to one of two conceptual schemes, namely the jobshop (typical of low volume, high product variety environments) and the transfer line (typical of high volume, extremely low product variety environments).

The cellular manufacturing concept was developed to exploit the advantages of both types of layout, and involves the pursuit of production activities in cells. A manufacturing cell is 'a group of processes designed to make a family of parts in a flexible way' (Black, 1991).

A jobshop layout results in the grouping of machinery according to process capability, i.e. a **process-based layout**. A transfer line, conversely, involves the grouping of machinery according to the processing requirements of a (single) product, i.e. a **product-based layout**. A cellular manufacturing layout requires the grouping of machinery according to the processing requirements of a family of parts, i.e. a **product-family layout**.

3.1.2 Types of cells

A wide variety of industrial cells exist, and are classified to distinguish on the basis of cell functionality. Cells may have the following functionality: **assembly**, **disassembly** and **manufacturing**.

Manufacturing cells may be further classified based on the design concept; the best-known design is based upon the principles of **group technology** (GT). The group technology concept seeks to 'partition a manufacturing system into cells, and part types to be manufactured into part families, based upon the similarity of part manufacturing characteristics' (Nagi, Harhalakis and Proth, 1990).

This chapter also discusses an alternative design concept which results in cells that are more flexible from a control standpoint than GT cells, namely **virtual cells**. Another interesting development is the idea of **fractal cells**.

3.1.3 Advantages and disadvantages of cellular manufacturing

The cellular manufacturing concept was born from a need to compromise the flexibility of the jobshop while retaining the production management simplicity associated with the transfer line layout in order to boost productivity in low-to-medium volume, low-to-medium part variety production environments.

The brief comparison of the manufacturing cell with the jobshop and flow line that follows is intended to highlight the advantages and disadvantages of cellular manufacturing:

1. **Flexibility** Manufacturing cells are less flexible in terms of the variability of routing than jobshops, yet are more flexible than flow lines. Routing flexibility is achieved via compromises in the mean machine utilization, since a cellular manufacturing environment typically requires the acquisition/maintenance of production capacity in excess of that associated with flow lines.
2. **Material flow system** Material handling is more efficient, and the material flow system is more streamlined in the case of a cellular manufacturing environment compared with a jobshops, and only marginally more complex (and less efficient) compared with transfer lines.
3. **Work-in-progress** Although the level of work-in-progress (WIP) inventory need not vary between the layouts, practices in industry have demonstrated that cellular manufacturing and transfer line layouts typically result in a lower level of WIP compared with jobshops. This is primarily facilitated by the fact that production management tasks are less complex in these environments, permitting increased control of excess inventory.
4. **Lead times** Lead times are typically short in the case of cellular manufacturing compared with jobshops (where considerable waiting and transportation time is incurred), and marginally longer compared with transfer lines. Shorter lead times reduce the need for unnecessary handling to intermediate stores/buffers, and obviate lost time resulting from product 'burial'.
5. **Production volume** Cellular manufacturing environments are suitable for low-to-medium volume production. Jobshops are ideally suited to prototyping and low volume production operations, while transfer lines are suited to medium-to-high volume production (depending upon the level of automation involved). The majority of production in most cellular manufacturing environments is in lot sizes of 50 or less.

3.1.4 JIT and cellular manufacturing

The links between cellular manufacturing and the just-in-time (JIT) manufacturing philosophy have important implications for material handling efficiency that form the basis for the justification and design of cellular manufacturing systems. We address these issues in a qualitative fashion before moving on to a discussion of the analytical tools used for the design of cellular manufacturing systems.

Golhar and Stamm (1991) identify four tenets comprising the JIT manufacturing philosophy: elimination of waste; employee involvement in decision-making; supplier participation; and total quality control.

The inter-relationship between JIT and cellular manufacturing is twofold: first, in many cases, a cellular manufacturing layout is developed with the express purpose of exploiting JIT methods for production control; and second, JIT considerations can be seen to play an important role in deriving optimization objectives for the cell design process.

In particular, the notion of the elimination of waste is a primary objective in the design of cells (this issue is discussed in Section 3.1.5). The importance of employee involvement (via decision-making) via human interfaces with cells cannot be overestimated: employee (as opposed to part) movements within cells play a critical role in determining the efficiency of cell operation. Moreover, in many cases, the sizing of manufacturing cells during the design stage is influenced by the size of worker teams. Team behavior research pegs the optimal number of workers in a cell at either two or anywhere between six and twelve.

3.1.5 Cellular manufacturing system design

Boctor (1991) lists three steps required for the implementation of cellular manufacturing concepts:

- economic and technical feasibility studies,
- design of the manufacturing system;
- system implementation.

We consider only the **design** of cellular manufacturing systems in this chapter.

Burbidge (1989) developed a hierarchical five-level analytical model to drive the design process called **Production Flow Analysis** (PFA). One of the major problems addressed by the PFA model is the solution of the **cell formation problem** (CFP). The CFP involves the grouping of parts into families and machines into cells, and the assignment of part families to machine cells.

In an ideal cellular manufacturing environment, raw material should be transported directly from dock to cell, and product (parts) should be manufactured completely within the cell and then transported to an assembly cell for assembly with other parts. This scenario supposes that products are produced exclusively in a single cell, and that no inter-cell transfers of parts is required. One of the primary objectives of the CFP therefore focuses explicitly on material handling efficiency by aiming to minimize the inter-cell transfer of parts. The solution of the CFP is thus a crucial part of the design process with regard to macro material flow system efficiency. We survey work on the CFP later in this section.

A secondary, micro material flow system design problem forming part of the PFA model is the **line formation problem** (LFP), which focuses on the layout of machines inside each cell. This problem is discussed in sections 3.2 (linear topology cell layouts) and 3.3 (network topology cell layouts). Whereas the efficiency of material handling can be improved by optimizing the LFP, typically larger gains (in terms of material handling time and costs saved) stand to be made by optimizing the CFP.

(a) Cell formation problem: Formulation

The design of a cellular manufacturing system has been structured in a variety of ways. Early work focused on the grouping of machines on the basis of part characteristics (part functionality, appearance, design features) classification codes (e.g. OPITZ codes). There now appears to be consensus in the literature (Gupta, 1991; Boctor, 1991) that the use of part characteristics for the development of part families and machine cells is inferior to the use of process information.

The CFP (and related problems) was studied extensively by Burbidge (1989). He and others developed a manual method to construct machine cells on the basis of the identification of key machines. This approach first identifies key machines in the facility, which serve as a foundation for each cell. The remaining machines are then iteratively assigned in decreasing 'order of importance' to the partial cells defined by the key machines. Burbidge defines a classification code ('SICGE' code) to classify machines in order of importance. Key machines are assigned an 'S' code, whereas general equipment (e.g. hand tools) is assigned an 'E' code. Machines are assigned to cells in order of 'decreasing' SICGE code (i.e. S → I → C → G → E). Burbidge's PFA model prompted research into quantitative methods to alleviate the computational burden of the manual resolution of the CFP.

Two of the most important goals of the CFP are:

G1: *Minimization of a measure of the inter-cell transfer volume.* This index corresponds directly to our desire to optimize (minimize) the volume of material flow (and thereby ensure the efficiency of material handling).

G2. *Maximization of a measure of individual machine utilization.* This goal has also been phrased in terms of the minimization of the number of duplicated machines needed to obtain mutually exclusive cells, i.e. a design that requires no inter-cell transfers. In general, this goal is employed to constrain capacity.

We now concisely state a simplistic version of the CFP (Boctor, 1991), and examine the realization of these goals therein:

Problem: Cell Formation Problem (CFP)
Instance: Given: matrix A ($a_{ij} \in A$) of dimension $M \times N$ of routing/production volume data; designer bound m on the maximum number of machines permitted in **any** cell.
Question: Minimize the value of Z in (3.1) such that (3.2) through (3.5) are satisfied. Read the solution to the CFP from the resulting X and Y matrices ($x_{ij} \in X$, and $y_{ij} \in Y$).

$$\text{(CFP)} \quad \text{Min}\, Z = \sum_{i=1}^{N} \sum_{j=1}^{M} a_{ij} \left(\sum_{k=1}^{C} |x_{ik} - y_{jk}| \right) \tag{3.1}$$

$$\text{s.t.} \sum_{k=1}^{C} x_{ik} = 1 \qquad i = 1, \ldots, N \tag{3.2}$$

$$\sum_{k=1}^{C} y_{jk} = 1 \qquad j = 1, \ldots, M \tag{3.3}$$

$$\sum_{i=1}^{N} x_{ik} \leqslant m \quad k = 1, \ldots, C \tag{3.4}$$

$$0 \leqslant x_{ik} \leqslant 1; \quad 0 \leqslant y_{jk} \leqslant 1; \quad x_{ik} \text{ and } y_{jk} \text{ integer } \forall i, j, k \tag{3.5}$$

where k = cell index; C = number of cells, C is a non-negative integer; M = number of parts, M is a non-negative integer; N = number of machines, N is a non-negative integer; m = maximum number of machines allowed per cell, m is a non-negative integer.

$$a_{ij} = \text{volume of part } j \text{ requiring processing on machine } i \tag{3.6}$$

$$x_{ik} = \begin{cases} 1 & \text{if machine } i \text{ is assigned to cell } k \\ 0 & \text{otherwise} \end{cases} \tag{3.7}$$

$$y_{jk} = \begin{cases} 1 & \text{if part } j \text{ is assigned to cell } k \\ 0 & \text{otherwise} \end{cases} \tag{3.8}$$

The above formulation of the CFP minimizes the number of part operations that are performed outside the cell to which a part is assigned (production volume considerations are included), and in general a_{ij} is a non-negative integer. Constraint (3.2) ensures that each machine is assigned to exactly one cell; constraint (3.3) ensures that each part type is assigned to exactly one cell; constraint (3.4) ensures that the total number of machines in any cell is less than some designer-specified (global) upper bound (note that it would be easy to introduce another system of constraints to handle a corresponding lower bound); constraint (3.5) provides the required integrality properties. In this formulation, goal G1 is realized via (3.1), and goal G2 is (somewhat indirectly) realized via constraint (3.4).

The CFP formulation is a variation of an assignment optimization problem which is known to be NP-hard (Rardin, 1992), suggesting that branch and bound, branch and cut (for small problems) and heuristic solution procedures are appropriate. Two basic types of formulations of the CFP have appeared in the literature:

1. *Heuristic formulations.* These typically rely (to a greater or lesser extent) on information contained in a (binary) part–machine incidence matrix A. This requires that the definition of $a_{ij} \in A$ as per (3.6) be specialized to:

$$a_{ij} = \begin{cases} 1 & \text{if part } j \text{ requires processing on machine } i \\ 0 & \text{otherwise} \end{cases} \tag{3.9}$$

In general, heuristic formulations include a wide variety of CFP formulations that are difficult to quantify, which are based upon rules of thumb, or which require optimization of a problem substantively remote from the CFP.
2. *Programming formulations.* These are defined in the usual mathematical programming sense.

Heuristic formulations

Heuristic formulations are motivated by a desire to exploit some heuristic algorithm. Typically, heuristic algorithms are based upon some intuitive notion that must be decoupled from the details of the CFP. Invariably, therefore, the heuristic formulation results in a simplification of the CFP, and implicitly/indirectly encodes (some of) the true objectives of the CFP. We survey some of the heuristic formulations proposed in the literature below.

The notion of a similarity coefficient has played an important role to-date in work on clustering algorithms. Early work relied on relatively simple derivations of similarity coefficients (e.g. Taboun, Sankaran and Bohle, 1991), where the single linkage similarity coefficient SM_{ik}, a measure of the similarity of parts i and k, is defined as:

$$SM_{ik} = \frac{\sum_{j=1}^{N} a_{ji} \cdot a_{jk}}{\sum_{j=1}^{N} (a_{ji} + a_{jk} - a_{ji} \cdot a_{jk})}, \quad i = 1, \ldots, M-1; \quad k = i+1, \ldots, M \tag{3.10}$$

These formulations were proposed to admit various permutation algorithms for the optimization of some measure of the similarity coefficients. The relationship between the similarity measure optimization and the CFP objective function is not clear – hence we term the formulation **heuristic**. Subsequent work (e.g. Chow and Hawaleshka, 1992) aimed at solving the 'machine chaining problem' (distance-based cluster-analysis – see Section 3.1.5b).

The clustering formulations are specifically structured to admit application of computationally inexpensive solution techniques, and consequently have spartan information requirements. Formulations based upon binary incidence matrices have inherent assumptions (Vengopal and Narendran, 1992), e.g. each part type is assumed to make identical capacity demands on each machine upon which it requires processing. These approaches suffer from several major information deficiencies that reduce their effectiveness in deriving good cellular manufacturing system designs, although various improvements have been proposed in the literature, including incorporation of operation sequence and production volume data (Gupta, 1991). These include:

1. Alternative routings are not accounted for. Multiple routings arise as a result of the functional similarity of workstations (Nagi, Harhalakis and Proth, 1990).
2. Production volume and machine capacity considerations are not accounted for (Vengopal and Narendran, 1992).
3. Cell size (and other management) constraints are not accounted for.
4. Key machines and bottleneck machines are not recognized.
5. The sequence and duration of operations is not accounted for (Logendran, 1991a).
6. Physical constraints that impact the ability to form certain groups of machines (e.g. immovable equipment and facility configuration) are not accounted for.
7. Tooling constraints are not considered (Taboun, Sankaran and Bohle 1991).
8. Potential upcoming part design changes are not accounted for.
9. Operator walking patterns are not accounted for (Chen, 1991, 1992).
10. Subcontracting opportunities are not accounted for (this could be effected by developing an appropriate cost model).

These problems have been partially addressed in formal programming models.

Programming formulations

Several programming formulations have appeared in the literature, (e.g. Boctor, 1991; Gunasingh and Lashkari, 1991; Logendran, 1991a, b; Nagi, Harhalakis and Proth, 1990). The main challenges facing researchers in this area appear to be: the development of formulations with special structure that can be exploited to improve computational efficiency; the efficient decomposition/synthesis of multiple optimization objectives; and the exploitation of developments in programming algorithms. We close this section by reviewing some of these developments.

Logendran (1991b) describes a series of models that are based upon Burbidge's idea of structuring cells around key machines. He describes three methods that are essentially based upon the idea of **dis**similarity coefficients for the identification of key machines. The reasoning behind this approach is that the identification of the most **dissimilar** machines will result in the most dissimilar cells with the least likelihood of inter-cell transfers. Logendran extends his research (1991a) by considering operation sequences and aspects of workstation utilization in an improved model.

Gunasingh and Lashkari (1991) propose two non-linear zero–one formulations of the CFP based upon tooling requirements, available tooling and processing times.

Variations (and properties thereof) of the model of Boctor – (3.1) through (3.5) – are discussed by Boctor (1991). Boctor proposes a linearized variation of (3.1), shows that (3.2)–(3.5) can be reduced, and that the resultant formulation realizes an integral solution.

Nagi, Harhalakis and Proth (1990) propose an iterative scheme comprised of a linear program (that is used to select optimal routings) and a two-phase heuristic that forms cells. This approach is novel in that capacity constraints are respected, and load distribution effects on cell design are explicitly accounted for.

(b) Cell formation problem: Solution

The decision version of the CFP is a hard computational problem (even a relaxed decision version of the part–machine matrix formulation is NP-complete–Miltenburg and Zhang, 1991). The most appropriate solution techniques used to 'solve' the CFP are typically heuristic, and are highly dependent upon the exploitation of special features of the formulation. In keeping with our discussion of **formulations** of the CFP, we discuss **solutions** (i.e. algorithms) of the CFP in two parts; namely solution procedures based upon heuristic formulations generally exploiting part–machine incidence matrices, and (pseudo) programming solutions.

Boctor (1991) classifies three basic types of solution procedure for the part–machine matrix formulation of the CFP:

1. **Matrix rearrangement procedures.** Typical examples of this type of procedure are: rank order clustering (ROC), the similarity coefficient (SC) method, the bond energy algorithm (BEA), and variations/modifications of these algorithms (Mittenburg and Zhang, 1991). These methods permute the column and rows of the $M \times N$ part–machine matrix A independently to optimize some heuristic measure of permutation efficiency. Figure 3.1 shows an initial part–machine matrix A, and a permuted matrix wherein clustering evident.
2. **Distance-based cluster-analysis procedures**. These methods arose as a result of work aimed at solving the 'machine chaining problem'. Typical methods in this class are based upon the development of a measure of association of every pairwise combination of machines to be clustered, and include: single linkage clustering (SLINK), total linkage clustering (CLINK), average linkage clustering (ALINK) (Gupta, 1991), and an algorithm by Chow and Hawaleshka (1992).
3. **Graph-decomposition techniques**. Typical examples of this approach in-

Parts \ Machines	10	20	30	40
10	1	0	1	0
20	0	1	0	0
30	1	0	1	0
40	0	1	1	1
50	1	0	1	0
60	0	1	0	1
70	1	0	1	0
80	0	0	0	1

UNCLUSTERED

Parts \ Machines	10	30	20	40
10	1	1	0	0
30	1	1	0	0
50	1	1	0	0
70	1	1	0	0
20	0	0	1	0
40	0	1	1	1
60	0	0	1	1
80	0	0	0	1

CLUSTERED

Fig. 3.1 Clustering in the part–machine incidence matrix.

clude the ideal seed non-hierarchical clustering (ISNC) algorithm and variations thereof (Mittenburg and Zhang, 1991).

We can add other heuristic techniques that use part–machine incidence matrices as their informational base to Boctor's classification, namely the work of Kusiak (1988), who developed a knowledge base of rules to guide solution of the CFP, the application of a variation of Lin and Kernighan's heuristic for graph partitioning (Askin and Chiu, 1990), and the application of genetic algorithms (GA) (Vengopal and Narendran, 1992).

The clustering solution techniques developed for solving the part–machine incidence matrix formulations of the CFP do not necessarily develop explicit solutions, e.g. the BEA method merely permutes the rows and columns of the part–machine matrix (the clustering task is left to the designer), and the SC method explicitly groups **either** the part or machine vectors (not both). Most of the heuristic solution methods group part and machine vectors separately, and then assign part families to machine cells in a staged procedure (this is the approach embodied in the ISNC algorithm). The independence of the part family and machine group formation in the heuristic formulations is a principal source of sub-optimality in heuristic cell design, and programming methods have been favored to solve these problems **simultaneously**.

Recent work has focused on the solution of programming formulations using heuristic programming, a term intended to represent a vast array of generally unrelated optimization techniques ranging from heuristic variations of optimal programming methods (e.g. column generation or pivot and complement procedures for linear programming formulations) through search techniques (e.g. local improving and tabu search) to simulated annealing SA) (Boctor, 1991). These methods are applied to improve the computational

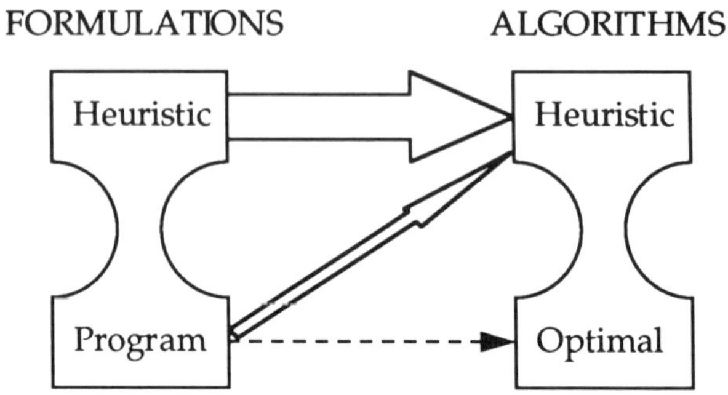

Fig. 3.2 Formulations vs. algorithms.

efficiency of the solution procedure, and invariably have no performance guarantees.

True programming techniques have also found application in heuristic settings via their incorporation in iterative schemes (e.g. Nagi, Harhalakis and Proth, 1990). In general, researchers have found computational efficiency a prohibiting factor for large mixed integer linear programs (Chen, 1991). Problem decomposition and relaxation appear to be worthy of further investigation; for example, Gunasingh and Lashkari (1991) decomposed an integer linear formulation of the CFP and solved Lagrangian relaxations of the subproblems.

In summary, the practicality of most formulations of the CFP is limited by the computational efficiency of the solution technique employed. The CFP problem is formally hard, and even heuristic formulations are generally computationally intractable in the optimization sense. Consequently, most solution techniques are based upon the application of heuristic **algorithms** (sub-optimal) to solve heuristic **formulations**. These ideas are captured in Fig. 3.2. Analysis in practice is based upon the application of one or more of the quantitative methods described to generate an initial design, which is then manually refined by engineering and management staff to account for additional constraints not included in the computer model. When a cell has been created (by whatever method) for a family of parts, the further problem of arranging the machines within the cell remains. This arrangement can have either a 'line' or 'network' orientation.

3.2 LINE ORIENTED CELLS

3.2.1 Line oriented cells and their advantages

A line oriented cell is a cell whose facility layout is linear. Although modern material handling equipment is intelligent and flexible enough for us to have a more complex layout, the linear layout of manufacturing facilities is still a popular design choice. Linear layouts exist in various forms such as a U-shape,

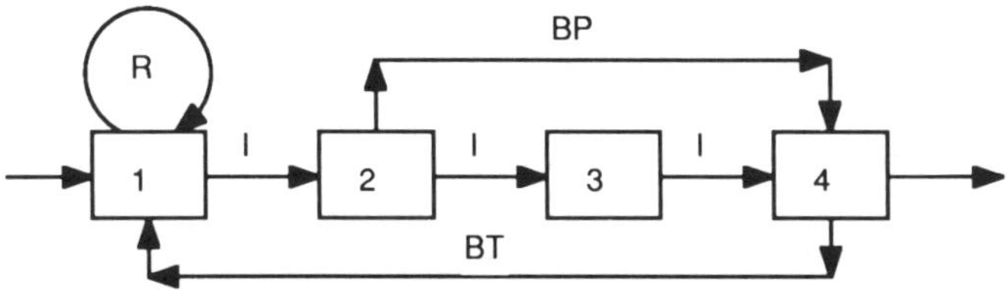

Fig. 3.3 Four different flow movements within a multi-product flow line. R = repeat; I = in sequence; BP = bypassing; BT = backtracking (Aneke and Carrie, 1986).

a straight line, and a circular loop. The form of a layout is determined by factors such as shape and area of the available space, material handling system used and linkage to other cells.

One main reason for the popularity of linear layouts is the simple material handling problems associated with them. Because the material flow is restricted by the structure of the flow line, the main factor contributing to the flow complexity in a flow line is the changing of flow directions. This can be realized by comparing the flow control problem of a single loop guide path to that of a network guide path for an automated guided vehicle (AGV). Apparently, due to the simpler structure of a single loop guide path, its flow control problem is easier. Although a linear layout may not be the optimal design choice, the desire for simplicity in the resultant layout and material handling problems makes it a popular choice.

3.2.2 Material flow analysis

Material flow analysis is a technique performed before establishing the physical layout. Its main purpose is to determine the logical position of each processing entity (machine or workstation) by analysing the flow sequences of products. In a multi-product flow line, four types of flow movements can be observed: repeat, in-sequence, bypassing and backtracking movements (Fig. 3.3). Among these flow movements, in-sequence movement is the most desirable because of its unidirectional movement. Backtracking movement is the least desirable since it complicates flow the most. A layout that can achieve more in-sequence movements usually has the advantage of smaller flow distance, easier material handling and more efficient production. On the other hand, a layout with a significant amount of backtracking movements usually results in a greater flow distance and a more difficult material handling problem. Therefore, the design of a layout where the flow movements are mostly in-sequence is the main goal of flow analysis.

3.2.3 Review of flow line design and analysis algorithms

Many flow line design algorithms have been proposed. They can be classified into two types. The first type is the single machine/workstation algorithm that deals with a problem scenario where only one machine of any type is allowed in a flow line. Most of the early flow line analysis techniques are of this type.

Some of the best known methods are summarized here. Noy (1957) proposed a sequence-demand method which was probably the first multi-product flow line analysis technique. Singleton (1962) presented a simplified approach by converting all the sequences of products to a common length scale, and determined the mean position of every processing entity. Moore (1962) discussed a technique called the sequence-demand method, which calculated the mean demand position of every operation. Hollier (1963) proposed several methods that were designed to achieve either one of the following two goals: maximizing the amount of in-sequence movements, or minimizing the amount of backtracking movements.

An alternative type of flow line analysis allows duplicate machines in a flow line. This type of problem is more realistic, since it is very common to have duplicate machines in an actual production environment. This type of problem was first studied by Carrie (1975). His procedure starts by building a line that contains a sufficient number of workstations so that the flow of every product can be accommodated without incurring any backtracking movements. The algorithm then eliminates uneconomical workstations and re-routes their operations to other workstations. Usually, many solutions can be generated after the elimination process. Finally, computer simulations are used to compare these solutions and select the best one. Since the elimination process depends very much on the designer's experience and personal preference, it will become inefficient once the number of possible choices becomes too large. Aneke and Carrie (1986) presented a method which constructs a flow line from both ends of the flow line. This method performs better in a situation where the products have similar workstation requirements at their early and later processing stages, with the differences mainly coming from intermediate processing stages. Like the earlier method of Carrie (1975), this method does not provide a clear-cut eliminating process for the unjustified workstations. Lee (1991) proposed a method that builds the flow line based on the sequence similarities of products. This algorithm starts by selecting the product that has the highest sequence similarity with the flow line constructed so far. It then modifies the flow line to accommodate the sequence of the newly selected product. Finally, the elimination of infeasible or uneconomic workstations is performed. An example of a solution devised using Lee's linear flow analysis can be seen in Fig. 3.4.

3.3 NETWORK ORIENTED CELLS

3.3.1 Network oriented cells and their advantages

A network oriented cell is one where the material flows are not confined to one rigid line. Instead, any processing entity in the cell can be directly

Product	Demand quantity	Operation sequence (in w.s. #)
A	20	2-3-4-6-8-9-7
B	10	14-2-3-4-5-10-11-12
C	15	2-4-6-8-9-13
D	10	1-2-3-5-11-12

(a)

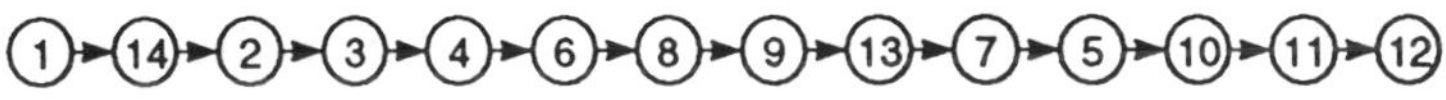

(b)

Fig. 3.4 (a) Product information for the illustrative example; (b) a linear layout configuration.

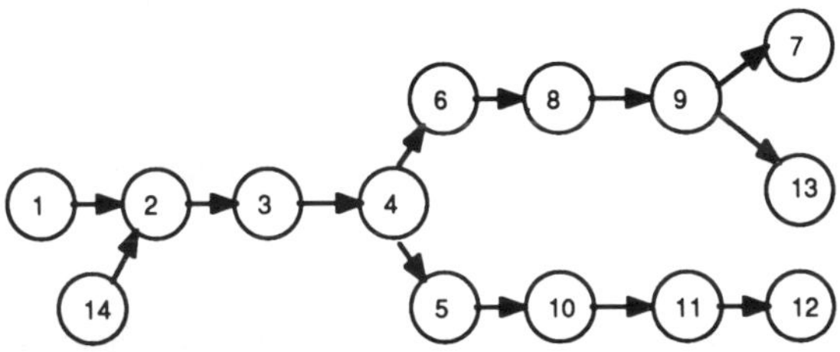

Fig. 3.5 A possible network configuration of the illustrative example.

connected to another entity if the flow between them is large enough. Since material handling is a non-value added activity, it is desirable to minimize the time spent on transportation. One way to achieve this is to reduce the distance among machines. Although a line layout may have the advantage of easier material handling problems, its rigid structure does not provide much routing flexibility. On the other hand, a network with branching flow can prevent some unnecessary bypassing movements that may occur in a linear flow line, and will result in a smaller total flow distance. In addition, the network model provides the routing alternatives and flexibility that are especially important in a flexible production environment. The advantages of the network configuration are illustrated by the following example. Using the product data from Fig. 3.4, a network configuration is constructed. As can be seen, the line configuration of the machines in Fig. 3.4b incurs extra flow distance compared with the network configuration in Fig. 3.5. Assume that the distance between any two adjacent nodes is one unit, and the products can be injected into the line at the first machine of their respective processing sequence and leave at their last required machine. In this case, the expected total flow distance (for all products) is 480 for the linear layout and 350 for the network layout. The better performance of the network design is due to its smaller number of bypass movements.

Another advantage of the network configuration is the flow independence among the different branches. Any blockage (due to a bottleneck machine, the breakdown of machines, or the breakdown of material handling equipment) in one branch will not affect other branches. Although designers can deduce network configurations from the results of linear flow line analysis, this deduction process will be very tedious when the operation sequence information is complex.

A typical network consists of nodes and directed arcs connecting the nodes. A node represents a processing entity that can be a machine, a manual/automatic workstation, or an integrated workstation containing several processing entities. The directed arc indicates the flow direction. A head node is a node without any preceding nodes, and a tail node is a node without any succeeding nodes. For example, in the network of Fig. 3.4, nodes 1 and 14 are head nodes and nodes 7, 12 and 13 are tail nodes. A path is a sequence of nodes that starts at a head node and ends at a tail node, e.g. 1–2–3–4–5–10–11–12 is a path.

3.3.2 Network oriented flow analysis

(a) Sequence similarity measure

A sequence similarity measure called the **compliant index** is used in the network oriented flow analysis. The compliant index between a product and a flow path is determined by the number of operations that appear in the same order in both the flow path and the operation sequence of the product. Two kinds of compliant indices can be obtained, depending on the direction of comparison. They are obtained by comparing the flow sequence of products with the sequene of the flow path from both directions, forward and backward. For example, for a flow path 1–2–3–2–4–5–6 and a product A with sequence 1–3–4–5–2, the comparison for both indices is illustrated in Fig. 3.6.

(b) Basic algorithm of network oriented flow analysis

A network oriented flow analysis was proposed by Ho (1991) for the network flow analysis problem. Similar to Lee's (1991) method; the network oriented flow analysis also considers the sequence similarity. At the first iteration of the proposed network oriented flow analysis, the processing sequence of the product with the highest demand quantity is used as the initial network. The subsequent iterations then update and built the network. There are three main steps in each of the subsequent iterations. The first step is the product selection, the selection of products depending on the demand quantity of products. A product with a larger demand quantity will have a higher priority of being selected. With this product selection rule, the major flows can be built into the network ahead of the minor ones. The second step selects the best flow path from the network by utilizing the compliant indices to

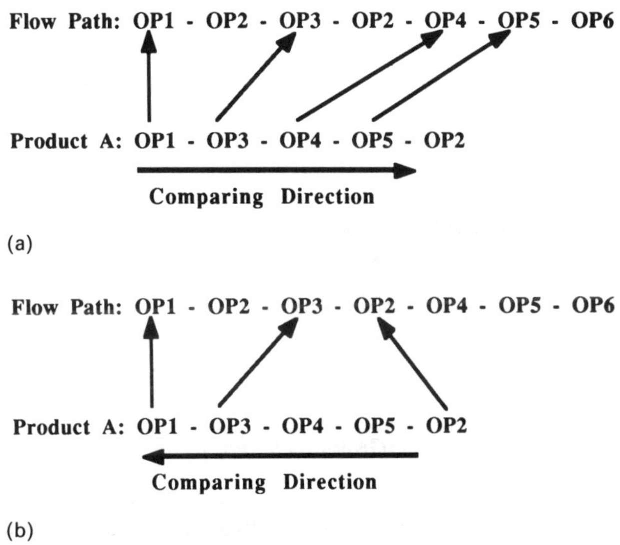

Fig. 3.6 Comparing process for obtaining (a) forward compliant index (=4) and (b) backward compliant index (=3).

Product	Demand	Sequence (in WS #)
1	150	3-2-4-5-6
2	100	3-2-5-6-4
3	50	3-2-1-4-6-8
4	200	1-2-4-6-7

WS	1	2	3	4	5	6	7	8
# Available	2	2	1	2	3	2	1	1

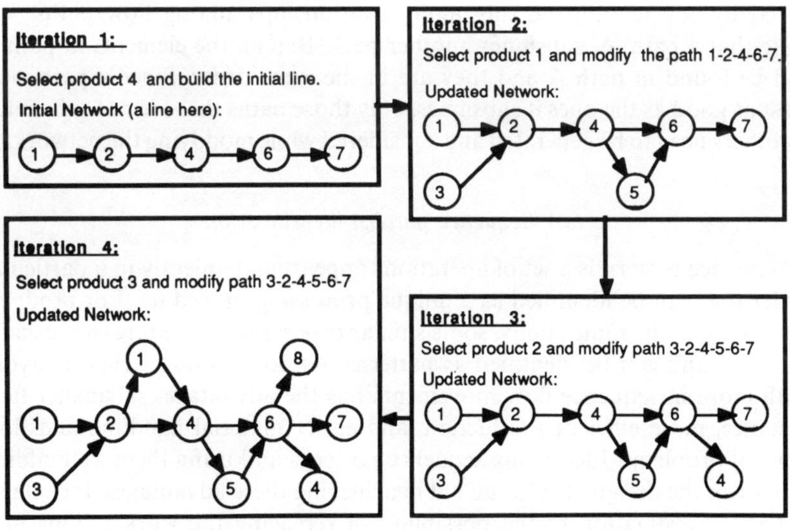

Fig. 3.7 Illustration of the network construction process.

accommodate the flow of the selected product. In the final step, the selected path is modified to accommodate the flow of the selected product. To accommodate those operations which cannot be located in the flow path, the modification procedure adds detouring or branching routes to the flow path. Figure 3.7 presents an illustration of the construction of a network.

3.3.3 Making network oriented flow analysis more efficient, and some extensions

The network flow analysis can become complex if the network is large; consequently, three accompanying procedures are proposed to assist the analysis process.

(a) Group technology

The application of group technology prior to network flow analysis has two advantages. First, a large problem can be divided into several smaller ones (groups or cells) which can be solved separately. Second, it is expected that the flow sequences of products belonging to the same group will be similar; as a result, the chance of branching in the network can be minimized. This is evident if the grouping technique considers the sequence similarities of the products. Coupling the grouping technology with the network oriented flow analysis algorithm can make the network analysis easier to perform.

(b) Elimination by subsumption

When selecting the best flow path for modification, some calculation can be saved by considering the subsuming relationships among flow paths. We state that a path, A, subsumes another path, B, if all the elements in path B can be found in path A and they are in the same order. Since a path is at least as good as the ones it subsumes, only those paths that are not subsumed by others need to be generated and considered while modifying the network.

(c) Sequence pattern identification

A sequence pattern is a set of operations appearing frequently in a particular order that can be identified as a unique processing procedure. For products belonging to the same family, some similar processing procedures are usually repeated and can be identified as patterns. As pointed out earlier, a layout with more in-sequence flow movements has the advantages of smaller flow distance, more efficient production, and easier material handling and flow control problems. Identifying sequence patterns and using them as 'building blocks' in the design of cells can aid in achieving these advantages. It can also aid the investigation of the possibility of replacing the workstations in a pattern with a smaller number of more intelligent and more integrated

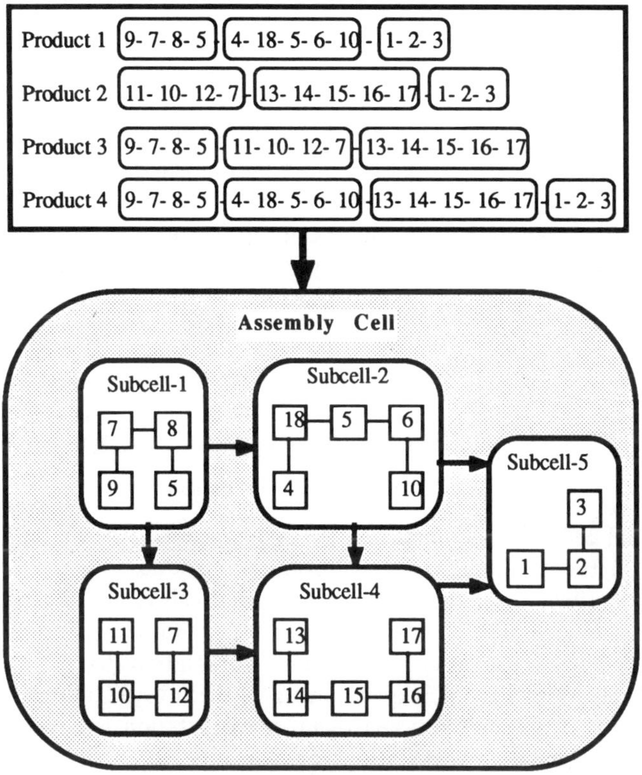

Fig. 3.8 Expository example of using sequence-patterns for the grouping of subcells and the layout design.

workstations, and/or the possibility of arranging them in a way such that operations can be performed more efficiently. In addition, the similarities of flow sequences within the patterns make the layout problems within them very easy.

The above ideas and advantages are illustrated by the example in Fig. 3.8. There are four different product types, which are assumed to be of the same family type and produced in the same cell. It is also assumed that the operations of the products are represented by workstation type numbers. There are five sequence patterns that can be identified from the operation sequences of these products. The workstations (operations) within each pattern are grouped into an operational module. Here, the whole manufacturing facility is referred to as a cell, while a subcell denotes the grouping of workstations (operations) within a pattern. We can apply the network oriented flow analysis and derive a network layout for these subcells. As can be seen, the subcells as well as the workstations within them can be arranged in a way such that both the material flows among and within the subcells become sequential. As a result, the layout problem within each subcell is straightforward.

A group technology cell has been previously defined as a set of workstations strongly inter-related and organized in such a way that the parts included in a given part family flow within this cell like as a flow line. The control and the physical configuration of workstations within a GT cell will not change in a foreseeable future. Difficulties with GT cells include what to do when demand of part-mix changes, and how much extra capacity will be required due to an improper balance of the workload. Traditional cellular manufacturing methods may not be appropriate if operation sequences or routings are prone to change with time. The following section discusses virtual cellular manufacturing systems, which are designed to cope with these problems.

3.4 VIRTUAL CELLULAR MANUFACTURING

3.4.1 The virtual cell concept

The virtual cellular manufacturing (VCM) concept to be described in this section has its roots in the virtual cell concept introduced by McLean, Bloom and Hopp (1986) in the early 1980s. A virtual cell is not identifiable as a fixed physical grouping of workstations, but as data files and processes in a controller. When a job order needs a set of workstations to be 'put together', a virtual cell controller takes over control of these workstations and makes communication possible between them. A workstation is either a member of a pool of available workstations or a member of a virtual cell. Major references on the virtual cell concept are Hocken and Nanzetta (1983), McLean and Brown (1986) and Simpson, Hocken and Albus (1982).

A VCM system does not promote opportunistic re-routing for each part after each operation. Instead, the route of a job (batch of parts) is defined at cell creation, and remains unchanged as long as the production goes as planned. In this manner, it is easier to plan the utilization of concurrent resources such as fixturing and tooling. Any unexpected event (breakdown, part-mix changes, etc.) can call for re-definition of one or more virtual cells, either directly or remotely affected by the unforeseen event.

3.4.2 Definition and terminology for VCM systems

Given the infancy of the VCM concept, it is appropriate to describe its operational characteristics and to define some terminology. The following list summarizes most of the peculiarities of a virtual cellular manufacturing system. Note that VCM systems are most appropriate for advanced manufacturing technology processors:

1. The workstations have the ability to switch between different types of part without significant setup time, providing tooling is available.
2. Most workstation types have replicates spread over the factory.

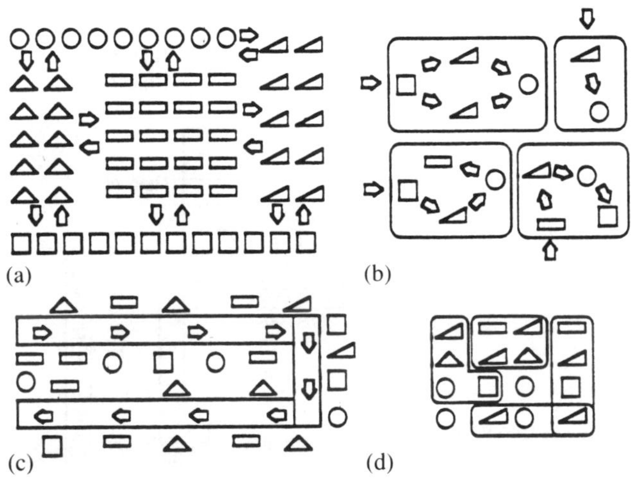

Fig. 3.9 Facilities layout. (a) process layout; (b) group technology cells; (c) product layout; (d) virtual cellular manufacturing.

3. The sequence of operations only specifies the sequence and type of workstations needed. Specific routes are only chosen during the virtual cell creation phase.
4. Once a virtual cell has been created for a batch order, every part in that order will be produced in that virtual cell unless some kind of unexpected event affects its production.

The physical configuration of a VCM system is not typical of any other manufacturing system. It is best to stress this difference by comparing its physical configuration with various manufacturing systems; process layout, product layout and group technology cell layout. Figure 3.9 shows material flow in process (a), product (b), classical GT cells (c), and virtual cellular (d) configurations. As stated above, certain manufacturing situations exist which will favor one layout over the other.

Except for VCM systems, the facility layout configuration models described above are available for, and are used in, contemporary factories. Looking ahead in time, however, trends indicate that world class manufacturing systems will have no more than 20 highly productive machines with automated material handling systems. These factories will operate in JIT mode under computer control. According to Solberg *et al.* (1985), systems having up to 10 workstations working around the clock will not be uncommon. These systems, although much smaller than current factories, will be much more productive. We believe that the virtual cellular layout strategy will become practical for scheduling parts to processors in this type of environment.

Figure 3.9(d) illustrates a virtual cellular layout which contains 16 workstations comprising four workstation types (differentiated symbolically by four different geometric shapes). This system currently has four active virtual

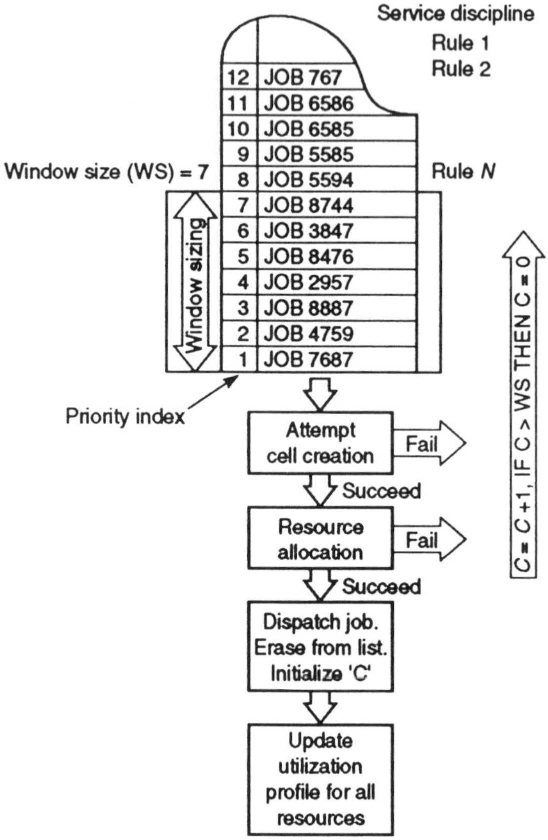

Fig. 3.10 Opportunistic scheduling.

cells and one shared workstation. Because virtual cells contain only the required workstations, the extra capacity required in a classical GT cell is no longer needed in a virtual cell system. Drolet, Montreuil and Moodie (1990) give more details on VCM layout.

3.4.3 An opportunistic scheduling strategy for virtual cellular systems

In this section, we propose a framework for scheduling and controlling a virtual cell-based system. The proposed framework contains four phases:

1. Job loading: prioritization and selection.
2. Virtual cell creation.
3. Resource allocation.
4. Dispatching of part units.

Figure 3.10 illustrates the overall procedure. Each step will now be presented in more detail.

(a) Job loading

Job loading consists of the priorization and selection of jobs to be introduced into the system. It uses opportunistic scheduling rules to produce a list of prioritized jobs. No scheduling rules are known to perform equally well in every situation, thus the job scheduler is provided with different scheduling rules for different situations. Following prioritization, the number of jobs to be considered for selection is reduced by means of window sizing. After selecting the most prioritized job in the window, the cell creation phase is attempted. If this phase fails, the controller repeats the process with the second most prioritized job in the window, followed by the third, and so on. This sequence is repeated when the capacity of any critical resource increases.

(b) Virtual cell creation

During virtual cell creation, we minimize the flow distance by determining an optimal set of workstations in a temporary production line (the virtual cell) for the selected job order. The outcome is the list of workstations that minimizes the material flow for that job order which is consistent with the requirement regarding the precedence and capacity constraints. Even though the details have been omitted, the general ideas encompassing this phase are portrayed in Fig. 3.11. Before proceeding, it is necessary to define the main elements represented in this figure.

The top-left frame illustrates a facility with 17 nodes, each representing a workstation. Each node is expressed in terms of the following convention: the numeral represents the type of workstation (e.g. milling, turning, inspection, etc.), and the character identifies each replication of a given workstation type. For example, node '1d' denotes replication 'd' of workstation type '1'. The frame itself symbolizes the outside walls of the facility. The axis system scales the layout and, thus, exact position of input/output workstations are shown. From such a layout we can calculate the inter-workstation distances (rectilinear or Euclidean); these measures, along with the utilization profile of resources, are useful during the virtual cell creation.

The second frame illustrates the set of nodes which can potentially be visited by the job, with their arcs. Each arc represents an achievable flow of parts between two nodes. Only the arcs connecting successive nodes restrained by the precedence constraint are shown.

After restructuring the second frame into the third view, we can easily distinguish the sequence of operations depicted by the precedence constraint of the job. The parts included in that job must go from a workstation of type 1 to a workstation of type 5, and then a workstation of type 3. The objective is to find the least expensive path from any workstation of type 1 to any workstation of type 3 via any workstation of type 5, while satisfying other constraints to be mentioned later. The output contains the list of workstations that minimize the material flow for that job, while being confined to the requirement regarding the precedence and capacity constraints.

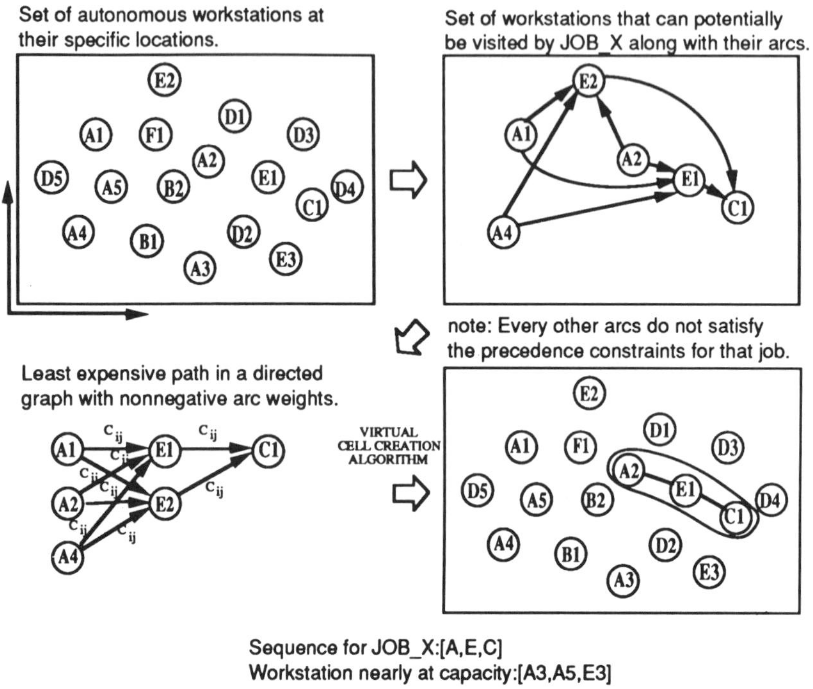

Fig. 3.11 Virtual cell creation.

The result of our example is shown in the last frame. The set of workstations displayed in this frame represents the virtual cell just created for JOB_X. Along with it, virtual cell controller will be created.

(c) Resource allocation

The third phase consists in assigning resources such as tools and fixtures to the newly created cell. The resources are delivered prior to activation of the virtual cell, and stay there for as long as the job is not finished.

(d) Dispatching

The final phase aims at dispatching the parts within, and occasionally between, each virtual cell. In this phase, the flow of parts and communication between cooperative workstations (members of the same virtual cell) is particularly important. The virtual cell controller has the logic to control the advance of processes within these workstations.

3.4.4 A linear programming model for VCM control

In this section, we propose an efficient linear programming model for creating and scheduling virtual cells. The solution satisfies the capacity constraints on all considered resources: workstations, fixtures and critical tools.

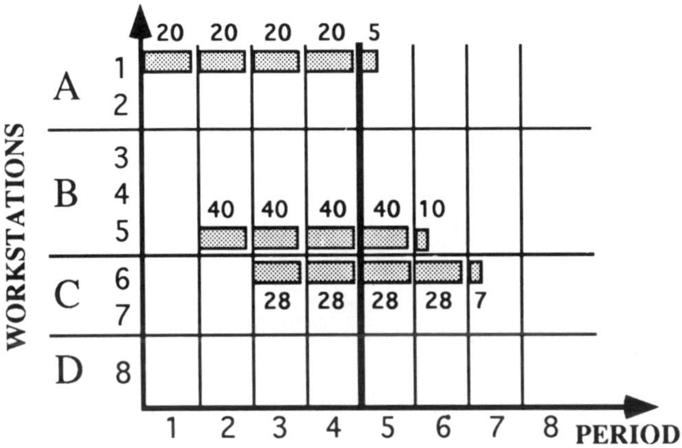

Fig. 3.12 Gantt chart illustrating variable $V_{3,156,1}$.

Let us now define the decisional variable V_{jrp}, which is critical to the understanding of the model. Variable V_{jrp} is the fraction of job j assigned to a virtual cell created at period p ($p \leqslant h$), and using resources on routing r. A Gantt chart which illustrates variable $V_{3,156,1}$ is represented in Fig. 3.12.

As shown, a job begins processing at period 1 using workstations A, B and C and its routing is $1 \to 5 \to 6$. The bottleneck workstation determines the maximum introduction rate of parts.

The model uses the following notation:

h = detailed scheduling horizon ($h \geqslant 0$)
H = aggregate scheduling horizon ($h < H$)
w = a resource type (e.g. workstation type, tool type, fixture type)
u = a unit of resource of some type w (a specific workstation, a tool, a fixture of some type w)
V_{jrp} = fraction of job j assigned to a virtual cell created at period p, assigned resources on routing r
W_{jp} = fraction of job j assigned to an aggregate virtual cell created at period p
N_j = number of parts in job j
D_r = distance travelled using routing r
AD_j = average distance travelled by job j using a routing r
F_j = flow time for job j (lifecycle of any virtual cell for job j)
L_{jp} = lateness of job j when processing begins in period p. In other words, $L_{jp} = \max\{0, p - (H - F_j)\}$
M = large cost incurred for each day of lateness on any job
P_{twjrp} = processing time required in period t on resource w when a virtual cell for job j using routing r and started in period p ($p \leqslant h$) has been created
P_{twjp} = processing time required in period t on resource w when an aggregate virtual cell for job j started in period p ($h < p \leqslant H$) has been created

C_{wut} = capacity of unit u of resource w in peiod t
AC_{wt} = aggregate capacity of all units of resource w in period t.

The model is now presented:

$$\text{Min} \sum_{p=0}^{h} \sum_{\forall j,r} V_{jrp}(N_j D_r + ML_{jp}) + \sum_{p=h+1}^{H} \sum_{\forall j} W_{jp}(N_j AD_j + ML_{jp}) \quad (3.11)$$

$$\text{s.t.} \quad \sum_{p=0}^{k} \sum_{\forall r} V_{jrp} + \sum_{p=h+1}^{H} W_{jp} = 1 \quad \forall j \quad (3.12)$$

$$\sum_{\forall j, p \leqslant h} P_{twjrp} V_{jrp} \leqslant C_{wut} \qquad \forall t: 0 \leqslant t \leqslant h, (w,u) \in r \quad (3.13)$$

$$\sum_{\forall j, \{p: h \leqslant p \leqslant H\}} [P_{twjrp} V_{jrp}] + \sum_{\forall j, \{p: h \leqslant p \leqslant H\}} P_{twjp} W_{jp} \leqslant AC_{wt}$$

$$\forall (w,u) \in r, t: h < t \leqslant H \quad (3.14)$$

$$V_{jrp} \geqslant 0 \quad \forall j, r, 0 \leqslant p \leqslant h \quad (3.15)$$

$$W_{jp} \geqslant 0 \quad \forall j, h < p \leqslant H \quad (3.16)$$

The first constraint (3.12) ensures that each job will be produced in its totality. There is one constraint for each job.

The first part of the constraint considers a certain number of periods in the future known as the **detailed horizon**. The detailed horizon begins at $t = 0$ and finishes at $t = h$. When variable $V_{jrp} = 1$, it represents the creation of a variable $V_{jrp} = 1$, it represents the creation of a virtual cell for job j at period p using the workstations on routing r. It is in fact a complete Gantt chart, as shown in Fig. 3.12.

The second part concentrates on the aggregate horizon. The aggregate horizon begins at $t = h + 1$ and finishes at $t = H$. When variable $W_{jp} = 1$, it represents the introduction of job j at period p. Here, specific routes r (by the same token specific virtual cells) are not considered.

Since variables are not restricted to be 0 or 1, they sometimes happen to be between 0 and 1, thus in these circumstances the rate of introduction and the size of the lots are adjusted proportionally. For example, if a variable V_{jrp} equals 0.3, and if the bottleneck limited the introduction rate to 20 units/period, then the rate would be adjusted to $20 \times 0.3 = 6$ units/period beginning at period t until an equivalent proportion of the lot has been introduced. Clearly, although the bottleneck fixes the boundaries for the introduction rate, the rate is really fixed by the scheduling model within these boundaries.

The second constraint (3.13) applies only to the detailed horizon. The constraint ensures that the demand will never exceed the capacity available at any workstation for any period during the detailed horizon. For this, no more than $n \times h$ constraints are required.

The third constraint (3.14) only applies to the **aggregate horizon**. The constraint ensures that the demand will not exceed the capacity available on any aggregate resource during any period t where $h < t \leqslant H$. An aggregate resource is a group of similar resources. For each group of resources w (e.g. similar workstations, fixtures, critical tools, etc.), there is one constraint for each period t where $h < t \leqslant H$. Only groups of similar resources and their overall capacity are considered in the aggregate planning horizon.

The objective function creates and schedules virtual cells in a way that minimizes the total distance flow for all considered jobs. This is so as long as the lateness factor equals 0; otherwise distance flow increases to permit the use of resources which are not as close.

Finally, it is possible to run the scheduling model overnight for generating an aggregate plan for the next daily production. In this case, it suffices to run the model with $h = 0$.

3.4.5 Conclusion

The concept of virtual cells seems promising, however its applicability relies on the development of scheduling algorithms or controllers capable of dealing with complex multi virtual cell interaction.

It is our opinion that a manufacturing system based upon virtual cells would maintain the low investment required for group technology cells while gaining some of the flexibility and efficiency associated with FMS.

A VCM system that uses a sequential scheduling framework would survive. Such a strategy, however, may not be very efficient, since the decision taken at time t may have a negative impact on the future. Careful planning is a must; a near optimal, linear programming-based scheduling algorithm, which permits planning, job splitting and variable rates of introduction over time, has been developed to assess this. An object oriented simulation environment which emulates a virtual cellular manufacturing system has been developed in Smalltalk. This environment is used in conjunction with the model. Results obtained so far are very encouraging: on most measures of performance, the VCM system performs 20% better than GT cell-based systems.

REFERENCES

Aneke, N.A.G. and Carrie A.S. (1986) A design technique for the layout of multi-product flowlines. *International Journal of Production Research*, **24**(3), 471–81.

Askin, R.G. and Chiu, K.S. (1990) A graph partitioning procedure for machine assignment and cell formation in group technology. *International Journal of Production Research*, **28**(8), 1555–72.

Black, J.T. (1991) *The Design of the Factory with a Future*, McGraw-Hill, New York.

Boctor, F.F. (1991) A linear formulation of the machine-part cell formation problem. *International Journal of Production Research*, **29**(2), 343–56.

Burbidge, J.L. (1989) *Production Flow Analysis*, Clarendon Press, Oxford.

Carrie, A.S. (1975) Layout of multi-product lines. *International Journal of Production Research*, **13**(6), 541–57.

Chen, H.-G. (1991) A mixed integer programming model for an operator cyclic walking pattern development. *Computers in Industrial Engineering*, **20**(1), 77–88.

Chen, H.-G. (1992) A Petri-net-based state-transition model for an optimal operator cyclic walking pattern. *International Journal of Production Research*, **30**(10), 2381–99.

Chow, W.S. and Hawaleshka, O. (1992) An efficient algorithm for solving machine chaining problem in cellular manufacturing. *Computers in Industrial Engineering*, **22**(1), 95–100.

Drolet, J.R., Montreuil, B. and Moodie, C.L. (1990) *Virtual Cellular Manufacturing Layout Planning*. Proceedings of the 1990 International Industrial Engineering Conference, San Francisco, CA, pp. 236–41.

Golhar, D.Y. and Stamm, C.L. (1991) "The just-in-time philosophy: A literature review. *International Journal of Productions Research*, **29**(4), 657–76.

Gunasingh, K.R. and Lashkari, R.S. (1991) Simultaneous grouping of parts and machines in cellular manufacturing systems – An integer programming approach. *Computers in Industrial Engineering*, **20**(1), 111–7.

Gupta, T. (1991) Clustering algorithms for the design of a cellular manufacturing system – An analysis of their performance. *Computers in Industrial Engineering*, **20**(4), 461–8.

Ho, Y.C. (1991) A Heuristic Sequence Pattern and Network Based Flow Analysis for Flexible Assembly Systems. Master's Thesis, Purdue University, West Lafayette, IN, USA.

Hocken, R.J. and Nanzetta, P. (1983) Research in automated manufacturing at NBS – An update on the National Bureau of Standards, Facility for Automation Research. *Manufacturing Engineering*, October.

Hollier, R.H. (1983) The layout of multi-product lines. *International Journal of Production Research*, **2**(2), 47–57.

Kusiak, A. (1988) "EXGT-S: A Knowledge-Based System for Group Technology. *International Journal of Production Research*, **26**(5), 887–904.

Lee, C.E.C. (1991) An Integrated Methodology for the Analysis and Design of Cellular Flexible Assembly Systems. PhD Dissertation, Purdue University, West Lafayette, IN, USA.

Logendran, R. (1991a) Impact of sequence of operations and layout of cells in cellular manufacturing. *International Journal of Production Record*, **29**(2), 375–90.

Logendran, R. (1991b) Effect of the identification of key machines in the cell formation problem of cellular manufacturing systems. *Computers in Industrial Engineering*, **20**(4), 439–49.

McLean, C.R., Bloom, H.M. and Hopp, T.H. (1986) *The Virtual Manufacturing Cell*. Proceedings of Fourth IFAC/IFIP Conference on Information Control Problems in Manufacturing Technology, Gaithersburg, MD.

McLean, C.R. and Brown, C.F. (1986) *The Automated Manufacturing Research Facility at the National Bureau of Standards*. Proceedings of the IFIP W.G. 5.7 Working Conference on New Technologies for Production Management Systems, Tokyo, Japan.

Miltenburg, J. and Zhang, W. (1991) A comparative evaluation of nine well-known algorithms for solving the cell formation problem in group technology. *Journal of Operations Management*, **10**(1), 44–72.

Moore, J.M. (1962) *Plant Layout and Design*, Macmillan, New York.

Nagi, R., Harhalakis, G. and Proth, J.-M. (1990) Multiple routings and capacity considerations in group technology applications. *International Journal of Production Research*, **28**(12), 2243–57.

Noy, P.C. (1957) Make the right plant layout – mathematical. *American Machinist*, March, 121–5.

Rardin, R.L. (1992) Discrete optimization, in *Handbook of Industrial Engineering*, (ed. G. Salvendy), Wiley Interscience, Chichester.

Singleton, W.T. (1962) Optimum sequencing of operations for batch production. *Work Study and Industrial Engineering*, 100–10.

Simpson, J.A., Hocken, R.J. and Albus, J.S. (1982) The automated manufacturing research facility of the National Bureau of Standards. *Journal of Manufacturing Systems*, **1**(1), 17–32.

Solberg, J.J., Anderson, D.C., Barash, M.M. *et al.* (1985) Factories of the Future: Defining the Target. A report of a research study conducted under National Science Foundation Grant MEA 8212074, Purdue University, West Lafayette, IN.

Taboun, S.M. Sankaran, S. and Bohle, S. (1991) Comparison and evaluation of similarity measures in group technology. *Computers in Industrial Engineering*, **20**(3), 343–53.

Vengopal, V. and Narendran, T.T. (1992) A genetic algorithm approach to the machine component grouping problem with multiple objectives. *Computers in Industrial Engineering*, **22**(4), 469–80.

CHAPTER 4

Unit load design and its impact on manufacturing systems performance

P.J. Egbelu

4.1 INTRODUCTION

When jobs are released to the production floor for processing in a manufacturing environment, the jobs must be moved between machines to satisfy all their processing needs. In most applications, due to the batch size of the job and other operational constraints, it is not possible to move the entire job as a single, whole unit; the job must be split into sublots called **unit loads**. Thus, it is the unit loads that move between machines instead of the whole job. Exceptions to this occur when the job is composed of a single workpiece, or the number of parts contained in the batch is small enough to justify handling the entire batch as a unit.

When a job is broken down into sublots for the purpose of handling, two or more sublots can be obtained from a batch, depending on the batch size. In material handling, the sublots are referred to as unit loads. A unit load as described is therefore a single component or groups of identical components arranged and restrained so that they can be picked up and moved between points as a single entity (Apple, 1972; Tanchoco and Agee, 1981). In manufacturing, a unit load may be a group of parts handled in a container, tote, box or pallet and moved as a unit. In a distribution system, a unit load may be a package, or several packages held together in a larger box or on a pallet. The concept behind the use of unit loads is to transport together multiple identical parts for the purpose of reducing the frequency of moves between points. By reducing the move frequency, the total move distance, and consequently handling cost, is reduced.

Simple as it is, the concept of unit loads is one of the most noteworthy concepts in manufacturing. It is one of the most economical approaches to the reduction of manufacturing costs, and to building efficiency into material handling. The impact of unit loads pervades several aspects of manufacturing, ranging from the raw material warehouse through production on the shop floor to finished product storage and warehouse. Because material handling is felt throughout the entire spectrum of manufacturing, and given that handling and storage of unit loads is a major component in material handling, it is not very difficult to see that any decision made regarding unit loads will undoubtedly affect several aspects of manufacturing. In particular, decisions on unit loads in a manufacturing system can be felt in the following areas:

- material handling equipment and capacity requirements;
- manpower requirements;
- machine requirements;
- buffer space requirements;
- aisle space requirements;
- storage equipment requirement;
- manufacturing lead time;
- manufacturing cost;
- material flow pattern;
- material tracking.

Each of the areas listed above is discussed briefly below.

4.1.1 Material handling equipment and capacity requirements

One of the input requirements for material handling system selection is the number of moves to be made. This is often provided in the form of a From-To matrix of material flow. The moves are equivalent to the number of unit load transfers to be made between stations. A high transfer volume is suggestive of conveyor type equipment. On the other hand, a low-to-medium volume flow is suggestive of mobile material handling equipment. The number of moves is directly related to unit load flows. Thus, by controlling the size of unit loads, the decision on what material handling equipment is required is influenced automatically. Furthermore, even within any given class of equipment, the capacity is also influenced by the dimensions and weight of the unit load.

In determining the fleet size requirements in cases where mobile handling equipment is recommended, a primary input into the decision process is the number of trips, and consequently the number of unit loads, to be moved between stations. The move frequencies coupled with the distances between stations is used in determining the fleet size requirements.

4.1.2 Manpower requirements

The issue of manpower requirements is directly related to fleet size requirements in situations where mobile transporters (e.g. industrial vehicles) are employed. Consider a situation where human operators are required to operate the transporters: the larger the fleet size, the larger will be the pull of operators required to operate the transporters, and *vice versa.*

4.1.3 Machine requirements

As discussed by Egbelu (1986, 1987b), the relationship between unit load design and machine requirements is not obvious to many people. However, what is obvious is that a machine can only process if there are parts available at the station. The availability of parts at a station is affected by the size of unit loads employed in transferring parts between stations. Unit load transfer is a part consolidation process. The smaller the unit loads, the more frequently parts can be moved and made available to succeeding machines, and thus reduce part starvation at succeeding stations. This will reduce machine idle time. On the other hand, if larger unit loads are used, it is conceivable that while the system waits to accumulate enough parts at a preceding station to build up a unit load, machines at succeeding stations may be starved of parts, and thus increase their idle time. Idle times reduce the production rate of machines. Thus, to meet the same production volume over a fixed period, many more machines would have to be available to process the jobs in order to compensate for the idle periods.

4.1.4 Aisle and buffer space requirements

A unit load may be transferred by stationary equipment such as a conveyor, or mobile equipment such as a vehicle. Consider the case where conveyor type equipment is used: the conveyor transport surface must be of a sufficient width to hold the unit load in order not to have excessive overhang. Excess overhang causes load instability. It is therefore obvious that larger and, consequently, wider unit loads must use wider conveyors. The plant floor area consumed by a conveyor segment is directly related to the conveyor's width and length.

In situations where mobile transporters are used, the minimum turning radius of a vehicle, and consequently the minimum aisle clearance required, is determined by the size of the vehicle and the length and width of the loads being handled. The longer and wider the load, the larger the aisle space required to allow for the free movements of vehicles and loads. Smaller loads are easier to manoeuver in narrow aisles. Similar reasoning applies to the buffer space requirement. To successfully place a load in a buffer, the buffer must be of a sufficient width to allow the load to be deposited. Again, this width requirements is directly related to unit load size.

4.1.5 Storage equipment

Equipment for the storage of unit loads must have a sufficient payload capacity to hold the weights that are to be placed on them. The larger and heavier the unit loads, the higher the payload rating requirements for the storage structure, and *vice versa*. Furthermore, the larger the unit loads, the larger will be the size of the storage cells or slots required to accommodate a load, and *vice versa*. To place a small load in a cell designed for large loads represents a poor utilization of space.

4.1.6 Production lead time

When poorly specified, unit loads can negatively influence the time taken to complete a job in a shop. Often overlooked, unit loads can be silent contributors to production delays, as illustrated in Fig. 4.1. The figure illustrates the throughput time for a production batch of 10 parts to be processed through two machine stations. The processing time per part at each station is five

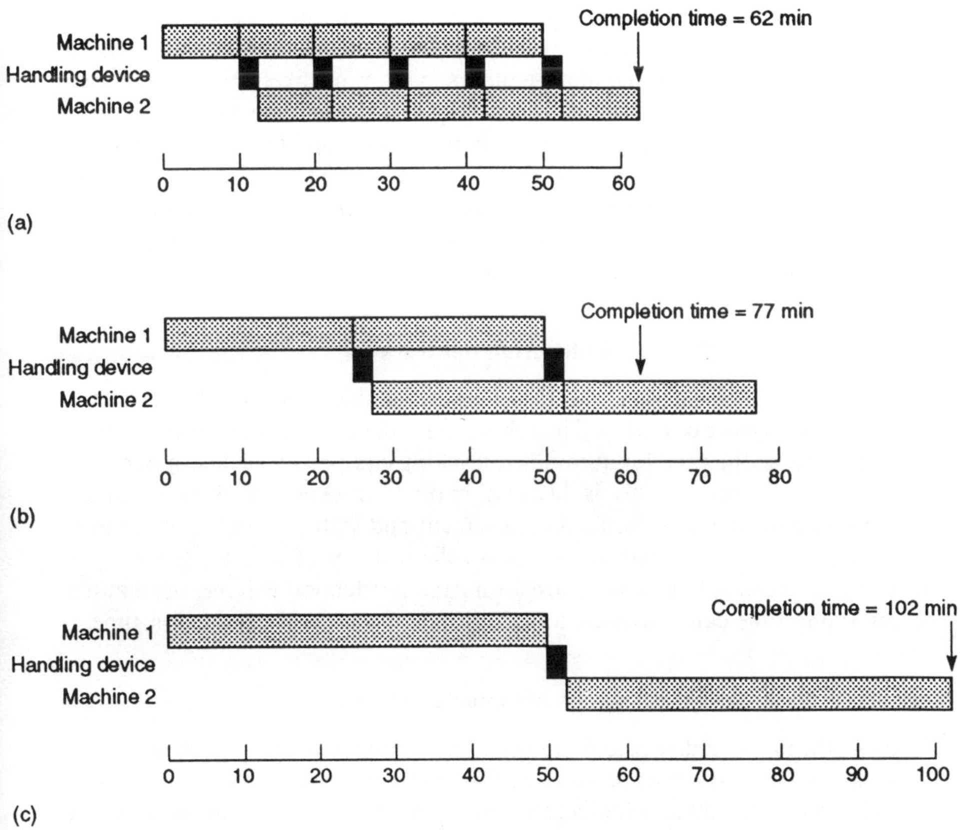

Fig. 4.1 Illustration of the effect of unit load size on completion time and handling frequency in the manufacture of a lot of 10 parts on two machines. Gantt charts based on handling (a) two parts per unit load; (b) five parts per unit load; (c) 10 parts per unit load.

minutes, and it takes two minutes to travel from station 1 to 2. Figure 4.1 simply shows the time required to complete the batch as the unit load size changes. For this particular situation, the optimal unit load size is to move two parts per load, as seen in the graph. Any other unit load size specification increases the completion time. For every product mix in manufacture in a given circumstance, there is a unit load size for each product that minimizes the overall completion time of all batches throughout the shop. It is desirable to find what the optimum unit load size is for each product during each production session.

4.1.7 Material flow pattern

Even though the shop routing for a job is known and may be fixed, the actual path followed in the shop for transporting the parts between stations is dependent upon the type of equipment used. The type of equipment selected is in direct response to the number of moves that would be made between stations. Larger unit load sizes call for fewer moves, and consequently a variable path; mobile handling equipment is more likely to be preferred, assuming that all other factors remain equal. On the other hand, the use of small unit load sizes increases the volume of flow, and thus justifies the use equipment suitable for high volume handling. Such equipment is more likely to be fixed path equipment such as conveyors. By selecting either mobile equipment (e.g. industrial vehicles) or stationary, fixed path equipment (e.g. roller conveyor), the pattern of material flow is automatically affected.

4.1.8 Material tracking

One of the benefits of material tracking is being able to account for the status of each workpiece or unit load in a shop. In moderate-to-high volume production systems, the need for automated tracking equipment, such as a bar code system, is essential. This is because manual tracking involves manually recording on standard forms the movement and status of unit loads or jobs as they progress through the shop. As the number of unit loads increases due to a decrease in unit load size for a fixed production volume, the manual recording time can constitute a high proportion of total production time.

4.1.9 Manufacturing cost

From the previous discussions, the economics of unit load handling is apparent. With the satisfaction of product quality and employee safety requirements, as absolute, manufacturing cost is the ultimate measure of manufacturing efficiency, since in the long run it is the most important factor that determines the competitiveness of a company. It is obvious that manufacturing cost is strongly influenced by decisions taken on unit load size policy. Many factors such as vehicle fleet size, machine and storage equipment requirements and manufacturing lead time all affect cost. This makes unit load size specification a candidate for cost saving opportunities. Figure 4.2 is a summary of the factors affected by unit load size specification, and also those factors which affect the specification of unit load size. By properly analysing the relationships between these factors and the size of unit loads, opportunities for cost savings can be identified.

4.2 CONTAINERS FOR FORMING UNIT LOADS

The time invested in constructing a unit load and the container or supporting structure used in forming a unit load depends upon the intended purpose of the unit load. Unit loads can be formed for internal handling within a shop, for storage and for distribution. These three basic areas of unit load application will be referred to in this chapter as **manufacturing unit loads**, **storage unit loads** and **distribution unit loads**, respectively.

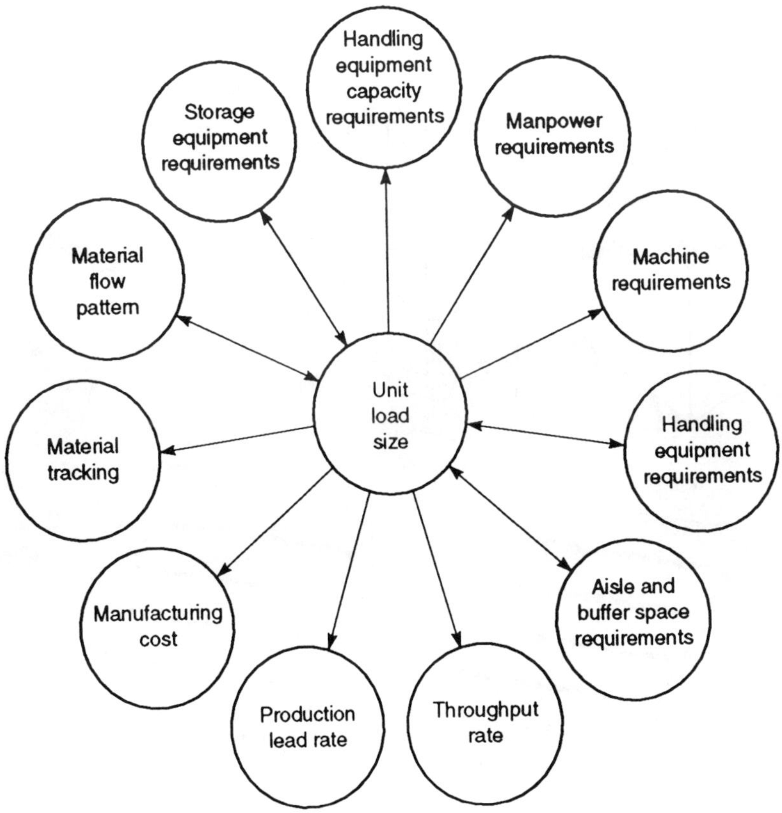

Fig. 4.2 Factors that affect and are affected by unit load size.

In manufacturing, raw materials enter the shop through the warehouse as storage and distribution unit loads from outside suppliers. Once in the shop, further transfers between stations are in the form of manufacturing unit loads. Manufacturing unit loads are formed using various forms of containers. The most common are tote boxes, tote pans, part bins, pallet boxes, skid boxes, racks and dump hoppers. These devices are as illustrated in Fig. 4.3.

Distribution unit loads are used for shipping materials or manufactured products to consumers or places reasonably removed from the plant. They often exist as pallet loads, and are constructed with the objectives of meeting customer specifications, maximizing cube space utilization, strength, stackability, ease of transfer, stability during shipment and the minimization of

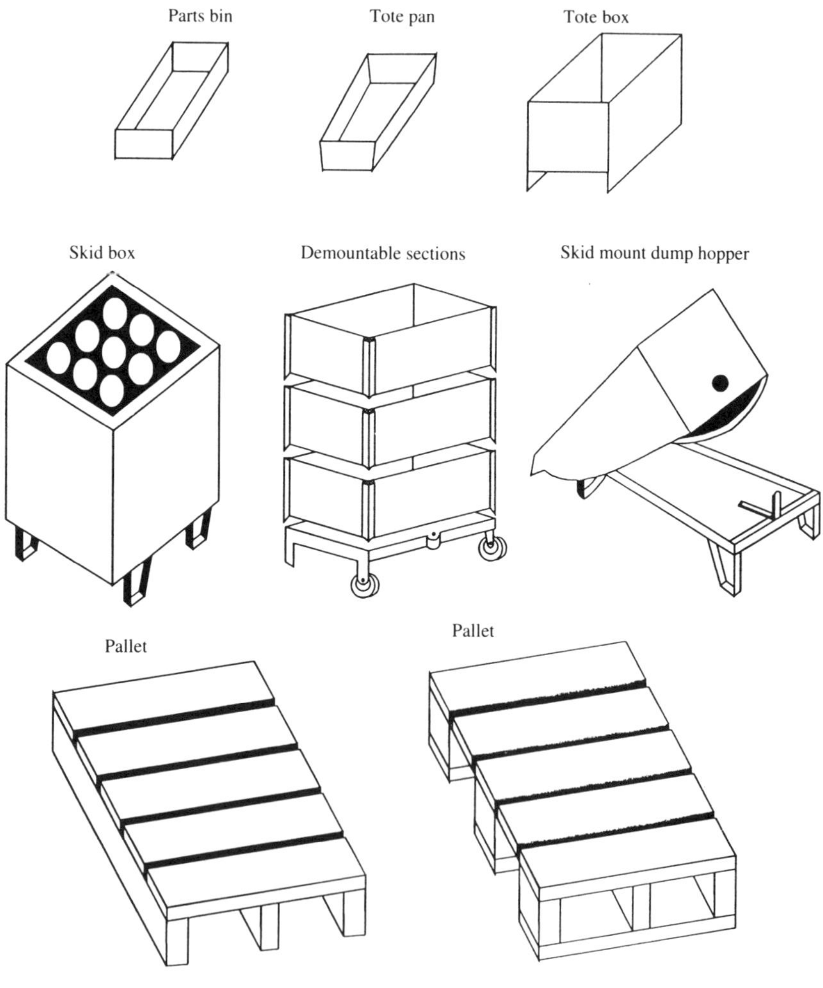

Fig. 4.3 Various forms of container.

shipping cost. Common structures most often used in constructing distribution unit loads are pallets and slip sheets, as shown in Fig. 4.3.

The third type of unit load is the storage unit load. Storage unit loads form the interface between manufacturing unit loads and distribution unit loads. In manufacturing, it is sometimes necessary to move work-in-progress (WIP) parts to storage to be held there until needed. When this is done, the items are usually stored using containers used for manufacturing unit loads. In a similar manner, completed parts are packaged and palletized on the shop floor for distribution. These pallet loads also go into storage until they are released for shipment. In a network of manufacturing facilities, what is shipped as finished product by one plant using a distribution unit load is received as raw material by another plant as a storage unit load. Thus, storage unit loads can be both manufacturing unit loads and distribution unit loads. Figure 4.4 is an illustration of the changing nature of unit loads throughout a network of manufacturing facilities.

Regardless of the type of unit load used, the common thread between them is to maximize material handling, storage and material tracking efficiency. In a manufacturing situation, all three types of unit loads may be in use simultaneously. For example, whereas items on the shop floor may be moved in tote pans or pallet boxes between workstations, the same set of containers may be used for storing in-process parts in WIP automated storage and retrieval systems or miniload systems. A given company may palletize packages of finished parts on pallets for shipment to customers while at the same time using a similar pallet to hold items for internal storage.

4.3 CONFIGURATION OF UNIT LOADS

Unit load configuration refers to the pattern in which the items in a unit load are arranged. Although different part types can be placed and handled together in a unit load, in practice this is not often done. The norm is to make all items in a unit load homogeneous. This implies that parts of different types are not mixed together in a unit load. Some of the benefits for having uniform items in a unit load include:

- ease of tracking unit items;
- ease of unitizing items;
- short item retrieval time – an operator does not have to search for the right item in the mix before locating a desired item since all items are similar;
- better cube space utilization in a unit load – when nonhomogeneous items are held together in a unit load, the efficiency of cube space utilization may be compromised due to improper mating of the items.

To improve cube space utilization in a unit load, the unitized items must be configured in an efficient manner. Again, the effort and care required to

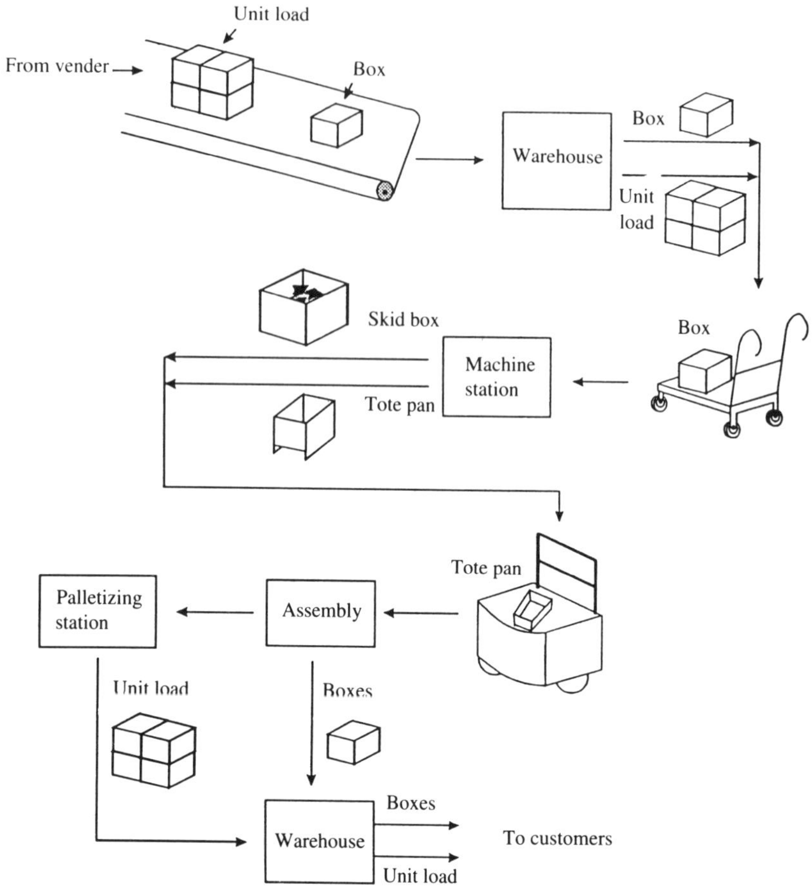

Fig. 4.4 Relationships between forms of unit loads in manufacturing.

configure a unit load depends on whether the unit load is to be used internally within the shop floor, placed in storage or shipped to an outside customer or another plant. Factors that affect the level of investment and effort required to configure a unit load are the distance the items are to be moved, the frequency of the moves and the methods of handling. Of the three types of unit loads, manufacturing unit loads require the least amount of time and effort to configure. In some cases, configuration simply involves throwing the items into the container in which the unit load is formed. No special effort is required in determining the exact loading pattern of unit items. Where the size of the parts prohibits the operator from casually dropping the items into the container, the items may be carefully placed in the container according to some orientation normally determined by intuition. Figure 4.5 is an illustration of typical manufacturing unit loads. The low effort invested

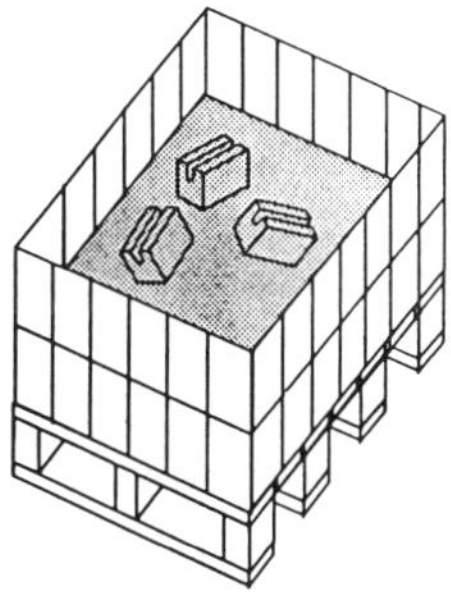

Fig. 4.5 Manufacturing/storage unit load.

in the configuration of the unit items is justified by the short distances the unit loads travel between machine stations, the high frequency of de-unitization and re-unitization at workstations, and the short life-span of unit loads.

Unlike manufacturing unit loads, distribution unit loads are configured according to some preselected pattern. Physical factors that determine the choice of a loading pattern include the dimensions of the container or pallet used, and the shape and dimensions of the items to be unitized. When the items to be unitized are of a regular shape, a geometric loading pattern can be readily determined for unitizing the items. However, when the items to be unitized are unstable, irregular or unstackable, special fixtures must be designed to account for these attributes. One approach to correct for irregularity, instability and unstackability is to pack the items into regular shaped containers, as in packaging situations. It is these regular containers or packages that are eventually formed into unit loads. When configured, a unit load formed on a pallet may appear as shown in Fig. 4.6.

Pallets are commonly used in forming distribution and storage unit loads. Because of their popularity of usage, pallets have been standardized (Apple,

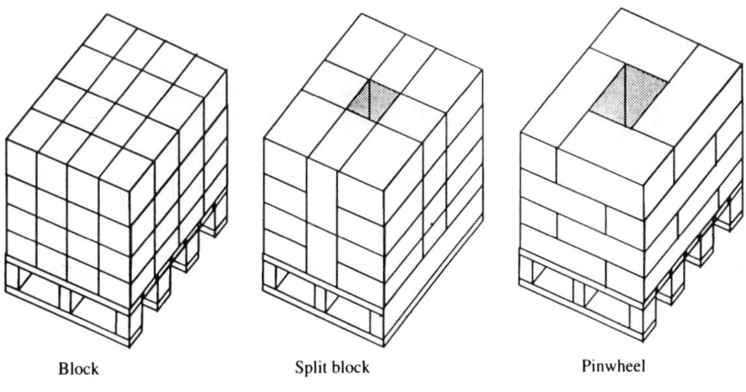

Fig. 4.6 Distribution/storage unit load.

1972) to facilitate intermodal handling from manufacturing, through the distribution pipeline, to the point of use. Standard pallet sizes are as given in Table 4.1 (Apple, 1972). The US Navy has also developed several configuration patterns for each standard pallet size based on the dimensions of the items to be loaded (i.e. palletized) on the pallet (Haynes, 1957; Warehouse Operations, 1966). In the development of the loading patterns, it is assumed that the items have regular shapes and rectangular faces. Although these patterns have been recommended, several algorithms that can generate the loading patterns for any combination of pallet and item dimensions have been developed. Among some of the algorithms reported in this area are those of Hodgson (1982), Han *et al.* (1989), Downsland (1987a,b) and Steudel (1979). Steudel developed an algorithm which, according to the results of his analysis, generates pallet loading patterns about 10% more efficient than those recommended by the US Navy study. These algorithms are essentially the application of the cutting stock problem. For detailed discussions of analytical models that focus on pallet pattern generation, interested readers are referred to the references given above. No further discussion on pallet loading is pursued in this chapter.

The remainder of this chapter uses the term 'unit load' without any special reference on how the items in the unit load are configured. Another key point in the chapter is the frequent use of the term 'unit load size'. Wherever this term is used, it should be understood to mean the number of items or parts contained in a unit load, unless otherwise stated.

Three basic models are presented in the rest of this chapter. The models are presented in sections 4.3, 4.4 and 4.5, respectively. In section 4.3, a simple model for selecting a unit load to maximize warehouse space utilization is presented. In section 4.4, a model extracted from Egbelu (1991) for specifying the optimal unit load size to minimize overall manufacturing time for a batch job is presented. The third model is extracted from Egbelu (1993), and is presented in section 4.5. This model addresses the problem of concurrent

Table 4.1 Standard pallet sizes

Pallet number	*Rectangular pallet sizes (inches)*	*Square pallet sizes (inches)*
1	24 × 32	36 × 36
2	32 × 40	42 × 42
3	36 × 42	48 × 48
4	32 × 48	
5	36 × 48	
6	40 × 48	
7	48 × 60	
8	48 × 72	
9	88 × 108	

specification of unit load and transport vehicle parameters in a manufacturing environment to minimize total manufacturing cost.

4.4 UNIT LOAD DESIGN FOR STORAGE

In warehousing facilities, items placed in storage may exist in pallet loads, tote pans and boxes, or in other forms of containers. In cases where pallet loads are stored, opportunity exists for space optimization through pallet size selection and loaded pallet height specification. The same opportunity exists for other types of storage containers. However, in this section, a procedure to optimize the height of a loaded pallet is described. It is emphasized that the pallet load height optimization problem is a subset of the more general problem of specifying the dimensions of a pallet load or unit load so as to optimize on some user-selected performance measure.

To understand the problem environment, consider the problem of loading regular-shaped boxes or packages on a pallet whose width and length are known. The dimensions of the boxes are also assumed known. If the warehouse maximum stacking height and the container crushing pressure are known, the optimal height of a loaded pallet to maximize warehouse space utilization can be easily determined. To model the problem, the following variables are defined:

H = warehouse maximum stacking height (inches);
h' = height of an empty pallet;
L = length of pallet;
W = width of pallet;
t = height of boxes to be palletized;
s = length of boxes to be palletized;
b = width of boxes to be palletized;
w = weight of a box;
p = crushing pressure of boxes;
t' = maximum allowable height of a unit load.

If items crush under pressure, stacking of loads directly on one another may not be possible if maximum utilization of warehouse cube space is to be realized. To maximize space utilization, the placement of unit loads on shelves is necessary. When shelving is used, the thickness of the horizontal shelf beams and the space allowance between a beam and the top of the unit load placed directly below it must be accounted for in optimizing the unit load height. Suppose d is the thickness of a shelf beam and g is the height of the space allowance between a beam and the top of the load directly below it. Then

$$h = h' + (d + g)X, \tag{4.1}$$

where $X = 1$ if shelving is used, and $X = 0$ otherwise.

From the variable definition, if the boxes are loaded on the pallet with the $s \times b$ side face down, the maximum number of layers of boxes that can be placed on a pallet is n, where

$$n = \min\left[\left(\frac{t' - h}{t}\right)^{-}, \left(\frac{psb}{w}\right)^{-} + 1\right], \tag{4.2}$$

where the notation $[x/y]^{-}$ represents the greatest integer less than or equal to x/y.

Therefore, if j is the number of layers of boxes on each pallet, then $1 \leqslant j \leqslant n$. The objective is to select the number of layers j' such that the warehouse space utilization is maximized. This is equivalent to selecting j' such that

$$F(j') = \max\left\{\left[\frac{H}{h + jt}\right]^{-} j\lambda_{sb}\right\}, \quad 1 \leqslant j \leqslant n, \tag{4.3}$$

where λ_{sb} is the number of boxes in one unit load layer if each box is placed with side $s \times b$ face down. If we set $F(s, b)$ to be the maximum number of boxes stored in one stack of pallet loads if side $s \times b$ boxes are placed face down, then

$$F(s, b) = F(j') \tag{4.4}$$

Similarly, analysis with box side $s \times t$ face down and side $b \times t$ face down must also be performed, unless a face is prohibited from being placed face down on the pallet. The overall optimum pallet load height is selected such that

$$F^* = \max\{F(l, b), F(l, t''), F(b, t'')\} \tag{4.5}$$

Note that if, for example, the $b \times t$ face is prohibited from being placed face down, then $F(b, t) = 0$. The general algorithm for unit load design to maximize warehouse space utilization for stackable loads is as summarized below:

1. Label the three distinct sides of the container types to be palletized. Let the sides be labeled 1, 2 and 3. Set $i = 0$ and $\theta = 0$.
2. Set $i = i + 1$.
3. With side i placed face down, apply the pallet layout pattern optimization algorithm to determine the layout of the first layer of boxes on the pallet.
4. Apply the procedure described to determine the maximum number of boxes $F(i)$ that can be stored on one stack of pallets if side i of boxes are placed face down.
5. If $F(i) > \theta$, set $I = i$ and $\theta = F(i)$. If $i < 3$, go to step (2).
6. Design unit load with side I placed face down. The maximum number of boxes stored per pallet stack is θ.

4.5 UNIT LOAD DESIGN FOR MANUFACTURING

Statistics indicate that about 78% of all manufacturing activities in the US fall under the classification of a batch production system (Grooves, 1987).

In batch production, parts are manufactured in lots that may range from a few parts, say less than 10, to many thousands. With improved manufacturing technology through advanced machine design, production planning and just-in-time (JIT) application, coupled with the need to lower inventory cost, the trend is towards smaller lot manufacture. One of the problems that must be addressed to ensure efficient manufacture of batch jobs is that of material handling between machines. For a given batch, a decision must be made as to whether the entire batch is to be moved as a unit between machines or broken down into sublots or unit loads for the purpose of handling. If the decision to move the batch in sublots is accepted, then the second stage decision of determining the optimal number of parts to handle as a unit load must be made.

As it is common with most manufacturing systems, a production stage visited by a batch job may consist of multiple, heterogeneous machines that differ in performance due to differences in age, make and model. Machines in a particular stage may occupy a particular section of the shop floor, or the stage designation may simply be logical. Although machines in a logical stage are functionally similar, they do not necessarily occupy a unique section of the shop floor. Machines within a logical stage may be located at different sections of a shop or different buildings in a manufacturing complex. Therefore, to move a unit load from one machine stage to another, a decision must be made as to which machine in the next production stage the unit load has to be sent if production efficiency is to be maintained. Such a decision must not only consider the production efficiency of the machines at the succeeding stage, but also the distances between the source machines and the potential succeeding machines.

As discussed section 4.1.6, when material handling effect is considered in production, the size of unit loads adopted in moving a batch through production does affect the completion time for the job. For this reason, when jobs are moved in unit loads through the shop, analysis needs to be performed to determine the best unit load size. In this section, a model for selecting the best unit load size for a batch job while in production to minimize the completion time is presented. The following assumptions are made in the model:

- parts in a unit load are identical;
- parts in a unit load constitute both a handling and processing unit or entity;
- once unitized, a unit load cannot later be broken down into smaller unit loads while in-process or combined with other unit loads to form a larger unit load;
- unit loads are transported between machines one at a time;
- all unit loads of a job are of equal size unless it is impossible to form a full load with the last unit load due to the batch size;
- once completed at a machine in a preceding stage, a unit load immediately proceeds to the succeeding production stage as soon as the material handling system becomes available;

- material transfer between every machine pair in adjacent production stages is possible subject to obeying the routing requirements of the jobs (if material transfer between any pair of machines is prohibited, the associated transfer time between them is set to infinity);
- the operation sequence required by a batch is known;
- each machine has a unit of material handling device that serves to move completed unit loads to their next destination.

The last assumption implies that each machine has a dedicated transportation device assigned to serve its needs in the case of a mobile system. For the case of a stationary system, this assumption implies machines in adjacent stages are all connected by the transportation system. The following variables are used in the model. Additional local variables are defined at different subsections of the chapter as the need arises. Only one batch job is considered in the model. For the case of multiple batch jobs, the interested reader is referred to Egbelu (1986, 1987b, 1990, 1991).

Q = number of parts contained in the batch (i.e. the batch size).

q = number of parts in a full unit load or sublot. This parameter is also referred to as the **unit load size.**

N_q = number of unit loads generated from a batch of size Q if a full unit load contains q parts.

$\bar{N}_q$ = the size of any partial unit load present within the set of the N_q unit loads generated from the batch size Q, given that a full unit load contains q parts.

P_{ij} = processing time per part on the jth machine of stage i.

Z_{ij} = setup time for machine j in stage i.

F_{ij} = preparatory time per unit load on machine j of stage i. This time is independent of unit load size.

$t_{ij(i+1)k}$ = travel time from machine j in stage i to machine k in stage $(i+1)$.

M = number of machine stages.

$\bar{M}(i)$ = set of machines in stage i.

$m_i = |\bar{M}(i)|$, the number of machines in the ith stage.

$A^n_{ij(i+1)k(q)}$ = the arrival time of the nth unit load at machine k in stage $(i+1)$, given that the unit load was sent to machine $k\in\bar{M}(i+1)$ from machine $j\in\bar{M}(i)$ and the size of the unit load is q.

$S^n_{ij(i+1)k(q)}$ = operation start time on the nth unit load on machine $k\in\bar{M}(i+1)$, given that unit load n was sent to machine $k\in\bar{M}(i+1)$ from machine $j\in\bar{M}(i)$ and the size of the unit load is q.

$C^n_{ij(i+1)k(q)}$ = the completion time of the nth unit load on machine $k\in\bar{M}(i+1)$, given that unit load n was sent to machine $k\in\bar{M}(i+1)$ from machine $j\in\bar{M}(i)$ and the size of the unit load is q.

$\bar{T}^n_{(i+1)k}$ = the earliest ready time of machine $k\in\bar{M}(i+1)$ after the arrival of unit load n at machine station $k\in\bar{M}(i+1)$. Note that $\bar{T}^n_{(i+1)k} = A^n_{ij(i+1)k(q)}$ if the machine is idle when unit load n arrives at station $k\in\bar{M}(i+1)$.

$R^n_{ij(i+1)(\ell,k)}$ = the earliest time the material handling device can pick up unit load n from machine $j \in \bar{M}(i)$ for delivery to machine $k \in \bar{M}(i+1)$, given that the unit load that preceded n on machine $j \in \bar{M}(i)$ was sent to machine $\ell \in \bar{M}(i+1)$.

4.5.1 Unit load and processing time submodel

Given that the batch size is Q and a full unit load size is $q \leqslant Q$, the number of unit loads, N_q obtainable from Q is given by (4.6):

$$N_q = [Q/q]^+, \quad 1 \leqslant q \leqslant Q. \tag{4.6}$$

If Q is not a perfect multiple of q, the last unit load will only be partially filled. The size $\bar{N}_q < q$ of any partial unit load is expressed as

$$\bar{N}_q = Q - (N_q - 1)q. \tag{4.7}$$

The total processing time $\bar{P}_{(i+1)k(q)}$ of a unit load of size q on machine $k \in \bar{M}(i+1)$ is defined as

$$\bar{P}_{(i+1)k(q)} = F_{(i+1)k} + qP_{(i+1)k}. \tag{4.8}$$

For a partial unit load, the corresponding processing time is given by

$$\bar{P}_{(i+1)k(\bar{N}_q)} = F_{(i+1)k + \bar{N}_q P_{(i+1)k}}. \tag{4.9}$$

Given that the nth unit load is processed on machine $k \in \bar{M}(i+1)$, the earliest operation start time $S^n_{ij(i+1)k(q)}$ of load n on machine $k \in \bar{M}(i+1)$ is defined by

$$S^n_{ij(i+1)k(q)} = \max\{A^n_{ij(i+1)k(q)}, \bar{T}^n_{(i+1)k}\}, \tag{4.10}$$

where machine $j \in \bar{M}(i)$ is the source of unit load n from the preceding stage, and $\bar{T}^n_{(i+1)k}$ is the completion time of the unit load that immediately preceded unit load n on machine $k \in \bar{M}(i+1)$. For $i = 0$, $A^n_{ij(i+1)k(q)} = 0$ for all $k \in \bar{M}(i+1)$. Also, if unit load n is the first to be processed on machine $k \in \bar{M}(i+1)$, then $\bar{T}^n_{(i+1)k} = Z_{(i+1)k}$. The completion time of unit load n on machine $k \in \bar{M}(i+1)$ is given by

$$C^n_{ij(i+1)k(q)} = \begin{cases} S^n_{ij(i+1)k(q)} + \bar{P}_{(i+1)K(q)}, & \text{for a full load} \\ S^n_{ij(i+1)k(q)} + \bar{P}_{(i+1)k(\bar{N}_q)}, & \text{for a partial load.} \end{cases} \tag{4.11}$$

The completion time $C^n_{ij(i+1)k(q)}$ will occasionally be denoted simply as $C^n_{\cdot\cdot(i+1)k(q)}$ when reference to the source machine $j \in \bar{M}(i)$ is unimportant.

4.5.2 Material handling submodel

Common material handling devices used for unit load transfer between machines in adjacent production stages can be classified broadly into two main categories:

- stationary or fixed position equipment (e.g. conveyors);
- mobile equipment (e.g. trucks, AGVs, etc.).

Under normal operating conditions, stationary equipment serving a machine can be assumed to be continuously available to transport a unit load to its next destination as soon as processing on the unit load is completed. This is not true with respect to a mobile equipment. For a mobile material handling system, the equipment must make a complete round trip between the source station and the destination station before the equipment becomes available again for another load pickup at the source station. For the sake of conciseness, only mobile handling systems will be considered in the model presented here. Models for stationary handling can be similarly developed. Readers interested in stationary handling systems are referred to Egbelu (1990, 1991). Although the material flow is unidirectional for the problem considered in this chapter, the movement of the material handling system itself is bidirectional. Because of the bidirectional assumption, the distance between pairs of machines in adjacent stages can be symmetric or nonsymmetric. Figure 4.7 depicts the type of production system modeled.

Given the operating characteristics of mobile material handling systems as described above, the arrival times of unit loads at a station and the ready times of the material handling system can be expressed algebraically. Given that unit load n was completed at time $C^n_{\cdot\cdot ij(q)}$ on machine $j \in \bar{M}(i)$, the arrival time of the load at station $k \in \bar{M}(i+1)$ is given by

$$A^n_{ij(i+1)k(q)} = \max\{C^n_{\cdot\cdot ij(q)}, R^n_{ij(i+1)(\ell,k)}\} + t_{ij(i+1)k}. \tag{4.12}$$

If unit load n was the first to be processed on machine $j \in \bar{M}(i)$, then $R^n_{ij(i+1)(\ell,k)} = 0$ since $\ell = 0$. From equation (4.12) it is obvious that the departure time of unit load n from station $j \in \bar{M}(i)$ is equal to $D^n_{ij(i+1)k(q)}$, where

$$D^n_{ij(i+1)k(q)} = \max\{C^n_{\cdot\cdot ij(q)}, R^n_{ij(i+1)(\ell,k)}\}. \tag{4.13}$$

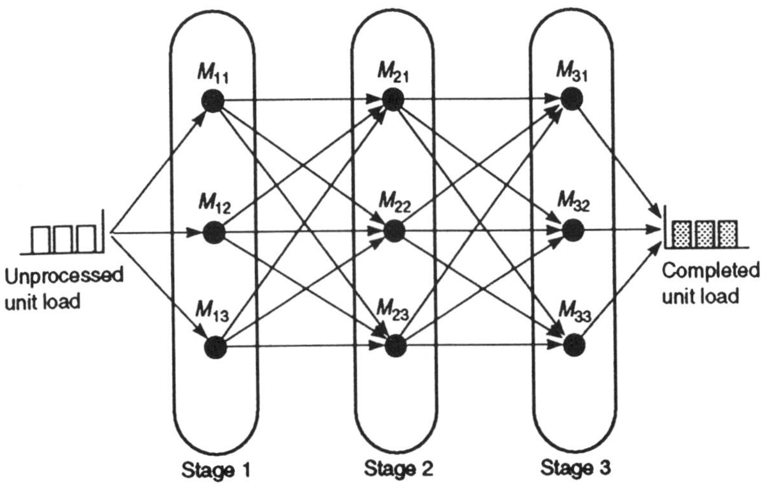

Fig. 4.7 Network flow representation of a batch job, where M_{hk} = the kth alternative machine in stage h.

Again, the ready time of the material handling system at station $j \in \bar{M}(i)$ after the pickup of unit load n from station $j \in \bar{M}(i)$ is given by

$$R^{m}_{ij(i+1)(k,\ell)} = \begin{cases} D^{n}_{ij(i+1)k(q)} + 2t_{ij(i+1)k}, & \text{for symmetric intermachine distances} \\ D^{n}_{ij(i+1)k} + t_{ij(i+1)k} + t_{(i+1)kij}, & \text{for nonsymmetric intermachine distances,} \end{cases} \tag{4.14}$$

where m refers to the unit load that will follow n on machine $j \in \bar{M}(i)$, and ℓ refers to any machine in stage $(i+1)$ to which load m may be routed.

4.5.3 Solution approach

For each unit load, a solution to the model will determine the following: the best machine routing; the arrival time at each machine visited; the operation start time; the operation completion time; and the departure time from each station visited.

From the relationships expressing the arrival time, operation start time and the completion time of a given unit load at a given machine station and stage, it is obvious that these parameters depend also upon the unit load departure time from the source machine at the preceding stage. The departure time itself is a function of the completion time and the material handling system ready time at the preceding machine. Thus, given a set of machines visited by a unit load, the completion time at each machine visited can be computed recursively using equations (4.10)–(4.13). The presence of the recursive relationships makes the use of dynamic programming (DP)-based heuristics very attractive as a solution approach to the problem. The application of the DP-based algorithm is also suggestive by the fact that the problem can be decomposed into stages and states. Each production stage corresponds to a DP stage, while the machines in each stage serve as states. Given that the number of sublots or unit loads to be processed is known, the problem of finding the shortest shop time for the batch is equivalent to that of finding the shortest completion time path for each unit load and picking the longest of these paths as the shop time.

The shortest path for each unit load is found using an algorithm similar to that commonly employed for project planning under the CPM (critical path method) or PERT (project evaluation and review technique) (Gido, 1985; Phillips *et al.*, 1976).

The best unit load size can be found by searching over the feasible range of unit load sizes for the batch. For each value of unit load size generated by the search routine, the number of unit loads from the batch size of Q is computed. Given the number of unit loads and the size of each load, the DP algorithm is invoked to return the minimum manufacturing time corresponding to that unit load size decision. Thereafter, a new search value for the unit load size is generated to invoke the dynamic programming algorithm again. For each search value of unit load size generated, the corresponding minimum manufacturing time for the entire batch is determined. The search value with the minimum overall batch manufacturing time becomes the adopted and the best unit load size decision.

Because the search on the unit load size is unidimensional and the batch completion time function is unimodal on unit load size, the one dimensional search algorithm known as the Golden Section (Beightler, Phillips and Wilde, 1979; Himmelblau, 1972) is used here. The DP-based heuristic algorithm is imbedded within the unidimensional search algorithm.

(a) Unidimensional search algorithm

The following additional variables are used in the algorithms:

$\bar{a}$ = size of the smallest unit load permitted in the shop.
$\bar{b}$ = size of the largest unit load permitted in the shop.
$S_{(q)}$ = the best schedule determined when a unit load of size q is used. The best schedule is defined as the schedule that yields the smallest determined completion time with a unit load size of q.
$T_{(q)}$ = the completion time associated with schedule $S_{(q)}$.

The detailed unidimensional search algorithm is as described below:

1. If $|\bar{M}(M)| > 1$, create a dummy last stage $M = M + 1$ and set the following parameters associated with the dummy stage as follows: $|\bar{M}(M)| = 1$; $P_{M1} = 0$; $Z_{M1} = 0$; $F_{M1} = 0$; and $t_{(M-1)jM1} = t_{M1(M-1)j} = 0$ for all $j \in \bar{M}(M-1)$.
2. Set $L = \bar{a}$ if $\bar{a}$ is given. Otherwise, set $L = 1$.
3. Set $U = \bar{b}$ if $\bar{b}$ is given. Otherwise, set $U = Q$.
4. Set $q_1 = [L + 0.38(U - L)]^+$, where $[x/y]^+$ is the lowest integer greater than or equal to x/y.
5. With $q = q_1$, calculate N_{q_1} and $\bar{N}_{q_1}$ according to equations (4.6) and (4.7).
6. Using parameters q_1, N_{q_1} and $\bar{N}_{q_1}$ calculated in steps (4) and (5), solve for the minimum completion time schedule using the DP-based algorithm (described in the next section). Let $S_{(q_1)}$ and $T_{(q_1)}$ be the output of the dynamic programming algorithm associated with the input parameters q_1, N_{q_1} and $\bar{N}_{q_1}$.
7. Set $q_2 = [L + 0.62(U - L)]^-$, where $[x/y]^-$ is the highest integer less than or equal to x/y. If $q_2 = q_1$, go to step 12.
8. With $q = q_2$, calculate N_{q_2} and $\bar{N}_{q_2}$ according to equations (4.6) and (4.7).
9. Using parameters q_2, N_{q_2} and $\bar{N}_{q_2}$, solve for the minimum completion time schedule as described in step 6. Again, let $S_{(q_2)}$ and $T_{(q_2)}$ be the output of the DP algorithm.
10. If $T_{(q_1)} \leqslant T_{(q_2)}$, set $U = q_2$, $q^* = q_1$, $T_{(q^*)} = T_{(q_1)}$, $S_{(q^*)} = S_{(q_1)}$, $N_{q^*} = N_{q_1}$ and $\bar{N}_{q^*} = \bar{N}_{q_1}$. Otherwise, set $L = q_1$, $q^* = q_2$, $T_{(q^*)} = T_{(q_2)}$, $S_{(q^*)} = S_{(q_2)}$, $N_{q^*} = N_{q_2}$ and $\bar{N}_{q^*} = \bar{N}_{q_2}$.
11. Go to step 4.
12. With the interval of uncertainty $[L, U]$ narrowed, evaluate all unit load sizes, $q = [L, U]$ and set $q^* = \bar{q}$, where $T_{(q^*)} = T_{(\bar{q})} \leqslant T_{(q)}$ for all $q = [L, U]$. The best solution corresponds to q^*, N_{q^*}, N_{q^*}, $S_{(q^*)}$ and $T_{(q^*)}$.

(b) Dynamic programming algorithm

The input into the DP-based heuristic algorithm are the parameters q_1, N_{q_1} and $\bar{N}_{q_1}$, or q_2, N_{q_2} and $\bar{N}_{q_2}$, defined in steps (4), (5), (7) and (8) of the unidimensional search algorithm. These parameters will simply be identified as q, N_q and $\bar{N}_q$ in this algorithm. The algorithm proceeds as follows:

1. Set $T_{\max} = 0$ and let the unit load number $n = 1$.
2. (a) Set the stage number $i = 1$.
   (b) Set $\bar{q}$ = the size of unit load n.
3. If $i \geqslant 2$, go to step (4). Otherwise, for each machine $j \in \bar{M}(i)$, compute:
   (a) $A^n_{00ij(\bar{q})} = 0$.
   (b) $S_{00ij(\bar{q})} = \bar{T}^n_{ij}$, where $\bar{T}^n_{ij} = Z_{ij}$ if machine $j \in \bar{M}(i)$ has not previously been assigned a unit load.
   (c) $\bar{P}_{ij(\bar{q})} = F_{ij} + \bar{q}P_{ij}$.

   $$C^n_{00ij(\bar{q})} = S^n_{00ij(\bar{q})} + \bar{P}_{ij(\bar{q})}.$$

   (d) Record the results computed in steps (a) through (c) and store in a working array against machine $j \in \bar{M}(i)$.
   (e) Go to step (6).
4. For each machine $k \in \bar{M}(i)$ and over all preceding machines $j \in \bar{M}(i-1)$, compute:
   (a) $D^n_{(i-1)jik(\bar{q})} = \max\{C^n_{\cdot\cdot(i-1)j(\bar{q})}, R^n_{(i-1)ji(\ell,k)}\}$.
   Note that $R^n_{(i-1)ij(\ell,k)} = 0$ if no unit load had been previously assigned to machine $j \in M(i-1)$. All machines start initially with no assignment. Machine ℓ is any machine in $\bar{M}(i)$, including the zero or dummy machine.
   (b) $A^n_{(i-1)jik(\bar{q})} = D^n_{(i-1)jik(\bar{q})} + t_{(i-1)jik}$.
   (c) $R^m_{(i-1)ji(k,\ell)} = A^n_{(i-1)jik(\bar{q})} + t_{ik(i-1)j}$, where $R^m_{(i-1)ji(k,\ell)}$ is the ready time of the material handling equipment at machine $j \in \bar{M}(i-1)$ and m is the unit load that will follow n immediately on machine $j \in \bar{M}(i-1)$. The index ℓ refers to any machine in $\bar{M}(i)$ that load m may be sent to.
   (d) $S^N_{(i-1)jik(\bar{q})} = \max\{A^n_{(i-1)jik(\bar{q})}, \bar{T}^n_{ik}\}$, where $\bar{T}^n_{ik} = Z_{ik}$ if no unit load had been previously assigned to machine $k \in \bar{M}(i)$.
   (e) $\bar{P}_{ik(\bar{q})} = F_{ik} + \bar{q}P_{ij}$.
   (f) $C^n_{(i-1)jik(\bar{q})} = S^n_{(i-1)jik(\bar{q})} + \bar{P}_{ik(\bar{q})}$.
5. (a) Determine the source machine $h \in \bar{M}(i-1)$ that yields the earliest completion time of unit load n on machine $k \in \bar{M}(i)$. Algebraically, this is equivalent to identifying the machine $h \in \bar{M}(i-1)$ such that

   $$C^n_{(i-1)hik(\bar{q})} = \min_{\forall j \in \bar{M}(i-1)} \{C^n_{(i-1)jik(\bar{q})}\}.$$ Break ties arbitrarily.

   (b) Record against machine $k \in \bar{M}(i)$ in a working array, the best source machine $h \in \bar{M}(i-1)$ and the associated parameters $R^m_{(i-1)hi(k,\ell)}$, $A^n_{(i-1)hik(\bar{q})}$, $S^n_{(i-1)hik(\bar{q})}$ and $C^n_{(i-1)hik(\bar{q})}$.
6. Set $i = i + 1$. If $i \leqslant M$, go to step (3).

7. Set $T_{max} = C^n_{(M-1)hM1}$ since there is only one machine in the last stage.
8. Trace back the earliest completion time path for unit load n. For each machine $k \in \bar{M}(i)$, $i = 1, 2, \ldots, M$ in the path, record in the schedule file $S_{(\bar{q})}$, the results $A^n_{(i-1)hik(\bar{q})}$, $S^n_{(i-1)hik(\bar{q})}$, $C^n_{(i-1)hjik(\bar{q})}$, h, and the material handling system ready time $R^m_{ik(i+1)(h',\ell)}$ computed at machine $h' \in \bar{M}(i+1)$ where $h' \in \bar{M}(i+1)$ follows machine $k \in \bar{M}(i+1)$ immediately on the path for unit load n.
9. Set $n = n + 1$. If $n \leqslant N_q$, go to step (2).
10. A complete schedule $S_{(\bar{q})}$ for all unit loads has been constructed. Set $T_{(\bar{q})} = T_{max}$ and return to the unidimensional search algorithm.

(c) Illustrative problem

To illustrate the unit load routing algorithm, it has been applied in the production planning of a job containing 100 pieces of identical parts. Five operations are to be performed on each part, with each operation requiring a separate production stage. The stages are numbered according to the operations. There are three machines each in stages 1, 4 and 5. Stages 2 and 3 each have four machines. Processing times P_{ij} per part on the machines are given in Table 4.2. Unit load preparatory times F_{ij} at machine stations are given in Table 4.3. Setup time requirements for the machines are summarized in Table 4.4. The data of Tables 4.2, 4.3 and 4.4 are drawn from the uniform distribution with parameters of $(a, b) = (0.5, 3.0)$, $(a, b) = (1.0, 3.5)$ and $(a, b) = (20, 50)$, respectively. Data on material handling times between pairs of machines in adjacent production stages are given in Table 4.5. It is assumed

Table 4.2 Part processing times

	Stage no.				
Machine	$i=1$	$i=2$	$i=3$	$i=4$	$i=5$
m_{i1}	1.98	2.64	2.12	1.98	2.35
m_{i2}	1.49	1.22	1.97	2.38	2.93
m_{i3}	2.42	0.77	2.27	1.53	2.00
m_{i4}	–	1.18	0.64	–	–

m_{ij} = jth machine in stage i.

Table 4.3 Unit load preparatory time at machine stations

	Stage no.				
Machine	$i=1$	$i=2$	$i=3$	$i=4$	$i=5$
m_{i1}	1.80	2.47	2.61	2.35	2.54
m_{i2}	1.92	3.30	3.07	2.56	2.68
m_{i3}	2.26	1.97	2.52	1.66	2.72
m_{i4}	–	1.02	2.05	–	–

Table 4.4 Setup times at machine stations

	Stage no.				
Machine	$i=1$	$i=2$	$i=3$	$i=4$	$i=5$
m_{i1}	39.01	48.65	29.46	41.15	21.06
m_{i2}	32.21	43.36	31.20	24.88	28.00
m_{i3}	20.90	46.35	26.42	41.09	49.00
m_{i4}	–	21.45	37.15	–	–

that the travel times between machines are symmetric (i.e. $t_{ij(i+1)k} = t_{(i+1)kij}$). Given the problem scenario described above, what are the best unit load size and unit load routing policies that will minimize the completion time of the job?

Using the algorithm, the earliest completion time for the job was determined to be 218.44 time units. Figure 4.8 is a summary of the material flow pattern and routing for the 17 unit loads. A full unit load contains six parts while the single partial load contained four parts. By tracing the machines visited by each unit load through all the stages, the routing for each unit load is uniquely identified. Figure 4.8 also gives the unit load sequencing on each machine. For example, the unit load sequencing on machine 4 in stage 2 is (1, 3, 6, 9, 12, 16), while the routing for unit load 9 is ($m_{11} \rightarrow m_{24} \rightarrow m_{34} \rightarrow m_{42} \rightarrow m_{51}$). The routing for all other unit loads are similarly read.

The model presented illustrates one way in which unit load design can be optimized to improve manufacturing efficiency. Models based on other design criteria such as cost and equipment requirements (machines and transporters) and configurations can be developed. The opportunity for cost reduction through unit load selection exists in all phases of material flow, as this chapter has attempted to demonstrate.

4.6 SIMULTANEOUS SPECIFICATION OF UNIT LOAD AND TRANSPORT VEHICLE PARAMETERS FOR MINIMUM COST MANUFACTURE

In this section, we present a procedure aimed at minimizing total manufacturing cost through the simultaneous specification of unit load and transporter parameters. Unlike the previous model in which the focus was only on the unit load size specification, in this section, the relationship between unit load size, transporter fleet size and payload capacity are jointly integrated into a decision model which, when solved, yields the best unit load size that optimizes on both the fleet size of mobile transporters to employ and the payload requirement of the transporters such that total manufacturing cost is minimized. Although the model assumes a job shop environment that uses automated guided vehicles (AGVs) as mobile transporters, it is generic enough to be

Table 4.5 Material handling time* between machines in adjacent stages

From machine	*To machine*													
	$i=1$				$i=2$				$i=3$			$i=4$		
	m_{21}	m_{22}	m_{23}	m_{24}	m_{31}	m_{32}	m_{33}	m_{34}	m_{41}	m_{42}	m_{43}	m_{51}	m_{52}	m_{53}
$m_{(i-1)1}$	6.43	9.14	10.67	6.26	9.57	11.64	5.18	6.27	10.52	6.44	11.80	10.96	7.49	5.78
$m_{(i-1)2}$	11.10	7.73	5.93	9.49	5.54	11.49	9.50	11.08	10.06	8.03	9.97	10.98	11.97	9.06
$m_{(i-1)3}$	5.22	6.57	11.55	9.21	7.94	10.85	8.39	9.34	6.60	5.88	7.68	11.99	10.34	11.19
$m_{(i-1)4}$	–	–	–	–	8.81	11.70	8.90	6.11	9.47	9.29	8.71	–	–	–

*Symmetric times for mobile handling systems.

m_{11}	3	6	9	12	16	
m_{12}	2	4	7	10	11	14
m_{13}	1	5	8	13	15	17

m_{21}	13	17				
m_{22}	5	7	11	15		
m_{23}	2	4	8	10	14	
m_{24}	1	3	6	9	12	16

m_{31}	2	5	8	11	15	
m_{32}	6	13				
m_{33}	4	10	14	17		
m_{34}	1	3	7	9	12	16

m_{41}	1	2	4	6	8	11	13	15
m_{42}	5	9	12	16				
m_{43}	3	7	10	14	17			

m_{51}	3	5	9	12	16			
m_{52}	7	10	14	17				
m_{53}	1	2	4	6	8	11	13	15

Fig. 4.8 Unit load routing and machine sequencing (mobile system).

used for any kind of manufacturing system involving multiple workstations and using any kind of mobile transporters other than AGVs.

Current procedures in the specification of vehicle payload capacity, and consequently the fleet size, implicitly assume that the unit load decision has already been made and is firm. The vehicle selection process is made to fit the unit load decision. Thus, the problem of the vehicle system designer is already constrained, therefore limiting his design flexibility. The lower the flexibility in design, the lower the opportunity for creativity and manufacturing cost reduction.

In a similar manner, the unit load system designer is limited by some implicit assumptions. Most unit load designers implicitly assume a given transporter payload capacity to guide their design decisions. Based on the assumed payload, the unit load size is obtained. In this respect, the unit load designer assumes an *a priori* decision on vehicle size, just as the vehicle system designer assumes an *a priori* decision on unit load size. Thus, the current design procedures of unit loads and transport systems are isolated, as illustrated in Fig. 4.9(a). What is required is a design procedure that clearly recognize the interdependency between the two problem areas, as shown in Fig. 4.9(b).

4.6.1 Development of cost models

As discussed in section 4.1 of this chapter, the unit load size decision affects total manufacturing cost through its effect on vehicle cost, overhead cost, inventory cost and pallet (container) cost. Development of a model which, when solved, will yield the best design parameters on unit load size, vehicle size and fleet size requires that the cost contributions from the various sources be integrated into a single design model. In the subsections that follow, cost models from the various sources of costs are developed. The models are eventually integrated to form the objective function of a decision model for optimizing on unit load and transporter parameters.

(a) Unit load cost model

A unit load as used in this chapter refers to a container, box, bin, pallet, etc. with its part contents. The parts in a unit load are assumed to be homogeneous. The size of the unit load for a given part is defined in terms of the number of parts in the unit load. The number of parts in a unit load is constrained by the volume and the weight limitations of the container. Thus, if q_{jk} is the unit load size for part j if container type k is used, then

$$q_{jk} = \min\left\{\left[\frac{\Psi_{jk} l_k b_k h_k}{v_j}\right]^-, \left[\frac{W_k}{w_j}\right]^-\right\}, \tag{4.15}$$

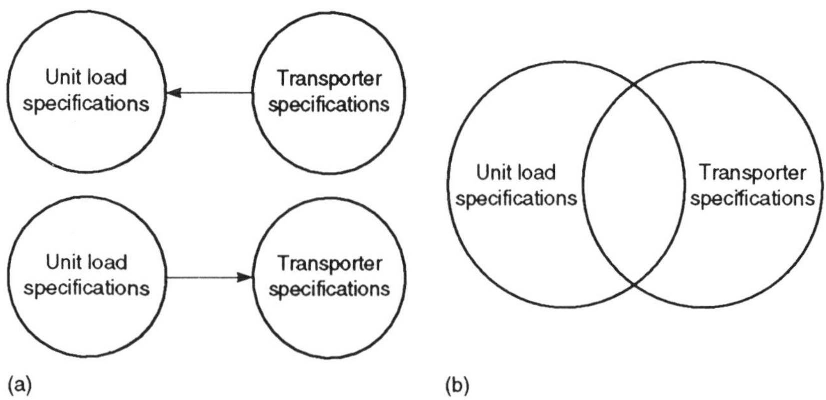

Fig. 4.9 Integrated design concept for unit loads and transporters.

where Ψ_{jk} = fill ratio for container type k with part type j; l_k, b_k, h_k = length, breadth and height of container k; W_k = weight limit that container type k can support; v_j, w_j = volume and weight per part of type j.

Again, the notation $[a/b]^-$ implies the integer portion of the ratio. The size $q_{jk} \geqslant 1$ for a valid unit load to form.

If container type k is adopted and part type j is manufactured in batch sizes of Q_j units, then the total number of unit loads of part j to be obtained per batch is given by U_{jk}, where

$$U_{jk} = [Q_j/q_{jk}]^+. \tag{4.16}$$

The relation $[a/b]^+$ refers to the next higher integer greater than a/b if the ratio is a noninteger. For ease of handling and storage, standardization of containers, whenever possible, is a desirable objective. To select an optimal container, a decision must be made from a set of K containers.

Containers can be designed for single or multiple use: the nature of operation dictates the desirable form. Associated with a container k is a cost. If the life expectancy and the usage frequency of a container is known, the expected cost per usage Z_k is a function of initial container cost c^k, life expectancy n^k, annual maintenance cost a^k, the number of use per year p^k, and the salvage value s^k of the container at the end of its economic life n^k, i.e.

$$Z_k = f(c^k, s^k, n^k, a^k, p^k)$$

For example, if the interest rate per dollar per year is i_0, then Z_k can be estimated as in equation (4.17) (Thuesen and Fabrycky, 1984; White, Agee and Case, 1977):

$$Z_k = [(c^k - s^k)(A/P, i_0, n^k) + s^k i_0 + a^k]/p^k, \tag{4.17}$$

where

$$(A/P, i_0, n^k) = \left[\frac{(1+i_0)^{n^k} - 1}{i_0(1+i_0)^{n^k}}\right].$$

Other methods of estimating container cost such as regression analysis may be employed.

(b) Calculation of vehicle cost

A driving force in selecting a vehicle for material handling purposes is the weight of the items or unit loads to be handled. The vehicle selected must have a rated capacity of no less than the weight of the heaviest load to be transported. Several factors influence the cost of an AGV. These include the electronic control system ϕ, guidance method g, rated capacity r, reliability t, accuracy a_c, and method of handling e – push, pull or on-board. Thus, if F_v represents the cost of a vehicle, then F_v can be represented as

$$F_v = f(\phi, g, r, t, a_c, e).$$

If the assumption is made that all vehicles under consideration have the same basic attributes except on the rated capacity, the cost per vehicle simplifies to

$$F_v = f(r).$$

The rated vehicle capacity r can be mapped to a container k that can support a weight of $W_k = r$. Thus, F_v can be described in terms of k and have F_v redefined to F_k (i.e. $F_k = F_v$). If the life expectancy of the vehicle γ_k, the salvage value δ_k at the end of its economic life, the annual maintenance cost π_k and the number of hours H of the operation annually are known, then Φ_k, the hourly cost of operating a vehicle with a rated capacity of $r = W_k$, is given by

$$\Phi_k = \frac{[(F_k - \delta_k)(A/P, i_0, \gamma_k) + \delta_k i_0 + \pi_k]}{H}. \tag{4.18}$$

(c) Inventory cost

As previously illustrated, unit load size affects the completion time of jobs, and consequently the work-in-process (WIP). Suppose $\bar{Q}_{jk}$ (unit loads) denotes the average WIP inventory for job j when container type k is used. Suppose J represents the number of part types, i'_0 the average daily holding cost per unit money, and C_w the average worth of a WIP part, then the total daily cost ξ_k due to WIP if container type k is

$$\xi_k = i'_0 C_w \sum_{j=1}^{J} q_{jk} \bar{Q}_{jk}. \tag{4.19}$$

(d) Overhead cost

Manufacturing overhead cost represents a constant source of cost that is incurred as long as the shop remains open. Some components of the overhead are insensitive to the volume of production. Factors contributing to overhead cost include supervision, air conditioning and heating, water and lighting, to mention a few. In this paper, the manufacturing overhead is assumed to be directly proportional to the time required to complete the batch of jobs. If T_k is the production makespan for the batch of jobs and Ω the hourly shop overhead rate, the total overhead σ_k incurred in manufacturing the jobs is

$$\sigma_k = \Omega T_k. \tag{4.20}$$

4.6.2 Development of integrated unit load and transporter parameter specification model

In the development of the unit load selection model, the following assumptions are made:

- once unitized the contents of a unit load remain together until completion;
- items in a unit load are treated as a single entity for both processing and handling purposes;

- the shop adopts a standard container; therefore, only one container needs to be adopted.
- only one AGV type can be adopted;
- a vehicle can transport one unit load at a time;
- the job mix is known and represents the job mix and workload profile of the shop;
- data on job routing, processing time and lot sizes are known;
- the machine configuration at each machine station is known.

In addition to the variables defined in earlier section, the following variables are used in the model:

T = total hours per day the shop operates;
N_k = number vehicles required if container type k is selected;
K = number of container types from which one is to be selected;
X_k = 1, if container type k is selected; 0, otherwise.

Given the cost submodels above and given that container type k is selected, the total production cost for a batch is as given by (4.21):

$$\text{Cost} = \sum_{j=1}^{J} Z_k U_{jk} + N_k \Phi_k T_k + \xi_k \left(\frac{T_k}{T} \right) + \Omega T_k,$$

or

$$\text{Cost} = \sum_{j=1}^{J} Z_k U_{jk} + N_k \Phi_k T_k + \Omega T_k + i'_0 \left(\frac{T_k}{T} \right) C_w \sum_{j=1}^{J} q_{jk} \bar{Q}_{jk}. \tag{4.21}$$

The variable T represents the number of hours the shop remains open per day. Thus, the selection model simplifies to solving the following problem:

Minimize

$$P = \sum_{k=1}^{K} \left[\left(\Phi_k N_k + \Omega + \frac{i'_0}{T} C_w \sum_{j=1}^{J} q_{jk} \bar{Q}_{jk} \right) T_k + \sum_{j=1}^{J} Z_k U_{jk} \right] X_k \tag{4.22}$$

such that

$$\sum_{k=1}^{K} X_k = 1 \tag{4.23}$$

$$\sum_{k=1}^{K} q_{jk} U_{jk} X_k = Q_j, \quad \forall_j \tag{4.24}$$

$$1 \leqslant q_{jk} \leqslant Q_j, \quad \forall_j \tag{4.25}$$

$$X_k = 0 \text{ or } 1 \tag{4.26}$$

$$q_{jk}, U_{jk} \text{ integer for } \forall_j, k.$$

The above mathematical program is unconstrained on T_k, the length of time required to complete the J jobs if container type k is used. In many production instances, the job completion time is subject to control. Suppose T_0 is an upper bound on T_k, then a constraint of the following form will be added:

$$\sum_{k=1}^{K} T_k X_k \leqslant T_o. \tag{4.27}$$

The above mathematical program is self-explanatory. Equation (4.22) is the objective function; (4.23) ensures that only one container type is selected; (4.24) ensures that exactly the batch size of a job is manufactured; (4.25) ensures that the unit load size of a job does not exceed the batch size; (4.26) is the non-negativity constraint on the binary variables X_k.

4.6.3 Solution methodology

The mathematical program presented above is a mixed integer program with decision variables on N_k, T_k, $\bar{Q}_{jk}$, U_{jk} and X_k. However, given a pallet type k, the values of W_k and X_k are automatically given. For example, if pallet type $k' = 3$ is selected, X_3 takes a value of 1 while the value of W_3 can be read directly from the matrix of input data on pallets. If W_3 is known, the number of unit loads U_{j3} associated with job j can be easily calculated according to equations (4.15) and (4.16). The number of vehicles N_3 required depends upon the total number of unit loads (i.e. $(\sum_{j=1}^{J} U_{j3})$) to be moved and the routing of the loads. Estimate of N_3 can be obtained according to the technique described by Egbelu (1987a). There is no closed form method known to the author which guarantees that the calculated value N_k is exact. Simulation is currently the most reliable method for determining N_k for any given application (Phillips, 1980). Since T_k and $\bar{Q}_{jk}$ depend upon N_k, they both have to be determined by simulation. With this difficulty, a simulation-based method for determining N_k, T_k and $\bar{Q}_{jk}$, $j = 1, 2, \ldots, J$, is proposed in this study. However, because simulation is an evaluation rather than an optimizing technique, the results from a given simulation run have a very limited interpretation. Multiple simulation runs followed by statistical analysis are required before the best unit load, and consequently AGV policy, is identified.

In view of the above discussions, an optimization based on searching over k, container type, is proposed in this chapter. For each container type k, several replications are run to estimate the total cost of the design as defined by the objective function. At each replication, a search is performed to find the best number of vehicles N_k. With N_k found, the corresponding T_k and $\bar{Q}_{jk}$ values can be obtained directly from the simulation output to compute the total cost. The replications are thereafter used to perform statistical analysis on the various container types by expected cost. The container with the highest ranking by minimum expected cost is selected as the preferred container k^*. Details regarding the algorithm are as presented below. The following additional variables are used in the algorithm:

$$Q(k) = \sum_{j=1}^{J} U_{jk}, \; k = 1, 2, \ldots, K;$$

R_k = number of replications required for container k;
Q^k = number of unit loads completed by the shop when container type k and N_k vehicles are used;
$P^*_{kk'}$ = the lowest total cost found in the k' replication for container type k;
$N^*_{kk'}$ = the number of AGVs used in the k' replication for container type k that yielded a total cost of $P^*_{kk'}$;
$T^*_{kk'}$ = production makespan corresponding to the design that yielded the total cost $P^*_{kk'}$, and using $N^*_{kk'}$ under the k' replication for container k.

The steps of the algorithm are as follows:

1. Identify the container types, $k = 1, 2, \ldots, K$ (see section on container choices for further discussion of this step).
2. Compute q_{jk}, U_{jk} and $Q(k)$ for $j = 1, 2, \ldots, J$; $k = 1, 2, \ldots, K$.
3. Initialize container, $k = 0$.
4. Set $k = k + 1$; for container type k, initialize the replication number, $k' = 1$.
5. Specify the number of replications R_k required for k.
6. Based on $Q(k)$ and the shop layout, estimate the number of vehicles $\bar{N}_k$ required if T_o is given. Otherwise, guess a starting value for $\bar{N}_k$. Set $N_{kk'} = \bar{N}_k$.
7. Set $P^*_{kk'} = +\infty$; $N^*_{kk'} = +\infty$.
8. Simulate the shop with the flow volume of $Q(k)$ and $N_{kk'}$ vehicles. Let $T(Q(k), N_{kk'})$ be the time required to complete the $Q(k)$ unit loads when $N_{kk'}$ vehicles are used.
9. Compute the total cost, $\bar{P}_{kk'}$ associated with the design of $Q(k)$ unit loads and $N_{kk'}$ vehicles, where

$$\bar{P}_{kk'} = \begin{cases} \infty, & \text{if } T(Q(k), N_{kk'}) > T_0 \\ \left(\Phi_k N_{kk'} + \Omega + \dfrac{t'_0}{T} C_w \displaystyle\sum_{j=1}^{J} q_{jk} \bar{Q}_{jk} \right) T(Q(k), N_{kk'}) + \displaystyle\sum_{j=1}^{J} Z_k U_{jk}, & \text{otherwise} \end{cases}$$

   Note that T_0 equals $+\infty$ if there is no upper bound on production makespan.
10. Compare $\bar{P}_{kk'}$ and $P^*_{kk'}$. If $\bar{P}_{kk'} < P^*_{kk'}$, update $P^*_{kk'} = \bar{P}_{kk'}$; $N^*_{kk'} = N_{kk'}$; and $T^*_{kk'} = T(Q(k), N_{kk'})$ or T_o if T_o is given.
11. Stopping condition reached (see discussion on stopping condition) for the k'th replication? If yes, go to step (12). Otherwise, reset $N_{kk'} = N_{kk'} + 1$ or $N_{kk'} - 1$ as may be desirable and go to step (8).
12. Output the results on $P^*_{kk'}$, $N^*_{kk'}$ and $T^*_{kk'}$.
13. Is $k' < R_k$? If yes, set $k' = k' + 1$ and go to step (7).
14. Is $k < K$? If yes, go to step (4).
15. Analyse the generated simulation output (see section on statistical analysis).
16. Select the contain k^* that has the highest ranking according to lowest cost.

(a) Algorithm implementation

The algorithm as described requires a search for $P^*_{kk'}$ for each iteration $k', k' = 1, 2, \ldots, R_k, k = 1, 2, \ldots, K$. Finding $P^*_{kk'}$ for each k' may require a search through several values of fleet size, $N_{kk'}$. Each search point requires a complete simulation of the shop. As a result, the number of simulation runs necessary to generate the data for all replications over all containers may be very prohibitive. If all the data is generated through one run of the algorithm, the computer time requested by the job may be very large. This lowers the job priority. To reduce the computer time requested for each run, the above algorithm can be implemented in segments. Two methods of segmentations readily present themselves. The first approach is to carry out the simulation for each replication and each container separately. If this approach is taken, a total of $R = \sum_{k=1}^{K} R_k$ separate computer runs will be made. The second approach is to have one long computer run for each container k. In this case, a total of K runs will be made. For the run corresponding to container k, R_k replications are required. The segmentation allows the simulation results to be generated in multiple sessions by the analyst. The segmentation will generally increase the total computer time used. However, the data collection process can be done in stages.

(b) Container Choices

For many applications, the set of containers to consider in the optimization is readily available. The set in such situation may simply consist of the containers currently in use at the plant. Container standardization only requires that a container be selected from the set.

On the other hand, if the analyst is starting on the basis of no knowledge as to which size containers to consider, the following algorithm is suggested to assist in generating the containers:

1. Let $\Gamma_1 = \max_{\forall j} \{w_j\}$, $\Gamma_2 = \min_{\forall j} \{w_j\}$, and $\Gamma_3 = \max_{\forall j} \{w_j Q_j\}$, where w_j and Q_j are the unit weight per part and batch size of part type j, respectively.
2. For the kth container, generate the maximum weight it can support according to the following relationship:

$$W_k = \begin{cases} \Gamma_1 + (k-1)\Gamma_2, & \text{for } k = 1, 2, \ldots, K-1 \\ \Gamma_3, & \text{for } k = K \end{cases}$$

In applying the algorithm, the analyst may choose to reduce the set of containers generated by eliminating some of the W_k values from consideration. Furthermore, safety reasons and equipment limitations may prohibit the consideration of weights W_k above a certain value. If an upper bound on unit load weight exists, then Γ_3 should be set equal to the upper bound.

(c) Stopping condition

The search for $N^*_{kk'}$ during the k'th replication is stopped if for some $N^*_{kk'}$, increasing or decreasing the number of vehicles to $N_{kk'} = N^*_{kk'} \pm 1$ yields a total manufacturing cost $P_{kk'} > P^*_{kk'}$. However, to guard against converging at a local minimum at any iteration, the search can be started with different values of $N_{kk'}$, and not necessarily with $N_{kk'} = \bar{N}_k$, the estimated fleet size when container k is used.

For the makespan constrained problem, the stopping condition is also satisfied if, regardless of the number of vehicles used, it is impossible to complete the jobs within the constrained time. Scenarios exhibiting this characteristic generally have zero marginal increase in throughput as the number of vehicles increases. Zero increase in throughput regardless of fleet size indicates a production system for which there is insufficient machine capacity given the unit load policy in force. For example, if a job requires two machines in sequence and the total processing time of the job on each machine is seven hours, if the entire batch is moved as a unit load, it is true that regardless of the number of transporters provided, it will be impossible to complete the job in an eight hour shift.

(d) Statistical analysis

The statistical component of the algorithm simply analyses the output data generated from the simulation experiments. This analysis is necessary to ensure that the lowest cost container among the containers tested is selected. Several ranking procedures exist. One commonly applied algorithm is that due to Dudewicz and Dalal (1975) and summarized by Law and Kelton (1982). The disadvantage of the algorithm, however, is that a large number of replications may be required. For a simulation-based optimization, the degree of computation required may be prohibitive. For this reason, we recommend a comparison procedure that is based on analysis of variance (ANOVA), followed by Duncan's Multiple Range test if still necessary after ANOVA (Walpole and Myers, 1985). The ANOVA method is well covered in most elementary statistics books. The hypothesis to be tested is:

$$H_0\colon u_1 = u_2 = u_3 \cdots = u_K$$

against

$$H_1\colon u_k \neq u_j \text{ for some } k \neq j$$

where u_k = mean cost of manufacturing if container type k is used. If the H_0 is accepted (i.e. all containers generate the same cost), further analysis is discontinued. In this case, the analyst can use any container without any major loss in cost trade-offs between containers. However, if the hypothesis fails (i.e. there is a difference in the mean costs generated by the containers), then Duncan's Multiple Range test is performed to determine which containers differ. Because the test ranks or groups treatment in the ascending order of their means, the container with the lowest ranking should be selected. Mathematically, this is equivalent to selecting the container with the rank 1 if

$$u_{[1]} \leqslant u_{[2]} \leqslant u_{[3]} \leqslant \cdots \leqslant u_{[K-1]} \leqslant u_{[K]},$$

where the notation $[r]$ signifies the container type with rank r from the set of K containers tested.

4.6.4 Example problem

In Egbelu (1993), an illustrative design scenario was described in which the model presented in this section was applied. Figure 4.10 shows the AGV layout involved in the design scenario. Data regarding the part type processed in the facility and the container and vehicle types considered are as summarized in Tables 4.6 and 4.7, respectively. As shown in Fig. 4.10, there are eight workcenters in the facility. The number of machines at workcenter 1, 2, 3, 4, 5, 6 and 7 are 1, 3, 3, 4, 4, 4 and 4, respectively. Workcenter 8 is a warehouse and therefore no machine is used there.

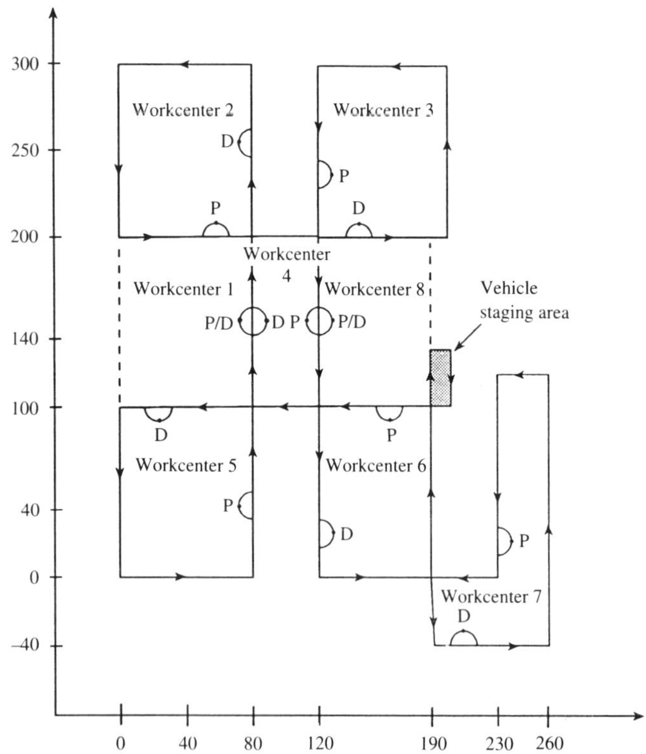

Fig. 4.10 Layout for the illustrative shop. D = load delivery station; P = load pickup station; → = direction of traffic flow.

Table 4.6 Part data

Part type	*Expected daily lot size*	*Wt/part (lb)*	*Vol/part (ft^3)*	*Routing*	*Processing time/part (min)*
1	50	375	1.5	1, 6, 7, 8	1.0[a], 11.25, 12.0[b]
2	90	625	2.0	1, 2, 5, 3, 8	1.0[a], 6.1, 7.2, 8.0[b]
3	100	450	2.0	1, 4, 2, 6, 8	1.0[a], 6.0, 5.0, 6.3[b]
4	70	850	3.5	1, 7, 5, 3, 4, 8	1.0[a], 10.6, 9.44, 5.2, 7.2[b]

[a]These are defined on a per unit load basis.
[b]No processing takes at the 8th station; this is a warehouse.

Table 4.7 Container and vehicle data

Container, k	W_k *(lb)*	*Volume*	*Cost/use* (\$)	*Cost/vehicle/hr* (\$)
1[a]	1000		4.00	9.00
2	1200	All	4.50	10.00
3	1500	containers	5.25	11.50
4	1700	are	5.75	12.50
5	1800	open	6.00	13.00
6	1900	pallets	6.25	13.50
7	2000		6.50	14.00
8	2300		7.25	15.50
9	2500		7.75	16.50

[a]Travel speed for AGVs = 150 ft/min; Shop overhead cost = \$270.00/hr; inventory carrying cost rate/day, $i_p^1 = 0.0004\%/\$/\text{day}$; $T_\sigma = 8$ hours.

Using the design approach presented, and running simulation replications for each candidate container as described in the solution methodology, the mean total manufacturing cost obtained and mean number of vehicles required for each container type are as given in Table 4.8. The simulation language AGVSim (Egbelu and Tanchoco, 1983) was used in performing the simulation. Based on the ANOVA result, container type 6 was selected as the design option that will result the overall minimum manufacturing cost.

4.7 CONCLUSION

Unit load design or size specification as an aspect of material handling can easily be overlooked because the impact of wrong a design on production system efficiency is not immediately apparent. The concept of moving items in sublots to minimize the frequency of handling is easily understood by people. However, how the decision to move a certain quantity of parts at a time affects other components of manufacturing is what is easily overlooked. It should be understood that unit load selection offers significant efficiency

Table 4.8 Result showing vehicle requirements and manufacturing cost by container type

Container (*k*)	*Payload capacity* (*lb*)	*Mean no. AGVs required*	*Mean total manufacturing cost*
1	1000	15	$4118.63
2	1200	15	4322.32
3	1500	12	4038.01
4	1700	9	3740.44
5	1800	8	3672.65
6	1900	8	3541.13
7	2000	8	3617.29
8	2300	8	3788.99
9	2500	8	3819.97

improvement opportunities to those who recognize its contributions. In a world where manufacturers shop for the same set technologies to enhance their competitiveness, it is the manager who recognizes the hidden powers of such seemingly uncomplex things as unit load selection that will reap the benefits of staying ahead of the game at little or no additional cost. Unit load size specification must be done logically and in an optimal manner to realize the full benefits of unitization.

REFERENCES

Apple, J.M. (1972) *Material Handling Systems Design*, John Wiley, New York.

Beightler, C.S., Phillips, D.T. and Wilde, D.J. (1979) *Foundation of Optimization*, Prentice-Hall, Englewood Cliffs, NJ.

Dowsland, K.A. (1987a) A combined database and algorithmic approach to the pallet loading problem. *Journal of Operational Research Society*, **38**(4), 341–5.

Dowsland, K.A. (1987b) Exact algorithm for pallet loading problem. *European Journal of Operational Research*. **31**, 78–84.

Dudewics, E.J. and Dalal, S.R. (1975) Allocation of measurements in ranking and selection with unequal variances. *Sankhya*, **B37**, 28–78.

Egbelu, P.J. and Tanchoco, J.M.A. (1983) *Designing the Operation of Automatic Guided Vehicle System Using AGVSim*. Proceedings 2nd International Conference on Automatic Guided Vehicles and IPA Conference, Stuttgart, Germany.

Egbelu, P.J. (1986) *Unit Load Design for Minimum Cost Manufacture in an Automated Guided Vehicle-Based Transport System*. Proceedings SME ULTRATECH Conference, **1**, Long Beach, CA, pp. 3.19–3.37.

Egbelu, P.J. (1987a) The use of non-simulation approaches in estimating vehicle requirements in an automated guided vehicle based transport system. *Material Flow*, **4**(3&4), 17–32.

Egbelu, P.J. (1987b) Integration of unit load selection in the design of an AGV-based automated manufacturing shop, in *Recent Developments in Production Research*, (ed A. Mital), Elsevier, Netherlands.

Egbelu, P.J. (1990) Batch production with unit load design and scheduling consideration, in *Progress in Material Handling and Logistics*, **2**, (eds J.A. White and I.W. Pence), Springer-Verlag, Berlin.

Egbelu, P.J. (1991) Batch production time in a multi-stage system with material handling consideration. *International Journal of Production Research*, **29**(4), 695–712.

Egbelu, P.J. (1993) Concurrent specification of unit load sizes and automated guided vehicle size in manufacturing system. *International Journal of Production Economics*, **29**, 49–64.

Gido, J. (1985) *An Introduction to Project Planning*, 2nd edn, Industrial Press.

Groover, M.P. (1987) *Automation, Production Systems, and Computer-Integrated Manufacturing*, Prentice-Hall, Englewood Cliffs, NJ.

Han, C.P., Knott, K. and Egbelu, P.J. (1989) A heuristic approach to the three dimensional cargo loading problem. *International Journal of Production Research*, **27**(5), 757–74.

Haynes, D.O. (1957) *Material Handling Equipment*, Clinton, Philadelphia.

Himmelblau, D.M. (1972) *Applied Nonlinear Programming*, McGraw-Hill, New York.

Hodgson, J.T. (1982) A combined approach to the pallet loading problem. *IEE Transactions*, **14**, 175–82.

Law, A.M. and Kelton, W.D. (1982) *Simulation Modeling and Analysis*, McGraw-Hill, New York.

Phillips, D.T. (1980) Simulation of material handling systems: When and which methodology. *Industrial Engineering*, **12**(8), 64–77.

Phillips, D.T., Ravindran, A. and Solberg, J.J. (1976) *Operations Research: Principles and Practice*, John Wiley, Chichester.

Steudel, H.J. (1979) Generating pallet loading pattern: A special case of the two-dimensional cutting stock problem. *Management Science*, **25**, 997–1004.

Tanchoco, J.M.A. and Agee, M.H. (1981) Plan unit loads to interact with all components of warehouse systems. *Industrial Engineering*, **13**(6), 36–48.

Thuesen, G.J. and Fabrycky, W.J. (1984) *Engineering Economy*, 6th edn, Prentice-Hall, Englewood Cliffs, NJ.

Walpole, R.E. and Myers, R.H. (1985) *Probability and Statistics for Engineers and Scientists*, 3rd edn, MacMillan, New York.

Warehouse Operations (1966) General Services Administration, US Government Printing Office, Washington, DC.

White, J.A., Agee, M.H. and Case, K.E. (1977) *Principles of Engineering Economic Analysis*, John Wiley, New York.

CHAPTER 5

Work-in-progress storage and handling capacity trade-offs in material flow design

C.J. Malmborg

5.1 INTRODUCTION

An important class of manufacturing facility design problems involves job shop systems where material flow is supported by discrete handling systems. Discrete handling describes systems that use a fixed number of vehicles (e.g. industrial trucks, AGVs, etc.) for transferring unit loads between workcenters. In designing job shop facilities using discrete materials handling systems, the designer must cost-effectively trade-off handling cpacity and the storage capacity used for buffering work in process unit loads. It is also necessary to effectively allocate the total buffer storage capacity among the individual workcenters of a facility. Treatment of the latter problem with respect to workcenter reliability issues has been well documented in the literature, particularly for flow shop systems with serial dependencies among workcenters. However, the problem is typically solved without regard to the interaction between storage requirements and the service level provided by a materials handling system. Buffer storage requirements in a job shop system can be sensitive to the relationship between handling demand and capacity, since slack capacity can be applied to minimize buffer queue fluctuations. The precise nature of this trade-off can be characterized in several ways, including the use of simulation (Ozden, 1988), Petri nets (Zeng, Wang and Jin, 1991) or analytical models (Malmborg, 1990; Maxwell and Muckstadt, 1982). For each of these approaches, it would be necessary to analyse a full set of plausible operating scenarios to effectively support the manufacturing system design process. Plausible operating scenarios would be defined with respect to factors which include:

- Production schedules (production quantities and part routings);
- handling parameters (vehicle fleet size, speed, routings, dispatching rules, etc.);

- storage parameters (buffer sizing and allocation between workcenters); and
- facility layout (travel distances between workcenters).

Simulation modeling would require coding and experimentation to provide feedback to the system designer as alternative design profiles are considered. Construction of stochastic Petri net models would be considerably easier than coding simulation models for all plausible operating scenarios in the early phases of the design process. Although not as diagnostic as simulation, Petri net models could also be used for the computation of basic performance measures that provide the designer with the feedback needed in the preliminary evaluation of intermediate design solutions (e.g. throughput and utilization measures, risks associated with deadlock situations, etc.). Effective use of Petri nets would require a protocol to allow the designer to perturb the graph representing the manufacturing system in an intelligent enumeration process. Thus, a major challenge lies in eliminating restraints on the designer's creativity that result from a reductance to discard existing nets to create new nets (representing completely new alternatives) as feedback is obtained. Although this is less of a problem with Petri nets than simulation, fixed costs in modeling of partial and/or intermediate solutions can significantly impede the earliest phases of the design process when flexibility is most crucial. Properly formulated analytical models can provide measures of output analysis that are comparable to those from stochastic Petri nets. They are also more easily imbedded in closed loop decision support systems that allow the designer to flexibly and effortlessly explore the design solution space. They can effectively remove inhibitions on the designer's creativity in convergence to a reduced set of promising candidate solutions that can be further screened using Petri net models (and eventually, validated through simulation). Analytical models also have the advantage of not requiring the designer to specify initial conditions for the model in deriving steady state results.

The purpose of this chapter is to provide a purely analytical construct for characterizing the trade-off between materials handling and buffer storage capacity in a job shop manufacturing system. The models describe the expected change in buffer storage requirements at the individual workcenters of a facility as a function of materials handling system parameters. The models can also be applied to predict the impact of buffer storage constraints on the production capacity of systems for which queue overflows influence workcenter output. Based exclusively on expected value calculations, the models provide a sensitivity analysis tool that is useful in block layout planning, storage space allocation, vehicle fleet sizing and cost estimation. Since the proposed models utilize flow rates between all combinations of workcenters in a facility, they are particularly applicable in batch manufacturing facility design problems.

The next section describes the analytical tools used to model the relationship between handling and storage capacity. The models approximate vehicle utilization based on unit load flow rates, routings and fleet size. This approximation is used to generate vehicle waiting times at each workcenter for a specific dispatching logic. These waiting times are used to generate queueing parameters that describe unit load buffering at individual workcenters. Buffer storage dynamics are used to generate storage dependent unit load flow rates within the facility. The third section presents an example problem through which the analytical models are illustrated and limited computational studies demonstrated. A summary and conclusions are provided in the final section.

5.2 ANALYTICAL MODELS OF STORAGE AND HANDLING CAPACITY TRADE-OFFS

Design parameters influencing the relationship between handling and storage requirements in the operation of a batch manufacturing system include the vehicle fleet size, facility layout and handling, and storage equipment parameters. Operating parameters influencing this relationship include the production schedule, vehicle efficiency, dispatching, part and vehicle routings and buffer storage disciplines. The models described below are designed to account for these factors in a single analytical construct which allows the user to instantaneously characterize storage and handling trade-offs as design and operating parameters are adjusted. The strategy of the models is to generate unit load flow rates based on a production schedule and vehicle travel times between workcenters for a given facility layout, vehicle routings and equipment parameters. Flow rates, travel times and fleet size are used to approximate vehicle utilization to estimate the volume of recirculation travel associated with alternative dispatching rules in a system. These results are exploited to characterize the service levels provided to individual workcenters by a materials handling system. Service levels experienced at workcenters directly influence work-in-progress storage requirements. In addition, storage policies (i.e. pooled or segregated storage of incoming and outgoing unit loads) directly impact total space requirements. If work-in-progress storage requirements exceed the space allocation at a given workcenter (in the absence of a secondary handling system), the output of the workcenter is affected. Depending upon the function of the workcenter, the production capacity of the manufacturing system may also be affected. The models described in this section can be used to quickly characterize these effects for alternative perturbations of the levels of design and operating parameters.

5.2.1 Description of model inputs

In a typical job shop system, a number of different products are manufactured with workcenters arranged in a process layout. Based on the routings and forecasted production levels for each product, a rate of unit load flow can be computed for each pairwise combination of workcenters in the facility. If we assume that unit loads throughout a system are uniform for handling purposes (e.g. a fully palletized system), we can characterize the handling load associated with a given production schedule in the form of a general material flow matrix. The actual material flow volume distance associated with this matrix is dependent upon the layout of the facility. Specifically, the travel distances between pairs of unit load transfer points are derived from vehicle routings with respect to the layout. Vehicle travel times are obtained through application of vehicle speeds and load transfer times. Given material flow rates and travel times between workcenters, the feasibility of material flow in a system is essentially determined by the available materials handling and work-in-progress storage capacities.

Handling capacity can be measured by the number of vehicles available and the dispatching rules used to control them. Vehicle dispatching refers to the rules applied to determine the sequence of handling transactions served, and the assignment of vehicles to transactions. Storage capacity in a system can be measured by the number of work-in-progress unit loads that can be stored. How this capacity is distributed in a system (i.e. the input and output unit load storage capacity of each individual workcenter) will also influence the effectiveness of this capacity. The notation summarized below is used in describing handling and storage capacity:

- W = the number of workcenters served by the materials handling system;
- $\mathbf{M}$ = the material flow matrix containing the elements m_{ij} for $i, j = 1, \ldots, W$, which represent the average number of (loaded vehicle) transactions per hour between workcenters i and j;
- T = the travel time matrix for the materials handling system containing the elements t_{ij} for $i, j = 1, \ldots, W$, which represent the vehicle travel time between workcenters i and j;
- N = the vehicle fleet size;
- e = the proportion of each hour that vehicles are available to service transactions (i.e. vehicle efficiency);
- b_{I1} = the vector of input queue buffer storage capacities for individual workcenters with elements b_{Ij} for $j = 1, \ldots, W$, which represent capacity in unit loads;
- b_{O2} = the vector of output queue buffer storage capacities for individual workcenters with elements b_{Oj} for $j = 1, \ldots, W$, which represent capacity in unit loads.

5.2.2 Modeling handling and storage relationships

If there are no limitations on storage capacity, the minimum possible materials handling workload for a facility (as measured in vehicle seconds per unit time) can be estimated as:

$$\phi = \frac{\left(\sum_{i=1}^{W}\sum_{j=1}^{W} m_{ij}t_{ij} + \sum_{i=1}^{W}\sum_{j=1}^{W} t_{ij}x_{ij}\right)}{\left(\sum_{i=1}^{W}\sum_{j=1}^{W} m_{ij}\right)}, \tag{6.1}$$

where the x_{ij} values are obtained from the solution to the linear transportation problem given by

$$\text{Min: } \sum_{i=1}^{W} \sum_{j=1}^{W} t_{ij} x_{ij} \tag{5.2}$$

$$\text{s.t. } \sum_{j \subset P1} x_{jk} = |h_k| \quad k \subset P2$$

$$\sum_{j \subset P2} x_{kj} = h_k \quad k \subset P1$$

$$x_{ij} \geqslant 0 \quad \forall_{ij}.$$

In the above formulation, $P1$ and $P2$ refer to the set of workcenters for which the h_k values are positive and negative, respectively, with

$$h_k = \sum_{j=1}^{W} m_{jk} - \sum_{j=1}^{W} m_{kj} \quad \text{for } k = 1, \ldots, W. \tag{5.3}$$

The logic of the formulation for ϕ in equation (5.1) is to compute the minimum vehicle minutes of travel necessary to meet the materials handling workload represented by **M** (assuming that buffer storage can be used to minimize empty travel associated with vehicle recirculation). The minimum volume of empty (recirculation) travel between workcenters is then added to this value by finding the travel minimizing assignment of net vehicle importing workcenters to net vehicle exporting workcenters. Since it assumes infinite storage capacity at each workcenter, equation (5.1) represents a lower bound on the material handling capacity needed to serve the material flow requirement in a system.

Storage capacity and work-in-progress inventory are typically scarce resources, thus it is not generally feasible to dispatch vehicles in a system to achieve the minimum theoretical vehicle resources required. Rather, vehicles must be continuously dispatched with transaction completions to prevent queue overflows at individual workcenters and provide acceptable waiting times for transaction demands. This results in more empty travel as vehicles respond to transaction requests from workcenters that may not have unit loads prepared to transfer to workcenters with active requests for transactions. For the extreme case of no storage capacity, the number of vehicles required in a system would approach the maximum number of handling transactions simultaneously active at one time.

In practical cases, a system will use a combination of vehicle initiated and workcenter initiated dispatching. Vehicle initiated dispatching rules are invoked when vehicles select from an active queue of materials handling transactions in a system. Examples of practical vehicle initiated dispatching rules include 'minimum outgoing remaining queue space', 'first come first serve', 'nearest workcenter', 'random workcenter', 'unit load shop arrival time', etc. Workcenter initiated dispatching rules are invoked during operating periods when the active queue of handling transactions is empty, and new transactions arriving to the system select from among the set of one or more idle vehicles. Practical examples of workcenter initiated dispatching rules include 'least utilized vehicle', 'longest idle vehicle', 'random vehicle', 'nearest vehicle', etc. Systems with more slack capacity tend to invoke workcenter initiated dispatching rules more often than systems with less slack capacity. However, all systems with handling capacity close to, but greater than, the average materials handling workload will use some combination of workcenter and vehicle initiated dispatching. The selection of dispatching rules in a system usually reflects a combination of objectives including resource conservation, workload leveling among vehicles, and service leveling among workcenters.

Therefore, dispatching influences the interaction between the handling and storage capacity by directly responding to the status of work-in-progress queues and determining the volume of vehicle resources consumed in recirculation travel. It follows that a key issue in modeling the relationship between handling and storage in a manufacturing system is the expected waiting time experienced by queued transactions. Although this value will vary with individual workcenters, an expected value for a system can be approximated for various dispatching scenarios. To obtain this value, assume a starting approximation of vehicle utilization. One possibility is to use

$$U' = \frac{\left\{\frac{1}{2}\left[\lambda\left(3\phi - 2\left(\sum_{i=1}^{W}\sum_{j=1}^{W} t_{ij}x_{ij}\right)\Big/\sum_{i=1}^{W}\sum_{j=1}^{W} m_{ij}\right)\right]\right\}}{60\,\mathrm{eN}}, \tag{5.4}$$

where

$$\lambda = \sum_{i=1}^{W}\sum_{j=1}^{W} m_{ij}$$

and U' represents the utilization level representing the average of utilization values associated with the minimum possible travel volume (from equation (5.1)), and that value associated with assuming that the empty travel volume exactly equals the loaded travel volume.

The initial estimate of utilization from equation (5.4) can be used to approximate the proportion of time that a system invokes workcenter initiated vs. vehicle initiated dispatching. For example, if a random vehicle/random workcenter (RV/RW) dispatching rule combination is applicable, we can approximate a refined estimate of the average empty travel time per transaction using

$$\omega = U' \sum_{i=1}^{W} p_i'\left(\sum_{j=1}^{W} p_j t_{ij}\right) + (1 - U') \sum_{i=1}^{W} p_i\left(\sum_{j=1}^{W} p_j' t_{ji}\right) \tag{5.5}$$

where p_i approximates the probability that a free vehicle is located at workcenter i, and p_j' estimates the probability that workcenter j is the source workcenter associated with a transaction request. The probability estimates are obtained directly from **M** using

$$p_i = \frac{\sum_{i=1}^{W} m_{ij}}{\sum_{i=1}^{W}\sum_{j=1}^{W} m_{ij}}, \quad i = 1,\ldots, W, \tag{5.6}$$

$$p'_j = \frac{\sum_{i=1}^{W} m_{ij}}{\sum_{j=1}^{W}\sum_{i=1}^{W} m_{ji}}, \quad j = 1,\ldots, W. \tag{5.7}$$

Using ω as a starting estimate of the average empty travel time per transaction, it is possible to apply a procedure described by Malmborg and Shen (1991a) to estimate empty travel time as a direct function of the number of vehicles in a materials handling system. In this procedure, a queueing model is used to obtain the probability distribution of the number of transactions in process as a function of the materials handling workload (λ), and the system average service rate (μ), where

$$\lambda = \sum_{i=1}^{W}\sum_{j=1}^{W} m_{ij} \quad \text{and} \quad \mu = \left[\frac{\left(\sum_{i=1}^{W}\sum_{j=1}^{W} m_{ij}t_{ij}\right)}{\lambda} + \omega\right]^{-1} \tag{5.8}$$

For example, if the M/M/N Poisson queueing model is applied, this probability distribution would be obtained as

$$P_0 = \left[\frac{\sum_{i=0}^{N-1} (N\lambda/\mu)^i}{i!} + \frac{\left(\frac{N\lambda}{\mu}\right)^N}{\left(N!\left(\frac{1-\lambda}{\mu}\right)\right)}\right]^{-1}$$

$$P_j = \frac{\left\{\left(\frac{N\lambda}{\mu}\right)^j P_0\right\}}{j!} \quad j = 1,\ldots, N$$

$$P_j = \frac{\left\{\left(\frac{N\lambda}{\mu}\right)^j P_0\right\}}{N!N^{j-N}} \quad j = N+1,\ldots,\infty, \tag{5.9}$$

where P_j denotes the probability of j materials handling transactions in the system.

The expected value of empty vehicle travel time as a function of N has separate components associated with vehicle and workcenter initiated dispatching. Both components are based on the specific dispatching rule used in a system. For example, the random workcenter or first come first served dispatching formulation of the expected value of empty vehicle travel time is given by

$$EV_V = \left(\sum_{l=N}^{\infty} P_l\right) \sum_{i=1}^{W} p_i' \sum_{k=1}^{W} p_k t_{ik}. \tag{5.10}$$

For vehicle initiated dispatching rules based on relationships between workcenters such as nearest workcenter or minimum remaining output queue space, the vehicle initiated component of the formulation is of the form

$$EV_V = \left(\sum_{l=N}^{\infty} P_l\right) \sum_{i=1}^{W} p_i \left[p_1' t_{i1} + \sum_{k=2}^{W} \left(\prod_{z=1}^{k} (1 - p_z') \right) p_k' t_{ik} \right], \tag{5.11}$$

where the ordering of terms on the index k in (5.11) is defined based on the specific dispatching rule used. For example, with nearest workcenter dispatching, the ordering on the index k would be defined with respect to the index i, i.e. $k = 1$ is associated with the nearest workcenter to workcenter i, $k = 2$ is associated with the second nearest workcenter, etc. The workcenter initiated component of the expected value of empty travel time is also rule-specific. For certain rules, the impact on the expected value is not directly dependent upon the number of vehicles available in a system. For example, with rules such as random vehicle, longest idle vehicle or least utilized vehicle, a reasonable approximation of empty travel time due to workcenter initiated dispatching is given by

$$EV_W = \left(\sum_{l=0}^{N-1} P_l\right) \sum_{j=1}^{W} p_j \sum_{k=1}^{W} p_k' t_{kj}. \tag{5.12}$$

However, the impact of other workcenter initiated dispatching rules on empty travel time is more directly influenced by the number of vehicles available in a system. For example, with rules such as nearest vehicle, the applicable formulation would be

$$EV_W = \sum_{j=1}^{W} p_j' \left[\sum_{k=1}^{W} \left\{ \sum_{l=0}^{N-1} p_l \left(\prod_{z=1}^{k-1} (1 - p_z)^{N-l} \right) (1 - (1 - p_k)^{N-l}) \right\} t_{kj} \right], \tag{5.13}$$

where the ordering of terms on the index k in (5.13) is defined by the workcenter initiated dispatching rule with respect to the index j in a manner analagous to that in (5.11). Using the appropriate subset of results from equations (5.10)–(5.13), the estimate of empty travel time per handling transaction is obtained by the sum $\tau = EV_V + EV_W$.

The logic of equation (5.11) is to sum the product of the probability that vehicle initiated dispatching is invoked on a randomly selected transaction, and the corresponding expected empty travel distance. (For problems where vehicle utilization exceeds 100%, the probability of vehicle initiated dispatching is 1.) The expected empty travel time portion of the formulation is given by the product of the probability that a vehicle becomes free at each workcenter i (i.e. p_i) and the corresponding travel time. The corresponding travel time consists of two separate terms. The first part of this term, $p'_1 t_{i1}$, represents the product of the probability that the 'first choice' workcenter (as defined by the dispatching rule) has a transaction pending and the expected travel time to the first choice workcenter. The second part of the expression takes each remaining workcenter in the preference order defined by the dispatching rule. It applies the travel time to this workcenter and the product defining the joint probability that the workcenter has a transaction request pending while other workcenters, of higher preference with respect to the dispatching rule, do not. The logic of equation (5.13) is to sum of the product of the probability that each workcenter initiates a request for a transaction p'_j and the corresponding expected empty vehicle travel time. The corresponding expected empty travel time is given by the sum of the probability that a vehicle is available at each workcenter, and the expected empty travel time to that workcenter. The expected empty travel time

$$\sum_{k=1}^{W}\left\{\sum_{l=0}^{N-1} p_l\left(\prod_{z=1}^{k-1}(1-p_z)^{N-l}\right)(1-(1-p_k)^{N-1})\right\}t_{kj}, \tag{5.14}$$

takes each workcenter in the preference ordering with respect to j that is defined by the dispatching rule, and adds the product of the joint probability that the workcenter has a vehicle available while higher preference workcenters do not, and the travel time to the workcenter from workcenter j. The product term for $k = 1$ is defined to equal 1.

The accuracy of equations (5.10)–(5.13) for predicting the expected value of empty travel time as a function of N has been validated through extensive simulation analyses Malmborg and Shen, 1991a, b). In the extreme cases of farthest vehicle/farthest workcenter (FV/FW) and nearest vehicle/nearest workcenter (NV/NW) dispatching, (5.11) and (5.13) have been shown to perform best when the initial estimates of ω are given by

$$\omega_{\mathrm{FV/FW}} = (U'/2)\left[\sum_{i=1}^{W} p_i\left(\sum_{j=1}^{W} p'_j t_{ji}\right) + \sum_{i=1}^{W} p'_i\left(\sum_{j=1}^{W} p_j t_{ij}\right)\right] + (1-U')\sum_{i=1}^{W} p_i \max_j(t_{ji}) \tag{5.15}$$

$$\omega_{\mathrm{NV/NW}} = (U'/2)\left[\sum_{i=1}^{W} p_i\left(\sum_{j=1}^{W} p'_j t_{ji}\right) + \sum_{i=1}^{W} p'_i\left(\sum_{j=1}^{W} p_j t_{ij}\right)\right] + (1-U')\sum_{i=1}^{W} p_i \min_j(t_{ji}) \tag{5.16}$$

respectively.

An estimate of the empty vehicle travel time based on the fleet size and dispatching rule can be used to measure the service level provided by the materials handling system, i.e. the expected waiting time following requests for transactions. The expected 'system wide' average waiting time could be estimated directly from the Poisson queueing model as

$$\beta = \frac{[(N(\lambda/\mu))^N P_0/N!(1-\lambda/\mu)]}{(N\mu-\lambda)} \tag{5.17}$$

where

$$P_0 = \left\{ \sum_{i=0}^{N-1} (N\lambda/\lambda)^i/i! + (N\lambda/\lambda)^N/N!(1-\lambda/\mu) \right\}^{-1} \tag{5.18}$$

where λ is defined in equation (5.8) and

$$\mu = \left[\frac{\left(\sum_{i=1}^{W} \sum_{j=1}^{W} m_{ji} t_{ij} \right)}{\lambda} + \tau \right]^{-1} \tag{5.19}$$

As the result above suggests, transaction waiting only occurs in the case of vehicle initiated dispatching.

When vehicle initiated dispatching is invoked, the actual transaction waiting time consists of β, plus the time necessary for vehicles to travel from the location where the last transaction was completed to the workcenter where they are dispatched. When workcenter initiated dispatching is invoked, actual waiting time consists only of travel time from the current location of the vehicle to the workcenter requesting the vehicle. For a given workcenter i, this time can be approximated as

$$\mu_{Oi} = \left[\left(\sum_{l=N}^{\infty} P_l \right) \left\{ \beta + \sum_{j=1}^{W} p'_j t_{ji} \right\} + \sum_{k=1}^{W} P_l \left(\sum_{l=0}^{N-1} P_l \left(\prod_{z=1}^{k-1} (1-p'_z)^{N-l} \right) (1-(1-p'_k)^{N-l}) \right) t_{ki} \right]^{-1} \tag{5.20}$$

where the Oi subscript denotes equation (5.20) as the 'service rate' for the output queue at workcenter i. In the second line of (5.20), dispatch travel time initiated by workcenter i, is given by an expected value defined for all possible materials handling system states. The expected value is the product of the probability that workcenter k is the source workcenter of the dispatched vehicle and the travel time to workcenter i. The index k in (5.20) is again defined with respect to the dispatching rule.

Using an estimate of the vehicle waiting time at individual workcenters, it is possible to model the impact of output queue storage constraints using a capacitated queueing model with service rate μ_{Oi} and arrival rate given by

$$\lambda_{Oi} = \sum_{j=1}^{W} m_{ij} \quad i = 1, \ldots, W. \tag{5.21}$$

The resultant probability distribution of the output queue system states at workcenter i would be given by

$$\pi_{Oi0} = (1-\lambda_i/\mu_i)/(1-(\lambda_i/\mu_i)_{bOi+1}) \tag{5.22}$$

$$\pi_{Oiq} = (\lambda_i/\mu_i)^q P_{Oi} \quad q = 1, \ldots, b_{Oi} \tag{5.23}$$

$$\pi_{Oiq} = 0 \quad q = b_{Oi} + 1, \ldots, \infty. \tag{5.24}$$

For situations where output queue overflows shut down workcenters, the results in (5.22)–(5.24) can be applied to modify the arrival rates to the input queues of workcenters that receive outputs of the affected workcenter as work-in-process. Specifically, the modified input queue arrival rate at workcenter i would be given as

$$\lambda_{Ii} = \sum_{j=1}^{W} (1 - \pi_{OjbOj}) m_{ji} \quad i = 1, \ldots, W. \tag{5.25}$$

The rate at which unit loads are removed from the input queue at workcenter i would similarly be obtained from

$$\mu_{Ii} = [(1 - \pi_{OibOi})\eta_i]^{-1} \quad i = 1, \ldots, W, \tag{5.26}$$

where η_i represents the standard processing time for unit loads in the input queue at workcenter i. The η_i terms for $i = 1, \ldots, W$ represent parameters associated with the processing function within each workcenter. For example, if a workcenter was a numerically controlled machining center, η_i would represent the machining cycle plus makeready time for each workpiece. The coefficient of η_i reflects the possibility that workcenter i is shut down as a result of an output queue overflow. A capacitated queueing model can again be applied to obtain the probability distribution of input queue states as

$$\pi_{IiO} = (1 - \lambda_i/\mu_i)/(1 - (\lambda_i/\mu_i)^{bIi+1}) \quad i = 1, \ldots, W \tag{5.27}$$

$$\pi_{Iiq} = (\lambda_i/\mu_i)^q p_{Oi} \quad q = 1, \ldots, b_{Ii} \tag{5.28}$$

$$\pi_{Iiq} = 0 \quad q = b_{Ii} + 1, \ldots, \infty. \tag{5.29}$$

5.3 APPLICATION OF THE HANDLING AND STORAGE CAPACITY MODELS

The procedures described in the previous section provide one basis for analytically modeling trade-offs in materials handling and storage capacity, as a function of the parameters defining a manufacturing system design. These models form the core of a general decision support system (DSS) for material flow system design that is currently under development. This DSS will allow the user to interactively generate graphs describing storage and handling capacity trade-offs such as:

- percentage change in work-in-progress storage space allocation vs. the minimum number of vehicles needed to support the material flow within the system (for a given percentage of lost production due to queue overflows);

- average storage space utilization by workcenter vs. the number of vehicles used in a system;
- average total space required as a function of the number of vehicles (for a given percentage of lost production due to queue overflows);
- average percentage of lost production due to queue overflows vs. percentage change in storage space allocation (for a given number of vehicles in the system);
- average percentage of lost production due to queue overflows vs. the number of vehicles used in the system (for a given storage space allocation).

The DSS is designed to allow the user to efficiently enumerate the design solution space by facilitating rapid modifications of parameter values including: storage space allocation, number of vehicles in the system, production volume forecasts, part routings, vehicle routings, the plant layout, vehicle dispatching rules, vehicle speed and efficiency, workcenter processing rates and workcenter critical path status (i.e. whether queue overflows influence the rate of total system output). In addition, the DSS is designed to provide the user with effective 'cuts' the output database to support meaningful adjustments in parameter values. For example, the user can generate a screen showing the probability distribution of input and output storage queue states at each workcenter to guide the reallocation of storage space among workcenters. Alternatively, the user can view the volume of vehicle travel along alternative flow paths and vehicle travel times within the facility to suggest modifications in a layout generated by the DSS through departmental location interchanges.

To illustrate an application of the system, consider the example facility with five workcenters, which is summrized in Fig. 5.1. The material flow matrix in Fig. 5.1 follows directly from production forecasts and part routings. The travel time matrix is obtained from the facility layout in Fig. 5.1, as well as vehicle routings, estimated travel speeds, load transfer times, etc. For the example problem with 14 vehicles to service the material flow requirements of the system and a vehicle efficiency factor of 90%, the key model outputs (including the transportation subproblem for vehicle recirculation) are summarized in Fig. 5.2. As the figure shows, the proposed system design would result in significant probabilities of output queue overflows at workcenters B and E, and significant probabilities of input queue overflows at workcenters A, C and D. This information would guide the user in reallocating work-in-progress storage capacity among the workcenters. It is also worth noting from the example outputs that vehicle utilization is relatively low. This information could lead the user to consider providing additional storage space within the facility in exchange for a significant reduction in the number of vehicles. At the core of the DSS is a 'CRAFT like' algorithm for generating alternative plant layouts of a system. Figures 5.3–5.6 illustrate this feature for four alternative layouts generated by the DSS. As the results indicate, most key model outputs are not extremely sensitive to variations in the proposed layout. This information could lead the user to conclude that

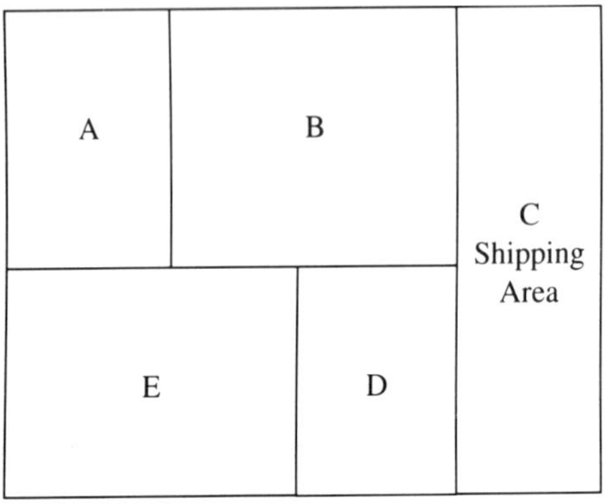

Travel Time Matrix T:
(in minutes)

	A	B	C	D	E
A	–	3	7	6	3
B	3	–	4	3	4
C	7	4	–	3	6
D	6	3	3	–	3
E	3	4	6	3	–

Veh. Speed: 500 ft./min.
Scale: 1" = 1000 ft.
Veh. Efficiency: e = .90
Veh. Dispatching: RV/RW

N=14 vehicles

Material Flow Matrix M:
(in unit loads per hour)

	A	B	C	D	E
A	–	4	6	0	2
B	5	–	9	4	6
C	0	0	–	0	0
D	3	4	6	–	7
E	5	3	4	1	–

$p_A=0.174$; $p_A'=0.188$
$p_B=0.174$; $p_B'=0.188$
$p_C=0.174$; $p_C'=0.188$
$p_D=0.174$; $p_D'=0.188$
$p_E=0.174$; $p_E'=0.188$

Workcenter Specific Parameters:

	Wkctr A:	Wkctr B:	Wkctr C:	Wkctr D:	Wkctr E:	
b_{Ij}	5	5	5	5	5	(Unit loads)
b_{oj}	5	5	5	5	5	(Unit loads)
η_i	15	20	25	15	15	(Unit loads)

Fig. 5.1 Summary of initial parameter values for the example problem.

selection of the final plant layout could be guided by qualitative criteria as opposed to material flow based measures.

5.4 SUMMARY AND CONCLUSIONS

A series of analytical models have been proposed to support tradeoffs in specifying the design and operating parameters of batch manufacturing facilities supported by discrete materials handling systems. The scope of these models includes production schedules, plant layout, storage capacity, vehicle fleet size, part and vehicle routings, vehicle dispatching and other key aspects of a manufacturing facility specification. Since these parameter values are rarely

$$\text{Min: } 3x_{AB}+7x_{AC}+6x_{AD}+3x_{AE}+3x_{BA}+4x_{BC}+3x_{BD}+4x_{BE}+7x_{CA}+4x_{CB}+$$
$$3x_{CD}+6x_{CE}+6x_{DA}+3x_{DB}+3x_{DC}+3x_{DE}+3x_{EA}+4x_{EB}+3x_{EC}+6x_{ED}$$

$$\begin{aligned}
\text{s.t.}\quad & x_{AB}+x_{AD} &&= 1\\
& x_{CB}+x_{CD} &&= 25\\
& x_{EB}+x_{ED} &&= 2\\
& x_{AB}+x_{CB}+x_{EB} &&= 13\\
& x_{AD}+x_{CD}+x_{ED} &&= 15\\
& x_{ij} &&\geq 0 \qquad i, j = A, B, C, D, E
\end{aligned}$$

$$x^* = \{x_{AB}^* = 1, x_{AD}^* = 0, x_{CB}^* = 12, x_{CD}^* = 13, x_{EB}^* = 0, x_{ED}^* = 2)$$

Model Outputs:

$\phi = 6.11$ (min.) $U' = 53.2\%$ $\omega = 3.60$ (min.) $\beta = 1.23$ (sec.)

Probability Dist. of the Number of Handling Transactions:

$P_0=0.001$ $P_1=0.005$ $P_2=0.019$ $P_3=0.046$ $P_4=0.083$ $P_5=0.120$
$P_6=0.144$ $P_7=0.148$ $P_8=0.134$ $P_9=0.107$ $P_{10}=0.077$ $P_{\geq 11}=0.115$

Workcenter Specific Parameters:

	Wkctr A:	Wkctr B:	Wkctr C:	Wkctr D:	Wkctr E:
μ_{oi}	60.9	22.6	10.0	11.5	18.3
λ_{oi}	24.0	15.0	0.0	5.0	11.0
π_{oio}	0.608	0.367	1.000	0.568	0.417
π_{oi1}	0.240	0.244	0.000	0.248	0.252
π_{oi2}	0.095	0.162	0.000	0.108	0.152
π_{oi3}	0.37	0.108	0.000	0.047	0.091
π_{oi4}	0.015	0.072	0.000	0.021	0.55
$\pi_{oi\geq 5}$	0.006	0.048	0.000	0.009	0.033
μ_{Ii}	14.9	19.0	25.0	14.9	14.5
λ_{Ii}	11.6	8.9	16.7	8.8	7.9
π_{Iio}	0.286	0.540	0.365	0.429	0.470
π_{Ii1}	0.222	0.251	0.244	0.253	0.255
π_{Ii2}	0.173	0.117	0.162	0.149	0.138
π_{Ii3}	0.134	0.054	0.108	0.088	0.075
π_{Ii4}	0.104	0.025	0.072	0.052	0.041
$\pi_{Ii\geq 5}$	0.081	0.012	0.048	0.030	0.022

Fig. 5.2 Model outputs for the example problem.

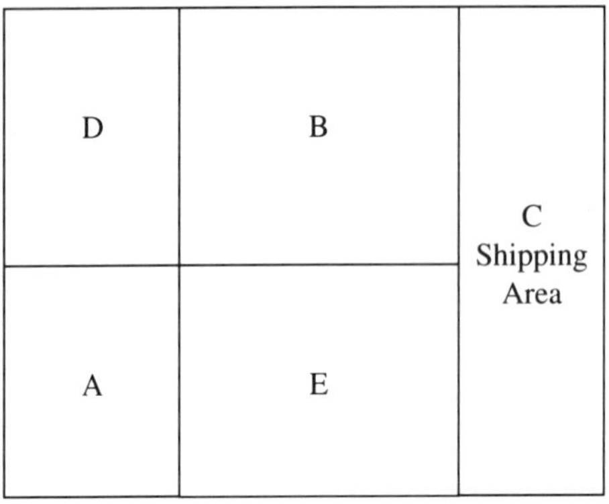

Travel Time Matrix T:
(in minutes)

	A	B	C	D	E
A	–	5	7	2	3
B	5	–	4	3	2
C	7	4	–	7	4
D	2	3	7	–	7
E	3	2	4	7	–

Veh. Speed: 500 ft./min.
Scale: 1" = 1000 ft.
Veh. Efficiency: e = .90
Veh. Dispatching: RV/RW

Model Outputs:

Min: $5x_{AB}+7x_{AC}+2x_{AD}+3x_{AE}+5x_{BA}+4x_{BC}+3x_{BD}+2x_{BE}+7x_{CA}+4x_{CB}+$
$7x_{CD}+4x_{CE}+2x_{DA}+3x_{DB}+7x_{DC}+7x_{DE}+3x_{EA}+2x_{EB}+4x_{EC}+7x_{ED}$

s.t.
$x_{AB}+x_{AD} = 1$
$x_{CB}+x_{CD} = 25$
$x_{EB}+x_{ED} = 2$
$x_{AB}+x_{CB}+x_{EB} = 13$
$x_{AD}+x_{CD}+x_{ED} = 15$
$x_{ij} \geq 0 \quad i, j = A, B, C, D, E$

$x^* = \{x_{AB}^*=0, x_{AD}^*=1, x_{CB}^*=11, x_{CD}^*=14, x_{EB}^*=2, x_{ED}^*=0\}$

$\phi = 6.20$ (min.) $U' = 53.5\%$ $\omega = 3.53$ (min.) $\beta = 1.12$ (sec.)

Probability Dist. of the Number of Handling Transactions:

$P_0=0.001$ $P_1=0.006$ $P_2=0.020$ $P_3=0.048$ $P_4=0.086$ $P_5=0.123$
$P_6=0.146$ $P_7=0.149$ $P_8=0.132$ $P_9=0.105$ $P_{10}=0.075$ $P_{\geq 11}=0.110$

Workcenter Specific Parameters:

	Wkctr A:	Wkctr B:	Wkctr C:	Wkctr D:	Wkctr E:
μ_{oi}	51.1	14.3	10.0	24.3	20.4
λ_{oi}	24.0	15.0	0.0	5.0	11.0
π_{oio}	0.536	0.148	1.000	0.794	0.473
π_{oi1}	0.252	0.155	0.000	0.163	0.255
π_{oi2}	0.118	0.162	0.000	0.034	0.137
π_{oi3}	0.056	0.170	0.000	0.007	0.074
π_{oi4}	0.026	0.178	0.000	0.001	0.040
$\pi_{oi\geq 5}$	0.012	0.187	0.000	0.000	0.021
μ_{Ii}	14.8	16.3	25.0	15.0	14.7
λ_{Ii}	11.0	8.9	16.1	8.2	7.6
π_{Iio}	0.311	0.467	0.383	0.467	0.494
π_{Ii1}	0.230	0.255	0.247	0.255	0.255
π_{Ii2}	0.170	0.139	0.159	0.139	0.131
π_{Ii3}	0.126	0.076	0.102	0.076	0.068
π_{Ii4}	0.093	0.041	0.066	0.041	0.035
$\pi_{Ii\geq 5}$	0.069	0.022	0.042	0.023	0.018

Fig. 5.3 Model outputs from revised layout 1.

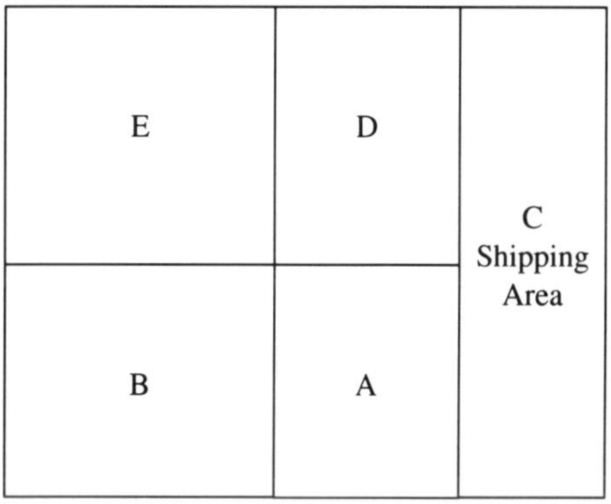

Travel Time Matrix T:
(in minutes)

	A	B	C	D	E
A	–	3	3	2	5
B	3	–	7	5	2
C	3	7	–	3	3
D	2	5	3	–	3
E	5	2	3	3	–

Veh. Speed: 500 ft./min.
Scale: 1" = 1000 ft.
Veh. Efficiency: e = .90
Veh. Dispatching: RV/RW

Model Outputs:

$$\text{Min: } 3x_{AB}+3x_{AC}+2x_{AD}+5x_{AE}+3x_{BA}+7x_{BC}+5x_{BD}+2x_{BE}+3x_{CA}+7x_{CB}+3x_{CD}+3x_{CE}+2x_{DA}+5x_{DB}+3x_{DC}+3x_{DE}+5x_{EA}+2x_{EB}+3x_{EC}+3x_{ED}$$

$$\begin{aligned}
\text{s.t.}\quad & x_{AB}+x_{AD} &&= 1\\
& x_{CB}+x_{CD} &&= 25\\
& x_{EB}+x_{ED} &&= 2\\
& x_{AB}+x_{CB}+x_{EB} &&= 13\\
& x_{AD}+x_{CD}+x_{ED} &&= 15\\
& x_{ij} &&\geq 0 \quad i, j = A, B, C, D, E
\end{aligned}$$

$$x^* = \{x_{AB}^*=0, x_{AD}^*=1, x_{AD}^*=0, x_{CB}^*=10, x_{CD}^*=15, x_{EB}^*=2, x_{ED}^*=0\}$$

$\phi = 5.81$ (min.) $\quad U' = 47.4\%$ $\quad \omega = 2.96$ (min.) $\quad \beta = 0.14$ (sec.)

Probability Dist. of the Number of Handling Transactions:

$P_0=0.002$	$P_1=0.015$	$P_2=0.044$	$P_3=0.089$	$P_4=0.133$	$P_5=0.160$
$P_6=0.161$	$P_7=0.138$	$P_8=0.104$	$P_9=0.069$	$P_{10}=0.042$	$P_{\geq 11}=0.043$

Workcenter Specific Parameters:

	Wkctr A:	Wkctr B:	Wkctr C:	Wkctr D:	Wkctr E:
μ_{oi}	120.0	21.0	18.3	25.4	13.3
λ_{oi}	24.0	15.0	0.0	5.0	11.0
π_{oio}	0.800	0.329	1.000	0.804	0.253
π_{oi1}	0.160	0.235	0.000	0.158	0.210
π_{oi2}	0.032	0.168	0.000	0.031	0.174
π_{oi3}	0.006	0.120	0.000	0.006	0.144
π_{oi4}	0.001	0.086	0.000	0.001	0.120
$\pi_{oi\geq 5}$	0.000	0.061	0.000	0.000	0.099
μ_{Ii}	15.0	18.8	25.0	15.0	13.5
λ_{Ii}	11.2	8.7	16.6	8.7	7.9
π_{Iio}	0.306	0.542	0.369	0.439	0.434
π_{Ii1}	0.229	0.251	0.244	0.253	0.253
π_{Ii2}	0.171	0.116	0.162	0.146	0.148
π_{Ii3}	0.128	0.054	0.107	0.084	0.086
π_{Ii4}	0.095	0.025	0.071	0.049	0.050
$\pi_{Ii\geq 5}$	0.071	0.012	0.047	0.028	0.029

Fig. 5.4 Model outputs from revised layout 2.

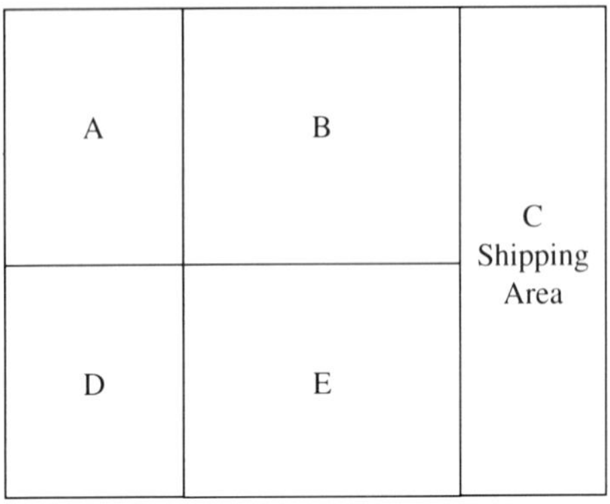

Travel Time Matrix T:
(in minutes)

	A	B	C	D	E
A	–	3	7	2	5
B	3	–	4	5	2
C	7	4	–	7	5
D	2	5	7	–	3
E	5	2	5	3	–

Veh. Speed: 500 ft./min.
Scale: 1" = 1000 ft.
Veh. Efficiency: e = .90
Veh. Dispatching: RV/RW

Model Outputs:

$$\text{Min: } 3x_{AB}+7x_{AC}+2x_{AD}+5x_{AE}+3x_{BA}+4x_{BC}+5x_{BD}+2x_{BE}+7x_{CA}+4x_{CB}+7x_{CD}+5x_{CE}+2x_{DA}+5x_{DB}+7x_{DC}+3x_{DE}+5x_{EA}+2x_{EB}+5x_{EC}+3x_{ED}$$

$$\begin{aligned} \text{s.t.}\quad & x_{AB}+x_{AD} = 1 \\ & x_{CB}+x_{CD} = 25 \\ & x_{EB}+x_{ED} = 2 \\ & x_{AB}+x_{CB}+x_{EB} = 13 \\ & x_{AD}+x_{CD}+x_{ED} = 15 \\ & x_{ij} \geq 0 \quad i, j = A, B, C, D, E \end{aligned}$$

$$x^* = \{x_{AB}^* = 0, x_{AD}^* = 1, x_{CB}^* = 13, x_{CD}^* = 12, x_{EB}^* = 0, x_{ED}^* = 2\}$$

$\phi = 6.87$ (min.) $U' = 56.0\%$ $\omega = 3.48$ (min.) $\beta = 1.05$ (sec.)

Probability Dist. of the Number of Handling Transactions:

$P_0=0.001$	$P_1=0.006$	$P_2=0.021$	$P_3=0.049$	$P_4=0.088$	$P_5=0.124$
$P_6=0.147$	$P_7=0.149$	$P_8=0.132$	$P_9=0.104$	$P_{10}=0.074$	$P_{\geq 11}=0.106$

Workcenter Specific Parameters:

	Wkctr A:	Wkctr B:	Wkctr C:	Wkctr D:	Wkctr E:
μ_{oi}	65.0	22.4	9.9	22.3	13.3
λ_{oi}	24.0	15.0	0.0	5.0	11.0
π_{oio}	0.632	0.364	1.000	0.776	0.255
π_{oi1}	0.234	0.243	0.000	0.174	0.210
π_{oi2}	0.086	0.163	0.000	0.039	0.174
π_{oi3}	0.032	0.109	0.000	0.009	0.144
π_{oi4}	0.012	0.073	0.000	0.002	0.119
$\pi_{oi\geq 5}$	0.004	0.049	0.000	0.000	0.098
μ_{Ii}	14.9	19.0	25.0	15.0	13.5
λ_{Ii}	11.3	8.7	16.6	8.7	7.9
π_{Iio}	0.301	0.549	0.368	0.437	0.434
π_{Ii1}	0.227	0.250	0.244	0.253	0.253
π_{Ii2}	0.171	0.114	0.162	0.147	0.147
π_{Ii3}	0.129	0.052	0.107	0.085	0.086
π_{Ii4}	0.097	0.024	0.071	0.049	0.050
$\pi_{Ii\geq 5}$	0.074	0.011	0.047	0.029	0.029

Fig. 5.5 Model outputs from revised layout 3.

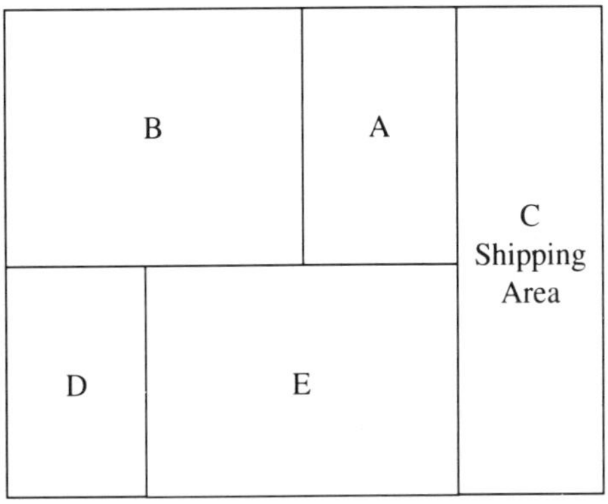

Travel Time Matrix T:
(in minutes)

	A	B	C	D	E
A	–	3	3	6	3
B	3	–	6	3	4
C	3	6	–	7	4
D	6	3	7	–	3
E	3	4	4	3	–

Veh. Speed: 500 ft./min.
Scale: 1" = 1000 ft.
Veh. Efficiency: e = .90
Veh. Dispatching: RV/RW

Model Outputs:

$$\text{Min: } 3x_{AB}+3x_{AC}+6x_{AD}+3x_{AE}+3x_{BA}+6x_{BC}+3x_{BD}+4x_{BE}+3x_{CA}+6x_{CB}+ 7x_{CD}+4x_{CE}+6x_{DA}+3x_{DB}+7x_{DC}+3x_{DE}+3x_{EA}+4x_{EB}+4x_{EC}+3x_{ED}$$

$$\begin{aligned} \text{s.t.} \quad & x_{AB}+x_{AD} &&= 1 \\ & x_{CB}+x_{CD} &&= 25 \\ & x_{EB}+x_{ED} &&= 2 \\ & x_{AB}+x_{CB}+x_{EB} &&= 13 \\ & x_{AD}+x_{CD}+x_{ED} &&= 15 \\ & x_{ij} &&\geq 0 \quad i, j = A, B, C, D, E \end{aligned}$$

$$x^* = \{x_{AB}^* = 1, x_{AD}^* = 0, x_{CB}^* = 12, x_{CD}^* = 13, x_{EB}^* = 0, x_{ED}^* = 2\}$$

$\phi = 6.95$ (min.) $U' = 53.0\%$ $\omega = 3.22$ (min.) $\beta = 0.41$ (sec.)

Probability Dist. of the Number of Handling Transactions:

$P_0 = 0.002$	$P_1 = 0.010$	$P_2 = 0.033$	$P_3 = 0.071$	$P_4 = 0.114$	$P_5 = 0.147$
$P_6 = 0.158$	$P_7 = 0.146$	$P_8 = 0.117$	$P_9 = 0.084$	$P_{10} = 0.054$	$P_{\geq 11} = 0.065$

Workcenter Specific Parameters:

	Wkctr A:	Wkctr B:	Wkctr C:	Wkctr D:	Wkctr E:
μ_{oi}	103.7	21.4	18.8	10.7	19.1
λ_{oi}	24.0	15.0	0.0	5.0	11.0
π_{oio}	0.769	0.340	1.000	0.538	0.440
π_{oi1}	0.178	0.238	0.000	0.252	0.254
π_{oi2}	0.041	0.167	0.000	0.118	0.146
π_{oi3}	0.010	0.117	0.000	0.055	0.084
π_{oi4}	0.002	0.082	0.000	0.026	0.048
$\pi_{oi\geq 5}$	0.001	0.057	0.000	0.012	0.028
μ_{Ii}	15.0	18.9	25.0	14.8	14.7
λ_{Ii}	11.6	8.9	16.7	8.7	7.9
π_{Iio}	0.290	0.533	0.365	0.428	0.472
π_{Ii1}	0.224	0.252	0.243	0.253	0.255
π_{Ii2}	0.172	0.119	0.163	0.149	0.137
π_{Ii3}	0.133	0.056	0.108	0.088	0.074
π_{Ii4}	0.102	0.027	0.072	0.052	0.040
$\pi_{Ii\geq 5}$	0.079	0.013	0.048	0.031	0.022

Fig. 5.6 Model outputs from revised layout 4.

known with certainty during the early phases of the facility design process, sensitivity studies based on a range of prospective parameter values are crucial to effective system design. The tools described in this chapter provide a purely analytical method for these types of sensitivity analyses. As a result, they provide the user with the flexibility to intelligently enumerate the design solution space without the need to overcome significant barriers associated with problem representation and model execution. The DSS under development as part of continuing research will make these tools directly accessible to the facility designer.

REFERENCES

Malmborg, C.J. (1990) A model for the design of zone control automated guided vehicle systems. *International Journal of Production Research*, **28**(10).

Malmborg, C.J. and Shen, Y.C. (1991a) Fleet size based modeling of vehicle dispatching effects in discrete materials handling systems. *IEEE Transactions on Systems, Man and Cybernetics* (in review, October).

Malmborg, C.J. and Shen, Y.C. (1991b) *Simulation Based Validation of Real Time Control Rules in Discrete Materials Handling Systems.* Proceedings 1991 IEEE International Conference on Systems, Man and Cybernetics, Charlottesville, VA, October.

Maxwell, W.L. and Muckstadt, J.A. (1982) Design of automatic guided vehicle systems. *IIE Transactions*, **14**(2), 114–24.

Ozden, M. (1988) A simulation study of multiple load carrying automatic guided vehicles in a flexible manufacturing system. *International Journal of Production Research*, **26**(8), 1353–66.

Zeng, L., Wang, H.P. and Jin, S. (1991) Conflict detection of automated guided vehicles: A Petri net approach. *International Journal of Production Research*, **29**(5), 865–79.

PART THREE

Alternative Material Flow Paths

CHAPTER 6

Reachability in material flow path design

K.H. Kim and J.M.A. Tanchoco

6.1 INTRODUCTION

The material flow path layout design is one of the most important variables in designing carrier-based material handling systems. The guide path layout design process includes the selection of lanes among candidate aisle spaces and determination of the flow directions of lanes.

Gaskins and Tanchoco (1987) first formulated the AGV flow path design problem as a zero–one integer programming model. The decision variable in their model is the direction of each unidirectional flow path segment. The computational efficiency of the first model was further improved by Kaspi and Tanchoco (1990), who applied a branch-and-bound procedure instead of solving a zero–one integer programming. Gaskins, Tanchoco and Taghaboni (1989) extended the previous model to allow multiple lanes in an aisle, and added an upper bound to the total amount of flow on each lane. Several other authors have addressed the AGV flow path problem (Venkataramanan and Wilson, 1991; Kouvelis *et al.*, 1992). A paper by Kim and Tanchoco (1993) gives a tighter bound for a branch-and-bound procedure based on the model developed by Kaspi and Tanchoco (1990), and shows some computational results.

All the above methods used the from–to matrix as input data in describing the intensities of material flows between pickup and delivery stations.

Although the above procedures minimized the total travel–distance–frequency, it did not guarantee that an empty vehicle which is departing a delivery station can get to an arbitrary pickup station requesting a delivery, a characteristic which defines the reachability of a flow path network. It results from the fact that the flow intensities of empty vehicles from delivery stations to pickup stations may not be considered explicitly in the from–to matrix, since the traffic behavior of empty vehicles is dynamic, uncertain and highly dependent on the operational control strategies implemented.

In this chapter, we discuss how to check the possible unreachability of the resulting flow path network by analysing the from–to matrix. A method is suggested which guarantees the reachability of the entire flow path network by modifying the from–to matrix. An implementation of this algorithm, coded in Fortran, is listed in the Appendix 6.A.

6.2 REACHABILITY AND CONNECTIVITY

The term, reachability, is clarified as follows:

Definition 1 A flow path network is said to be reachable if there is at least one path from each of the delivery stations to every pickup station.

Now, we introduce a useful concept from graph theory – the strong connectivity of a graph, which means that there is a path from each node to every other node.

Definition 2 Let $G = (V, G)$ be a directed graph. We can partition V into equivalence classes V_i, $1 \leqslant i \leqslant r$, such that nodes v and w are equivalent if and only if there is a path from v to w and a path from w to v. Let E_i, $1 \leqslant i \leqslant r$, be the set of arcs connecting the pairs of nodes in V_i. The graphs $G_i = (V_i, E_i)$ are called the **strongly connected components** of G. A graph is said to be **strongly connected** if it has only one strongly connected component (Aho, Hopcroft and Ullman, 1974).

Note that no G_i is a proper subgraph of a strongly connected subgraph of G (Tarjan, 1972), and the strongly connected components are uniquely determined for a given graph by the properties of the equivalence class.

It is possible to convert the from–to matrix to a graph using pickup/delivery stations as nodes and a flow requirement from station i to station j as an arc from node i to node j. We define this new graph as the P/D graph. As an illustration, consider the flow path layout network in Fig. 6.1(a) and the from–to matrix given in Table 6.1. The corresponding P/D graph is shown in Fig. 6.1(b). Then the following property holds:

Property 1 If the P/D graph constructed from a from–to matrix is strongly connected, any flow path network, which provides at least a path for every positive element in the from–to matrix, is reachable.

Proof Select an arbitrary pair of delivery and pickup stations, node i and node j. Then from the connectivity of the P/D graph, there is a path from node i to node j on the P/D graph which can be expressed as $i \rightarrow i_1 \rightarrow i_2 \rightarrow \cdots \rightarrow i_k \rightarrow j$ without loss of generality, which means there is a positive flow requirement from each preceding node to the immediate succeeding node. From the assumption on the flow path network, the above statement implies that there is at least a path on the flow path network from each preceding node

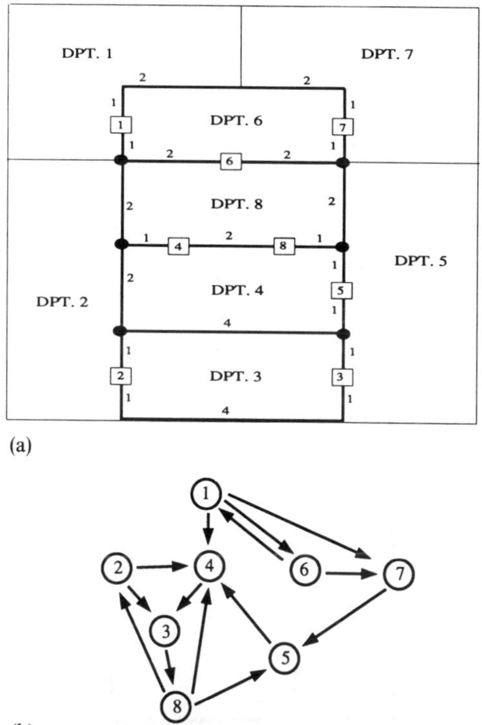

Fig. 6.1 (a) Initial layout (□ = P/D stations; —— = AGV lanes – numeric value on each lane is its length); and (b) P/D graph for the example.

Table 6.1 From – to flow matrix between P/D stations

	To							
From	1	2	3	4	5	6	7	8
1	–			39		5	10	
2		–	5	12				
3			–					6
4			13	–				
5				15	–			
6	90					–	5	
7					16		–	
8		30		6	10			–

to the immediate succeeding node. Therefore, there is at least a path from node i to node j on the flow path network. Q.E.D.

All the published algorithms for designing flow path networks provide at least one path for every positive element in the from–to matrix, and try to minimize the travel distance at the same time. Thus, we only have to concern

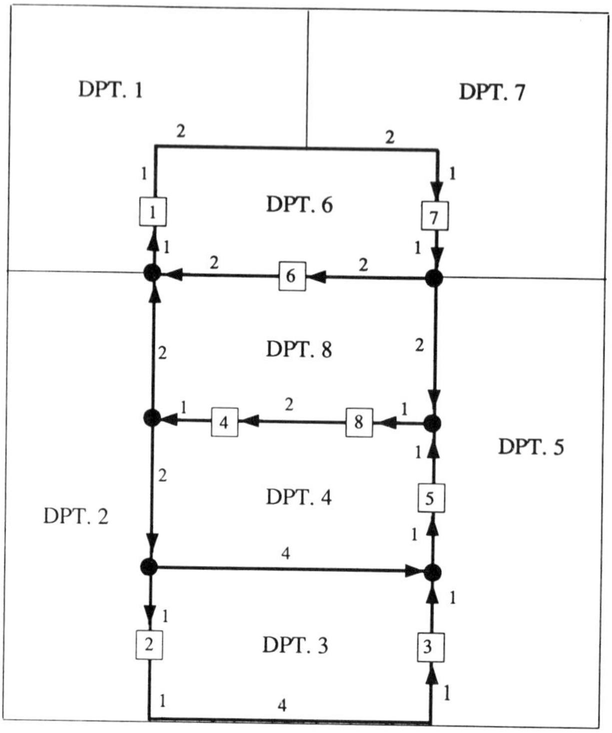

Fig. 6.2 Example of an unreachable flow path. → = AGV lanes; □ = P/D stations.

ourselves with the connectivity of the P/D graph instead of being concerned about the reachability of every candidate for the optimal flow path network during the design procedure. Note that, even if the P/D graph is not strongly connected, it does not mean that the resulting flow path network is not reachable. Instead, it only means that its reachability can not be guaranteed. Figure 6.2 illustrates an unreachable but optimal flow path network for the problem of Fig. 6.1 and Table 6.1, resulting from the flow path design algorithm suggested by Kaspi and Tanchoco (1990).

6.3 STRONGLY CONNECTED COMPONENTS OF THE P/D GRAPH

In this section, a procedure is introduced to check whether a given P/D graph is strongly connected or not, and to find out strongly connected components in case the whole graph is not strongly connected. Detailed explanations may also be found in Aho, Hopcroft and Ullman (1974).

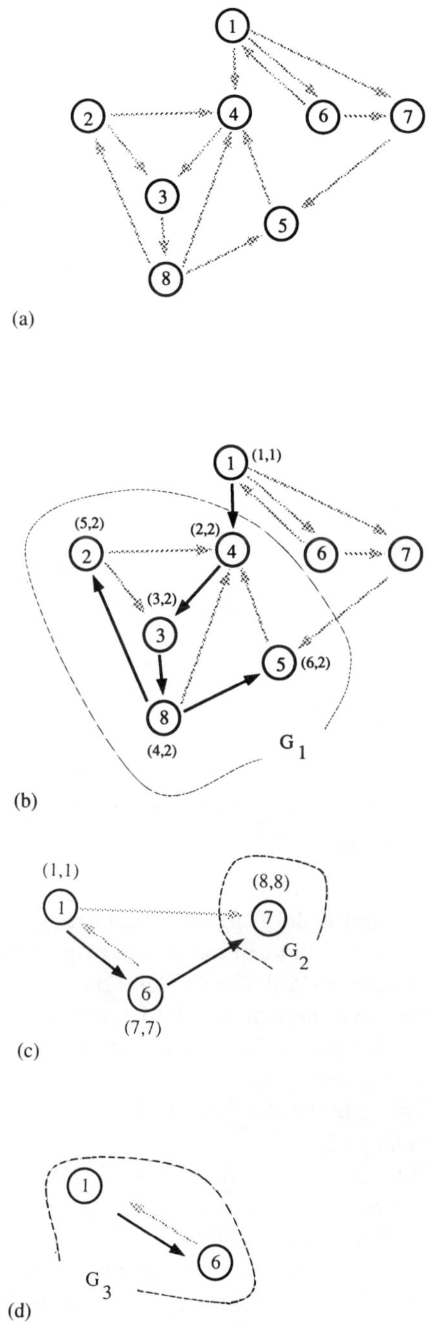

Fig. 6.3 FISCC procedure. (a) Initial P/D graph; (b) after finding the first SCC; (c) after finding the second SCC; (d) after finding the third SCC (→ = arcs on depth-first search tree; (m, n) are such that $m = VS(i)$, $n = SVS(i)$.

The depth-first search procedure is used to find the strongly connected components (SCC) of a P/D graph.

Let us introduce some notations and definitions:

$VS(i)$: visiting sequence number of node i in the search procedure

r_k: the root of an SCC which is found in the kth place. The root of a SCC is defined as the lowest-numbered (VS) common ancestor of nodes in the SCC. Once r_k is identified, we can identify G_k by collecting those nodes which are descendants of r_k but are in none of $G_1, G_2, \ldots, G_{k-1}$, while $G_1, G_2, \ldots, G_{k-1}$ are SCCs which have already been identified before r_k is found

$SVS(i)$: the smallest VS value of neighbors which have arcs from node i or its descendants. Formally stated,

$$SVS(i) = \min(\{VS(i)\} \cup \{VS(j) \mid \text{there is a cross arc or a back arc from } i \text{ or a descendant of } i \text{ to } j\})$$

$VS(i)$ and $SVS(i)$ are illustrated in Fig. 6.3. The significance of $SVS(i)$ is that if we encounter a node whose $VS(i) = SVS(i)$ and whose descendants are all searched, then we can conclude that the node is a root node

$PA(i)$: parent node number of node i

$NEWCH(i)$: child node number of node i being considered and visited for the first time

$OLDCH(i)$: child node number of node i which is being considered now but has already been visited.

The algorithm to find SCCs, called the Find-SCC (FISCC) procedure, may be summarized as follows:

Step 0: (**Initialize**) $IS = 0$. Go to step (1).

Step 1: (**Choose the initial node**) Choose an arbitrary initial node from the graph. If no node remains in the graph, stop. Otherwise, $IS = IS + 1$, $VS(i) = IS$, $SVS(i) = VS(i)$. Go to step (2).

Step 2. (**Consider the next child**) If no child remains, go to step (5). If the next child is the node which has already been visited (old child), go to step (4). Otherwise, go to step (3).

Step 3. (**New child**) Move to the child's node. $IS = IS + 1$, $VS(i) = IS$, $SVS(i) = VS(i)$. Go to step (2).

Step 4. (**Old child**) $SVS(i) = \min\{SVS(i), VS \text{ (old child)}\}$. Go to step (2).

Step 5. (**Find root and SCC**) If $VS(i) > SVS(i)$, go to step (6). Otherwise, this node is a root. List members of the SCC (the root and its descendants). Remove the listed nodes from the graph. Go to step (6).

Step 6: (**Backtracking**) Check if there is a parent node of the current node. If not, go to step (1). Otherwise, move to the parent node and set $SVS(i) = \min\{SVS(i), SVS \text{ (child just visited)}\}$. Go to step (2).

Table 6.2 and Fig. 6.3 illustrates the FISCC procedure for the example of Table 6.1.

Table 6.2 FISCC procedure

i	*VS*	*SVS*	*PA*	*NEWCH*	*OLDCH*	*Node*
1	1	1	–	4	–*	Initial node
1	1	1	–	4	–	Node 4 is
4	2	2	1		–	new child
						of node 1
1	1	1	–	4	–	Node 3 is
4	2	2	1	3	–	new child
3	3	3	4		–	of node 4
1	1	1	–	4	–	Node 8 is
4	2	2	1	3	–	new child
3	3	3	4	8	–	of node 3
8	4	4	3			
1	1	1	–	4	–	Node 2 is
4	2	2	1	3	–	new child
3	3	3	4	8	–	of node 8
8	4	4	3	2		
2	5	5	8	–		
1	1	1	–	4	–	Node 3 is
4	2	2	1	3	–	old child
3	3	3	4	8	–	of node 2
8	4	4	3	2		
2	5	3	8	–	3	Revise *SVS*(2)
1	1	1	–	4	–	Node 4 is
4	2	2	1	3	–	old child
3	3	3	4	8	–	of node 2
8	4	4	3	2		
2	5	2	8	–	4	Revise *SVS*(2)
1	1	1	–	4	–	Backtracking
4	2	2	1	3	–	to node 8
3	3	3	4	8	–	
8	4	2	3	2	4	Revise *SVS*(8)
2	5	2	8	–	–	
1	1	1	–	4	–	Node 5 is
4	2	2	1	3	–	new child
3	3	3	4	8	–	of node 8
8	4	2	3	5	–	
2	5	2	8	–	–	
5	6	6	8	–		
1	1	1	–	4	–	Node 4 is
4	2	2	1	3	–	old child
3	3	3	4	8	–	of node 5
8	4	2	3	5	–	
2	5	2	8	–	–	Revise *SVS*(5)
5	6	2	8	–	4	
1	1	1	–	4	–	Backtracking
4	2	2	1	3	–	to node 8
3	3	3	4	8	–	

*No more nodes to consider.

(*continued*)

Table 6.2 (*Contd.*)

i	*VS*	*SVS*	*PA*	*NEWCH*	*OLDCH*	*Node*
8	4	2	3	–	–	Revise *SVS*(8)
2	5	2	8	–	–	
5	6	2	8	–	–	
1	1	1	–	4	–	Backtracking
4	2	2	1	3	–	to node 3
3	3	2	4	–	–	
8	4	2	3	–	–	Revise *SVS*(3)
2	5	2	8	–	–	
5	6	2	8	–	–	
1	1	1	–	4	–	Backtracking
4	2	2	1	–	–	to node 4
3	3	2	4	–	–	$VS(4) = SVS(4)$
8	4	2	3	–	–	
2	5	2	8	–	–	$r_1 = 4$
5	6	2	8	–	–	$G_1 = \{4, 3, 8, 2, 5\}$

6.4 MODIFICATION OF FROM–TO MATRIX

Suppose that a P/D graph may be partitioned into multiple SCCs. The next problem is how to modify the from–to matrix to make the P/D graph a strongly connected graph.

If we represent a SCC (G_i) as node i and draw an arc from node i to node j when there exists at least an arc from a P/D station in G_i to a P/D station in G_j, we can construct a new graph which we call a 'G-graph'.

Then, the P/D graph is strongly connected if and only if the G-graph is strongly connected. We suggest a method to make the G-graph strongly connected below.

Note that a G-graph is a directed acyclic graph, since if the graph has a cycle, all nodes on the cycle might have been condensed into a single SCC in the FISCC procedure.

Then, we can construct the following three subsets of SCCs:

$F_1 = \{G_{1i} | \text{SCC which has no incoming arc from other SCCs}\}$
$F_2 = \{G_{2i} | \text{SCC which has both incoming arc from other SCCs and outgoing arc to other SCCs}\}$
$F_3 = \{G_{3i} | \text{SCC which has no outgoing arc to other SCCs}\}$

Note that an SCC with no incoming and outgoing arc belongs to both F_1 and F_3.

Property 2 The lower bound on the number of additional arcs required to make a G-graph a strongly connected graph is equal to $\max\{N_1, N_3\}$, where N_i is the number of elements in F_i.

Proof A necessary condition for a P/D graph to be strongly connected is that every SCC has at least an incoming arc and an outgoing arc. Thus, the conclusion holds. Q.E.D.

Now we propose a method called the Make-It-Strongly-Connected (MISC) procedure to make a G-graph strongly connected, as follows:

Step 1: Find out a pair of SCCs, $G_{1i} \in F_1$ and $G_{3j} \in F_3$, for which there is no path from G_{1i} to G_{3j}. Add an arc from G_{3j} to G_{1i}. Delete G_{3j} and G_{1i} from F_3 and F_1, respectively. Repeat this step unitl no other pairs can be found.

Step 2: Add arcs from elements of F_3 to elements of F_1 in the manner which satisfies that (1) if F_1 has more elements than F_3, at least one added arc has its tail on each element in F_3 and exactly one added arc has its head on each element of F_1; (2) if F_3 has more elements than F_1, at least one added arc has its head on each element of F_1 and exactly one added arc has its tail on each element of F_3.

Property 3 The G-graph with arcs added by the MISC procedure is strongly connected. The procedure requires the minimum number of added arcs which will make the G-graph strongly connected.

Proof Let us consider the resulting graph from step (1), and we maintain the elements of F_1, F_2 and F_3 as in the resulting graph from step (1) in the following explanation. Then, the following assertions hold for the graph:

1. It is a directed acyclic graph. This assertion follows from the fact that the initial G-graph is a directed acyclic graph, and we do not make any cycle in step (1).
2. There is at least a path from every element in F_1 to every element in F_3. This assertion comes from the fact that there remains no (G_{1i}, G_{3j}) pair which has no path from G_{1i} to G_{3j} as a result of step (1).
3. For each G_{2i}, we can always find at least a path from an element of F_1 to an element of F_3 which passes through the G_{2i}. This assertion comes from the fact that the graph has no cycle (assertion (1)).

Then let us consider the final graph, but still maintain the elements of F_1, F_2 and F_3 resultant from step (1) in the following explanation:

4. There is a path from each G_{1i} to every other node. This assertion is validated by the following assertions (a), (b) and (c):
   (a) From (2), there is a path from every G_{1i} to every G_{3j}.
   (b) Since every G_{1k}, for $k \neq i$, has at least an arc from an element of F_3 by step (2), considering (2), there is a path from every G_{1i} to every other G_{1k}.
   (c) From (3) and (b), we can conclude that there is a path from every G_{1i} to every G_{2k}.

5. Every G_{2j} has at least a path to a G_{3k} by (3) and G_{3k} has a path to at least a G_{1i} by step (2), which means that every G_{2j} and G_{3k} has a path to every other node by the assertion (4).

From (4) and (5), we can conclude that the resulting graph from step (2) is a strongly connected graph. From the fact that the number of arcs added exactly coincides with the lower bound of Property 2, the conclusion holds.

Once we identify pairs of (G_{1i}, G_{3j}) on which arcs should be added, we may choose any one arbitrary node from each of G_{1i} and G_{3j} to connect a corresponding pair of SCCs. A different procedure to make a graph strongly connected is discussed in the paper by Eswaran and Tarjan (1976).

6.4.1 Illustration

Figure 6.4 illustrates MISC procedure for an example with six SCCs.

In the example of Fig. 6.3, there is a need to add an arc from one node among 2, 3, 4, 5, 8 to one between 1, 6 to make the whole P/D graph strongly connected.

If we add infinitesimal values to the zeros in the from–to matrix corresponding to the added arcs in the P/D graph, the flow path network constructed will become reachable only if the algorithm used provides at least one path for every positive element in the from–to matrix. Furthermore, since infinitesimal values are added, the optimality of the final solution will not be affected.

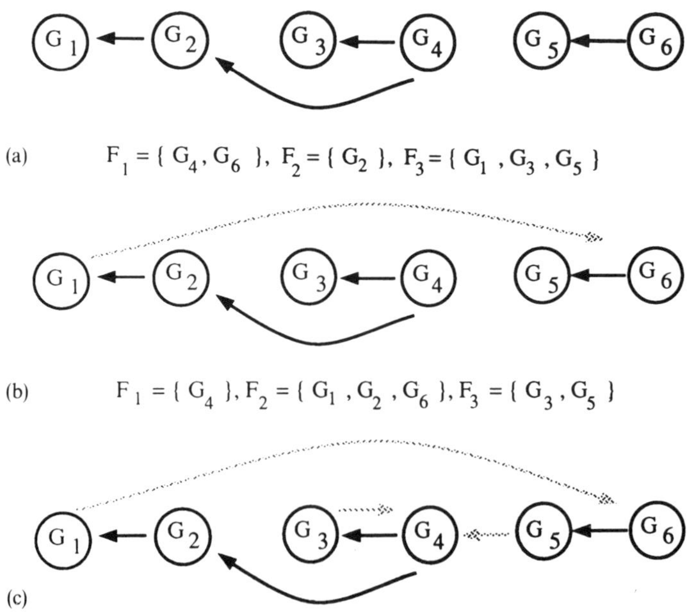

Fig. 6.4 MISC procedure. (a) Initial G-graph; (b) after step (1); (c) after step (2).

Table 6.3 illustrates a modified from–to matrix for the example of Table 6.1, and Fig. 6.5 shows the optimal flow path designed using the modified from–to matrix of Table 6.3 and the initial layout of Fig. 6.1.

Table 6.3 Modified from – to flow matrix between P/D stations

	To							
From	1	2	3	4	5	6	7	8
1	–			39		5	10	
2		–	5	12				
3			–					6
4	ε		13	–				
5				15	–			
6	90					–	5	
7					16		–	
8		30		6	10			–

ε ≪ 1.

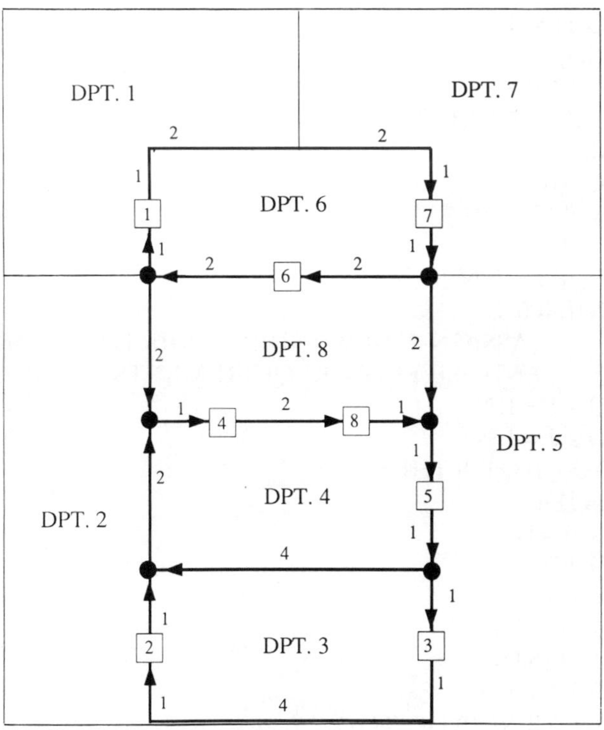

Fig. 6.5 Example of the optimal and reachable flow path. → = AGV lanes; □ = P/D stations.

APPENDIX 6.A FLOW REACH: MAKE FROM–TO MATRIX REACHABLE*

```
      DIMENSION TITLE(60)
      DIMENSION Q(50,50)
      DIMENSION QQ(50,50), NN(50), IVS(50), ISVS(50), IFA(50),
     1NEWSN(50), IRT(10), ISCC(50), G(10,10), R(10,10), F(10)
CC MODIFY TRANSPORT REQUIREMENT MATRIX TO
CC    GUARANTEE REACHABILITY OF EMPTY CARRIER FROM
CC    DELIVERY STATION TO PICK-UP STATION
CC                N: NUMBER OF INITIAL NODES IN FROM TO
CC                CHART
CC                L: NUMBER OF FLOW REQUIREMENTS
CC                Q(I,J): FLOW REQUIREMENT FROM NODE I
CC                  TO NODE J
      KK = 0
      READ(5,1111)(TITLE(I), I = 1,60)
1111  FORMAT(60A1)
      READ(5,*)N,L
      DO 1100 I = 1,N
      DO 1100 J = 1,N
1100  Q(I,J) = 0.
      DO 1000 K = 1,L
1000  READ(5,*)I,J,Q(I,J)
      INS = 0
      ISC = 0
      I1 = 0
      IS = 0
      DO 10 I = 1,N
10    NN(I) = 0
CC            ASSIGN A NEW INDEX(NN(I)) TO THE NODES
CC            WHOSE FLOW REQUIREMENTS ARE POSITIVE
      DO 1 I = 1,N
      DO 2 J = 1,N
      IF(Q(I,J).GT.0) THEN
      I1 = I1 + 1
      NN(I) = I1
      GO TO 1
      END IF
2     CONTINUE
1     CONTINUE
      DO 3 J = 1,N
      IF(NN(J).NE.0) GO TO 3
```

```
      DO 5 I = 1, N
      IF (Q(I, J).GT.0) THEN
      I1 = I1 + 1
      NN (J) = I1
      GO TO 3
      END IF
5     CONTINUE
3     CONTINUE
CC          I1: TOTAL NUMBER OF NODES ACTUALLY
CC            CONSIDERED
      DO 7 I = 1, I1
      DO 7 J = 1, I1
7     QQ (I, J) = 0.
CC          REVISE THE INDEX OF Q(I, J)
      DO 6 I = 1, N
      DO 6 J = 1, N
      IF (Q(I, J).GT.0) THEN
      QQ (NN(I), NN(J)) = Q(I, J)
      END IF
6     CONTINUE
CC          IVS(I): SEQUENCE NUMBER OF NODE I
CC          ISVS(I): THE SMALLEST VISITING SEQUENCE OF
CC            NEIGHBORS
CC          IFA(I): THE FATHER NODE OF NODE I
CC          NEWSN(I): THE NODE NUMBER OF NEW SON
CC            BEING CONSIDERED
CC          IS: INDEX TO ASSIGN SEQUENCE NUMBER
      DO 11 I = 1, I1
      IVS(I) = 0
      ISVS(I) = 0
      IFA(I) = 0
11    NEWSN(I) = 0
CC          SELECT A NEW NODE UNASSIGNED
19    DO 15 I = 1, I1
      IF (IVS(I).EQ.0) THEN
      INN = I
      GO TO 12
      END IF
15    CONTINUE
12    IS = IS + 1
      IVS (INN) = IS
      ISVS (INN) = IS
13    CONTINUE
      WRITE (6, *) 'INN = ', INN, 'SON = ', NEWSN (INN), 'IFA = ',
        IFA (INN), 'IVS = ', IVS (INN), 'ISVS = ', ISVS (INN)
```

```
CC              IF THERE IS NO SON LEFT
        IF(NEWSN(INN).EQ.11) GO TO 14
        DO 111 J=NEWSN(INN)+1,I1
CC              FIND OUT THE NEXT SON
        IF(QQ(INN,J).GT.0.AND.IVS(J).NE.999) THEN
        NEWSN(INN)=J
CC              IF THE SON IS OLD SON
        IF(IVS(J).GE.1) THEN
        ISVS(INN) = AMIN1(ISVS(J),ISVS(INN))
        GO TO 13
CC              IF THE SON IS NEW SON
        ELSE
        IFA(J) = INN
        NEWSN(INN) = J
        INN = J
        GO TO 12
        END IF
        END IF
111     CONTINUE
CC              BACKTRACKING
14      JNN = IFA(INN)
CC              IF THIS IS THE TOP NODE OR ROOT NODE
        IF(JNN.EQ.0.OR.IVS(INN).EQ.ISVS(INN)) GO TO 100
        ISVS(JNN) = AMIN1(ISVS(INN),ISVS(JNN))
        INN = JNN
        GO TO 13
CC              ISC: INDEX OF STRONGLY CONNECTED
CC                COMPONENT (SCC)
CC              IRT(ISC): THE ROOT OF SCC(ISC)
CC              ISCC(I): ISC INDEX OF NODE I
100     ISC = ISC + 1
        IRT(ISC) = INN
        IVSINN = IVS(INN)
        WRITE(6,*)'SCC NUMBER', ISC
CC              WRITE THE MEMBERS OF A NEW SCC
CC              INS: NO. OF NODES WHOSE MEMBERSHIP
CC                ARE ASSIGNED
        DO 110 K = IVS(INN), IS
        DO 120 I = 1,I1
        IF(IVS(I).EQ.K) THEN
        ISCC(I) = ISC
        DO 130 J = 1,N
        IF(NN(J).EQ.I) THEN
        WRITE(6,*) 'NODE =',J
        INS = INS + 1
        END IF
```

```
130     CONTINUE
        IVS(I) = 999
        END IF
120     CONTINUE
110     CONTINUE
        IS = IVSINN - 1
CC              IF ALL THE SCC ARE IDENTIFIED
        IF(JNN.EQ.0.AND.INS.EQ.I1) GO TO 200
CC              IF SOME OTHER NODES ARE REMAINED
        IF(JNN.EQ.0.AND.INS.NE.I1) GO TO 19
                BACKTRACKING
        INN = IFA(INN)
        GO TO 13
CC              TO INITIALIZE G MATRICES
CC              G: G-MATRIX
CC              R: REACHABILITY MATRIX OF SCC
CC              F: INDEX OF F TO WHICH A SCC BELONGS
200     IF(ISC.EQ.1) GO TO 3000
        DO 210 I = 1, ISC
        DO 210 J = 1, ISC
        G(I, J) = 0.
        IF(I.EQ.J) THEN
          G(I, J) = 1.
        END IF
210     CONTINUE
CC              BUILD A MATRIX
        DO 220 I = 1, ISC - 1
        DO 230 J = I + 1, ISC
        DO 240 II = 1, I1
        If(ISCC(II).EQ.I) THEN
          IX = II
          DO 250 JJ = J1, I1
          IF(ISCC(JJ).EQ.J) THEN
            IY = JJ
            IF(QQ(IY, IX).GT.0.) THEN
              G(J, I) = 1.
              GO TO 230
            END IF
          END IF
250       CONTINUE
        END IF
240     CONTINUE
230     CONTINUE
220     CONTINUE
CC              ******************STEP 1******************
CC              TO FIND THE INDEX OF F FOR EACH SCC
```

```
288   CALL FINDF(G, F, ISC)
CC          DERIVE REACHABILITY MATRIX
      CALL REACH(G, R, ISC)
      ICHECK = 0
      DO 260 I = 1, ISC
      IF(F(I).EQ.1..OR.F(I).EQ.2.) GO TO 260
      DO 270 J = 1, ISC
      IF(F(J).EQ.3..OR.F(J).EQ.2.) GO TO 270
      IF(R(J,I).EQ.0.) THEN
        DO 280 K = 1, N
        DO(NN(K).EQ.IRT(I)) IX = K
280     IF(NN(K).EQ.IRT(J)) IY = K
        Q(IX, IY) = 0.001
        WRITE(6,*) 'ADD EPSILON TO Q(',IX,',',IY,') – STEP 1'
        G(I,J) = 1.
        ICHECK = 1
        GO TO 288
      END IF
270   CONTINUE
26    CONTINUE
      IF(ICHECK.EQ.0) GO TO 290
      GO TO 288
CC          *****************STEP 2*****************
290   ICHECK = 0
      DO 300 I = 1, ISC
      IF(F(I).EQ.3.) THEN
        DO 310 K = 1, N
310     IF(NN(K).EQ.IRT(I)) IX = K
        ICHECK = 1
        F(I) = 2.
        GO TO 301
      END IF
300   CONTINUE
301   DO 320 J = 1, ISC
      IF(J).EQ.1.) THEN
        DO 330 K = 1, N
330     IF(NN(K).EQ.IRT(J)) IY = K
        ICHECK = 1
        F(J) = 2.
        GO TO 302
      END IF
320   CONTINUE
302   IF(ICHECK.EQ.0) GO TO 3000
      Q(IX, IY) = 0.001
      kk = kk + 1
```

```
      WRITE(6,*) 'ADD EPSILON TO Q(',IX',',IY,') – STEP 2'
      GO TO 290
3000  WRITE(7,1111) (TITLE(I),I = 1,60)
      kk = kk + 1
      WRITE(7,*) N,KK
      DO 3001 I = 1,N
      DO 3001 J = 1,N
      IF(Q(I,J).LE.0.) GO TO 3001
      WRITE(7,*) I,J,Q(I,J)
3001  CONTINUE
      WRITE(7,*) N + 1,N + 1, "–99."
      STOP
      END
CC             SUBROUTINE TO ASSIGN INDEX OF F
CC             WHICH EACH SCC BELONGS TO
      SUBROUTINE FINDF(A,F,ISC)
      DIMENSION A(10,10), F(10)
      DO 1 I = 1,ISC
      IIN = 0
      IOUT = 0
      DO 2 J = 1,ISC
      IF(I.EQ.J) GO TO 2
      IF(A(I,J).EQ.1.) IOUT = 1
2     CONTINUE
      DO 3 K = 1,ISC
      IF(I.EQ.K) GO TO 3
      IF(A(K,I).EQ.1.) IIN = 1
3     CONTINUE
      IF(IOUT.EQ.1.AND.IIN.EQ.0) F(I) = 1
      IF(IOUT.EQ.0.AND.IIN.EQ.1) F(I) = 2
      IF(IOUT.EQ.0.AND.IIN.EQ.1) F(I) = 3
      IF(IOUT.EQ.1.AND.IIN.EQ.0) F(I) = 4
1     CONTINUE
      RETURN
      END
CC             TO CHECK REACHABILITY FROM
CC             NODE I TO NODE J.
CC             B(I,J) = 1 IF THERE IS A PATH
CC             NODE I TO NODE J
      SUBROUTINE REACH(A,B,ISC)
      DIMENSION A(10,10), B(10,10), C(10,10), D(10,10)
      DO 10 I = 1, ISC
      DO 10 J = 1, ISC
      B(I,J) = A(I,J)
      C(I,J) = A(I,J)
```

```
10      CONTINUE
        DO 1 II = 1, ISC
        DO 11 I = 1, ISC
        DO 11 J = 1, ISC
11      D(I, J) = 0.
        DO 2 I = 1, ISC
        Do 3 J = 1, ISC
        IF(I.EQ.J) GO TO 3
        DO 4 K = 1, ISC
4       D(I, J) = D(I, J) + C(I, K)*A(K, J)
3       CONTINUE
2       CONTINUE
        DO 6 I = 1, ISC
        DO 6 J = 1, ISC
6       C(I, J) = 0.
        DO 5 I = 1, ISC
        DO 5 J = 1, ISC
        IF(I.EQ.J) GO TO 5
        If(D(I, J).GT.0.) THEN
        C(I, J) = 1.
        B(I, J) = 1.
        END IF
5       CONTINUE
1       CONTINUE
        RETURN
        END
```

REFERENCES

Aho, A.V., Hopcroft, J.E. and Ullman, J.D. (1974) *The Design and Analysis of Computer Algorithms*, Addison Wesley, Reading, MA, pp. 172–95.

Eswaran, K.P. and Tarjan, R.E. (1976) Augmentation problems. *SIAM Journal of Computing*, **5**(4), 653–65.

Gaskins, R.J. and Tanchoco, J.M.A. (1987) Flow path design for automated guided vehicle systems. *Internationl Journal of Production Research*, **25**(5), 667–76.

Gaskins, R.J. Tanchoco, J.M.A. and Taghaboni, F. (1989) Virtual flow paths for free-ranging automated guided vehicle systems. *International Journal of Production Research*, **27**(1), 91–100.

Kaspi, M. and Tanchoco, J.M.A. (1990) Optimal flow path design of unidirectional AGVS systems. *International Journal of Production Research*, **28**(6), 1023–30.

Kim, K.H. and Tanchoco, J.M.A. (1993) Economical design of material flow paths. *International Journal of Production Research*, **41**(6), 1387–1407.

Tarjan, R. (1972) Depth-first search and linear graph algorithms. *SIAM Journal of Computing*, **1**(2), 146–60.

CHAPTER 7

Single-loop guide paths for AGVs

J.M.A. Tanchoco and D. Sinriech

7.1 INTRODUCTION

The increasing use of flexible manufacturing systems, in general, and the widespread use of robots and other general purpose computer controlled machines, in particular, has placed a greater demand on the performance of material handling systems. The variety of products and parts produced in those systems, each of which requires that a different series of operations be performed, results in fluctuating transport requirements.

While their initial applications have been in warehousing, there is growing use of Automated Guided Vehicles (AGVs) in the manufacturing area. The reason is that to exploit the capabilities of those flexible manufacturing systems, it is equally necessary to use a flexible material handling system.

The design of an AGV system is comprised of two stages. The first is the design of the guide path and the location of the pickup/delivery stations, and the second is the design of a supervisory controller that will manage the transport system.

7.1.1 The guide path design

Designing the vehicle's guide path can be approached in different ways, based on the initial assumptions made. The main differences are:

1. What has to be designed:
   (a) Design of the facility layout, guide path and pickup/delivery station location.
   (b) Design of the guide path and the pickup/delivery station location based on an existing facility layout.

The material in this chapter is obtained from Tanchoco, J.M.A. and Sinriech, D. (1992).

   (c) Design of the guide path based on an existing pickup/delivery stations and facility layout.
2. The objective of the model:
   (a) To minimize the flow in the system based on the material flow between departments.
   (b) To minimize the time/distance the AGV has to travel to complete its assignments.

Gaskins and Tanchoco (1987) assumed that the facility layout and pickup/delivery stations exist and formulated the guide path layout problem as a zero–one integer programming model. Usher, Evans and Wilhelm (1988) assumed that the facility layout is given, and proposed a procedure to locate the pickup/delivery stations. The proposed procedure consists of two phases. Phase I uses the model developed by Gaskins and Tanchoco (1987) in finding the direction of flow. Phase II uses a heuristic method for improving the system's performance by changing the location of the pickup/delivery stations. This procedure is very tedious when we consider a large number of departments, and still does not give an optimal solution. The reason is that pickup and delivery stations' location and guide path design have to be determined simultaneously, and not separately. Kaspi and Tanchoco (1990) assumed that the pickup/delivery station location and facility are given, and developed a procedure that is based on a zero–one integer programming model and a branch-and-bound method. This approach reduces the computation time to get the best guide path design because not all of the possibilities are enumerated and optimality is guaranteed. More recently, Goetz and Egbelu (1990) suggested a method for determining the location of pickup/delivery stations based on a finite set of available sites. The Gaskins and Tanchoco model was also used as the base model.

7.1.2 The controller design

The control of systems described in the previous subsection is a complicated task because of the dynamic nature of the manufacturing system. That means decisions have to made online, and they have to be fast and accurate. An approach to this problem and the complexity of the decisions that have to be taken are discussed in Taghaboni and Tanchoco (1988). The decisions that have to be made online are:

- which AGV to send to pickup/drop a load? This is referred to as a **dispatching decision**;
- to which machine should the AGV be sent? This is the first part of the routing decision. That means defining the destination;
- which route should the AGV take when making a pickup/delivery? This is the second part of the routing decision. This part defines the best way to get to the selected destination;
- determining the arrival and departure times of the vehicles at each segment

and intersection to ensure collision-free journeys. This was referred to as the **scheduling decision**;

- routing and dispatching vehicles to pickup/deliver jobs in such a way that **shop locking** will be prevented. This means that some look-ahead procedure has to be implemented.

It is clear how complex the control problem is. One approach is to address the control problem associated with conventional AGV guide paths and apply greater intelligence to the controller function, as suggested by Taghaboni and Tanchoco (1988). A second approach is to consider a less complex guide path design. One such example is the zoning configuration commonly found in dial-a-ride transportation systems, as discussed in Stein (1978). Bozer and Srinivasan (1989) also suggest the zoning method for AGVs, but no procedure is given as to how to partition the guide path into non-overlapping zones. Another way to reduce the complexity of the controller is to design a single-loop guide, as we propose in this chapter.

In a paper published in 1984, Egbelu and Tanchoco (1982) suggested the single-loop concept. They define a loop as a fixed sequence of departments that vehicles visit, and called it the **sequential dispatching strategy**. They state that the motivation for using such a configuration is the simplicity of the traffic control algorithms and the elimination of the shop locking possibility. The low performance they obtained is due to the fact that the loop they chose and the location of the pickup/delivery stations were not optimal, and the extra vehicle capacity that was needed for the single-loop was not met. Haines (1985) suggested a control algorithm for a loop material handling system. In this study, a distinction is made between a serial loop pattern, a single recirculating loop pattern and multiple recirculating loop pattern. The benefits and drawbacks of every option are discussed. However, there was no procedure given for designing those patterns. Although not motivated by the guide path design problem, Bartholdi and Platzman (1989) use the single-loop assumption to derive analytical results for the AGV control problem. The performance calculation they use is based on the procedure developed by Fuller (1972), which deals with scheduling the order of reading information from a drum memory.

The above approaches were made as one step towards a better system understanding. This study, contrary to those others, views the single-loop as the final guide path configuration.

7.2 A PROCEDURE TO FIND AN OPTIMAL SINGLE-LOOP DESIGN

The goal of the procedure is to find the 'best' single loop guide path and to locate the pickup/delivery stations along the loop. To get the 'best' loop, the one that will minimize the flow of parts in the system, a solution procedure **OSL** has been developed. The procedure contains the following parts:

1. An integer programming formulation to find an initial valid* loop to start the enumeration procedure, One possible approach is to use the Model VSLP (Valid Single Loop Problem), as suggested by Sinriech (1990). Additional constraints to avoid solutions with disjoint graphs are incorporated in this chapter. The complete model is discussed in section 7.2.2.
2. An enumeration procedure **FASL** that starts with a given single-loop guide path and finds all the possible valid single-loop guide paths.
3. A set of three rules: **CAN, ANIN** and **FNIN**. These rules are used to reduce the number of loops that have to be considered as candidates to the final guide path by eliminating loops with inferior performance capabilities.
4. An integer programming formulation to find the optimal location of the pickup/delivery stations along a given loop based on a from–to material flow matrix. The value of a given solution is the number of units that need to flow from station to station multiplied by the distance traveled. The formulation will be called the SLSLP (Single Loop Station Location Problem). The solution to SLSLP sets directions to the arcs that belong to the single loop.
5. A lower bound calculation procedure for the value of a given solution (that means a lower bound to the SLSLP formulation). The reason for that procedure is the fast results it gives compared to the much slower results from the SLSLP model.

7.2.1 General assumptions

The assumption is that the department layout exists and cannot be changed, but the positions of the pickup/delivery points are the system variables that have to be defined. The layout is represented as a network $G=(N, A)$, where N is the set of nodes which represent convergences and divergences (i.e. aisle intersections) and A is a set of arcs which represent the existing path segments (aisles) that connect the nodes (this can be thought of as an undirected graph).

Theorem 7.1 If a layout which can be represented as an undirected graph is (1) a connected graph, and (2) every vertex in the graph has a degree greater than one, then there exists at least one subgraph which forms a single loop.

Proof In general every point on the graph can be connected to any other point on the graph, especially every pickup station to every delivery station, because the graph is a connected graph. In extreme cases, when two departments have only one node in common, that node has to be divided into two adjacent nodes so a single-loop can be formed, as shown in Fig. 7.1.

*By definition, a valid loop contains at least one arc of each department in the layout, although pickup and delivery stations are only located at nodes, as explained later.

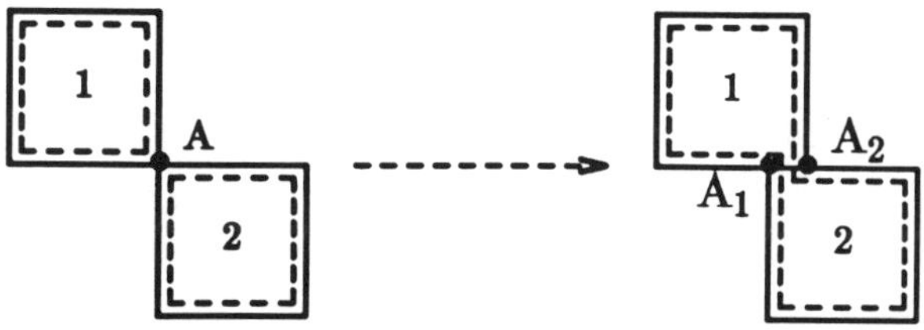

Fig. 7.1 Dividing a common node.

Fig. 7.2 Eliminating tree-shape guide paths.

The second part of the theorem prevents having a tree shape guide path. Every node has an incoming arc and an outgoing arc, as shown in Fig. 7.2.

In this study, the problem of finding the guide path is limited to undirectional single-loop guide paths. The distances calculated between departments are path distances: that means movement between departments is restricted to the defined aisles in the layout. The parts process plan and job mix is given, and will determine the from–to material flow matrix. The material handling system will be comprised of AGVs and single lane guide paths. This formulation is an NP-hard problem, but when solving problems containing around 15 departments, the solution is obtained in reasonable time.

7.2.2 Model *VSLP*

VSLP (Valid Single Loop Problem) finds a valid single-loop guide path. By definition, a valid loop contains at least one arc of each department in the layout. This problem is similar to the Traveling Salesman Problem (TSP). However, there are some major differences:

- The number of cities that are included in the TSP tour is fixed, while in the VSL problem it is not known in advance how many vertices will be included in forming a valid single-loop. However, there is an upper and a lower bound. These are the number of vertices N and the number of departments M, respectively.
- In the TSP the tour is started from a vertex labeled 0. This point is known and fixed. In the VSL problem, fixing a specific vertex as the starting point forces that point to be included in the loop. This limits the set of solutions due to the additional constraint added to the problem.

- The direction of the arcs in the TSP is essential. Even if the distances are symmetric ($d_{ij} = d_{ji}$) the direction is needed to determine the order of the tour. In the VSL problem, the distances are always symmetric and the solution is a set of arcs that form a closed loop.

(a) Notation

Departments $i, j = 1, \ldots, M$
Nodes $n, k = 1, \ldots, N.$
Arcs $m = 1, \ldots, p,$ two connected nodes (n, k), define an arc

$$x_{nk} = \begin{cases} 1 & \text{if arc } (n,k) \text{ is included in the guide path} \\ 0 & \text{otherwise} \end{cases}$$

$$z_n = \begin{cases} 1 & \text{if node } n \text{ is included in the guide path} \\ 0 & \text{otherwise} \end{cases}$$

A = the set of all arcs in the network.
S_i = the set of arcs that form department i.
Q_n = the set of arcs that are incident to node n.
d_{nk} = the length of arc (n, k).

$$\bigcup_{i-1}^{M} S_i = A.$$

$$S_i \cap S_j = \text{common arcs between departments } i, j.$$

(b) The objective function

The objective in this case is to find the shortest length single-loop guide path layout. Thus, the objective is

$$\text{minimize} \sum_{(n,k) \in A} d_{nk} x_{nk}.$$

(c) The constraints

There are six types of constraints in VSLP:

1. To have a valid single-loop, at least one arc of every department has to be in the final guide path:

$$\sum_{(n,k) \in S_i} x_{nk} \geqslant 1 \quad i = 1, \ldots, M.$$

   There is a constraint of this type for each department.
2. If a node is included in the guide path, that node has to have two and only two incident arcs. Otherwise, no arcs are connected to that node:

$$\sum_{(n,k) \in Q_n} x_{nk} = 2z_n \quad n = 1, \ldots, N.$$

   There is a constraint of this type for each node in the layout, excluding the start node which is duplicated.

3. Only one incident arc to an included node can be an incoming arc:

$$\sum_{n \in Q_k} x_{nk} \leqslant 1 \quad k = 1, \ldots, N.$$

   There is a constraint of this type for each node in the layout, excluding the node which is added as a duplicate node to ensure a closed loop.
4. Only one incident arc to an included node can be an outgoing arc:

$$\sum_{n \in Q_k} x_{kn} \leqslant 1 \quad k = 1, \ldots, N.$$

   There is a constraint of this type for each node in the layout, excluding the node which was duplicated to ensure a closed loop.
5. To eliminate having a disjoint loop solution, the constraint derived by Miller, Tucker and Zemlin (1960) is used. They suggest that one of the nodes in the layout be duplicated. Therefore, the duplicate node will be included automatically in the loop. y_n and y_k are dummy variables which are unrestricted real numbers:

$$y_n - y_k + N x_{nk} \leqslant N - 1 \quad n, k \in A, \quad n \neq k.$$

   There is a constraint of this type for each arc in the layout.
6. $$x_{nk}, x_{kn} = 0 \text{ or } 1 \quad \forall n, k \in A.$$

The above formulation does not ensure the optimality of the solution to the shortest length single-loop guide path. The optimality depends upon which node in the layout is duplicated.

7.2.3 The enumeration procedure *FASL* (Find All Single Loops)

This procedure finds all the single-loop guide paths in a given layout. The maximum possible number of single-loops in a layout can be found by the formula:

$$\sum_{i=1}^{M} \binom{\mathrm{M}}{\mathrm{i}},$$

because the different loops are a combination of departments inside and outside the loop. Not all of the combinations are valid single-loops, therefore this is the largest number of possible single-loops. The formula can be improved slightly, but it is only to show the large number of possible loops for even a medium size layout. This number is used to determine the size of the arrays used in the coded **FASL** procedure. The actual number of valid single-loops is much lower.

This procedure starts from a valid loop found by solving VSLP. The procedure keeps adding departments into the loop until the loop contains the largest number of departments and is still a valid loop, i.e. at least two nodes of every department are included in the single loop. The second phase is to take out departments from the loop while checking that the new loop has not been previously considered and is a valid loop, until the smallest number of departments that comprise a valid loop is reached. If a dead-end is reached, the algorithm traverses back to a point where a new loop can be found. This goes on until no new loop can be found. The procedure is formally described below:

begin
1. *set* i to 1
2. *put* the departments included in the loop given by the solution in the list $\{ID_i\}$. the rest of the departments (outside the loop) in the list $\{OD_i\}$.
3. *if* $\{OD_i\}$ is empty.
4.   *go* to 13.
    else
5.   *create* a new loop by adding the first department in $\{OD_i\}$ to the existing loop.
6.   *remove* that department from $\{OD_i\}$.
7.   *if* the new loop already exists in $\{ID_i\}$ or the new loop is not valid.
8.   *go* to 3.
        else
9.   *increment* i by one.
10.   *add* the new loop to $\{ID_i\}$.
11.   *create* the new $\{OD_i\}$ from all departments outside the loop.
12.   *go* to 3.
        end if
    end if
13. *put* the departments included in loop i from $\{ID_i\}$ in $\{OD_i\}$.
14. *if* $\{OD_i\}$ is empty.
15.   *go* to next stage†.
    else
16.   *create* a new loop by removing the first department in $\{OD_i\}$ from the loop.
17.   *remove* the department from $\{OD_i\}$.
18.   *if* the new loop already exists in $\{ID_i\}$ or the new loop is not valid.
19.   *go* to 14.
        else
20.   *increment* i by one.
21.   *add* the new loop to $\{ID_i\}$.
22.   *create* the new $\{OD_i\}$ from all the inside departments.
23.   *go* to 14.
        end if
    end if

† In every stage there are two parts, expansion and contraction. When both are done the next stage begins until no new departments are found.

7.2.4 Eliminating inferior loops by dominance

The set of single loops that is found by the enumeration procedure becomes very large as the number of departments in the layout increases. Therefore, this set has to be reduced to a manageable size. To reduce the number of loops, three rules were developed. These will eliminate loops whose performance is dominated by other loops. Only the remaining loops will be considered by the algorithm to find the best loop.

Rule 1: CAN (Contain All Nodes)

If $loop_1$ contains all the nodes $loop_2$ contains in the same order, and $loop_2$ contains all the departments $loop_1$ contains, and $loop_1$ and $loop_2$ are not the same loop, then the set of solutions achieved by using $loop_1$ dominates the set of solutions achieved using $loop_2$.

Proof Based on Theorem 7.2 (discussed in section 7.2.5), stations will be located only at nodes. Therefore, $loop_1$ which contains all nodes $loop_2$ contains in the same order, plus additional ones, achieves each solution $loop_2$ can, and more. This means that the set of solutions using $loop_1$ dominates the set of solutions using $loop_2$.

In the example shown in Fig. 7.3, $loop_1$ contains nodes 1, 4, 8, 6, 7, 3, 2, and $loop_2$ contains nodes 1, 4, 8, 6, 7, 3. Therefore, the solution using $loop_2$ cannot include positioning a station at node 2, while this is possible using $loop_1$.

Rule 2: ANIN (All Nodes and Inferior Nodes)

If $loop_2$ contains all the nodes of $loop_1$ in the same order, and $loop_2$ contains all the departments $loop_1$ contains, and $loop_2$ is longer, then the set of

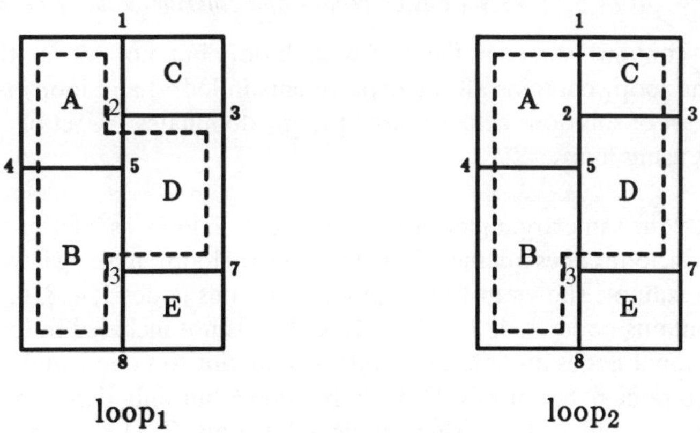

Fig. 7.3 Rule 1 examples.

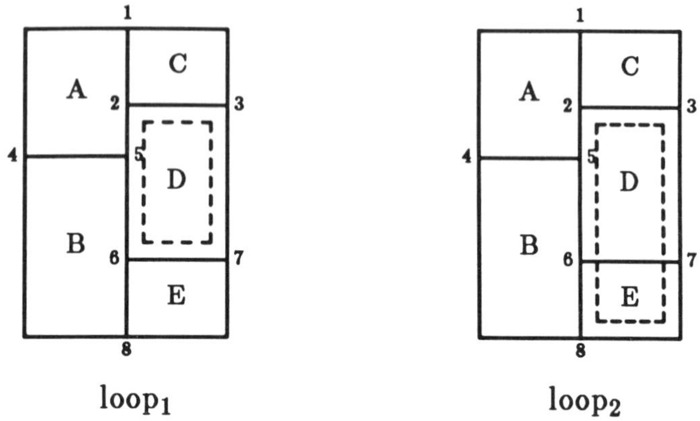

Fig. 7.4 Rule 2 example.

solutions achieved using loop_1 dominates the set of solutions achieved using loop_2.

Proof Although loop_2 contains additional nodes, those nodes are inferior because all the departments that share the new node also share the old node which is adjacent to it, and because of the additional nodes the travel distance between departments is longer. Therefore, loop_2 is inferior to loop_1.

In the example shown in Fig. 7.4, loop_1 contains nodes 2, 5, 6, 7, 3 and loop_2 contains nodes 2, 5, 6, 8, 7, 3. The additional node 8 is inferior to node 6 because B, D, E share node 6, but only departments B, E share node 8. By locating a pickup or delivery station of a department at node 8, no improvement will be achieved as compared to locating the department at node 6. The distance that parts have to travel is increased because of the additional two arcs (6–8 and 8–7) that were added.

Rule 3: FNIN (Fewer Nodes and Inferior Nodes)

If loop_2 contains some of the nodes in loop_1, but not all, in the same order, and loop_1 contains all the departments in loop_2, and loop_2 is longer, then the set of solutions achieved using loop_1 dominates the set of solutions acheived using loop_2.

Proof By the same argument presented in Rule 2, loop_2 is inferior to loop_1. In addition, loop_2 does not include all the nodes of loop_1, making it worse.

In the example shown in Fig. 7.5, loop_1 contains nodes 1, 4, 8, 6, 5, 2 and loop_2 contains nodes 1, 4, 8, 6, 7, 3, 2. Node 5 is not included in loop_2, and the additional nodes are inferior. Node 3 is inferior to node 2 and node 7 is inferior to node 6, because B, D, E share node 6 but only departments D, E share node 7, and A, C, D share node 2 but only C, D share node 3. By locating a pickup or delivery station of a department at nodes 7 or 3, no

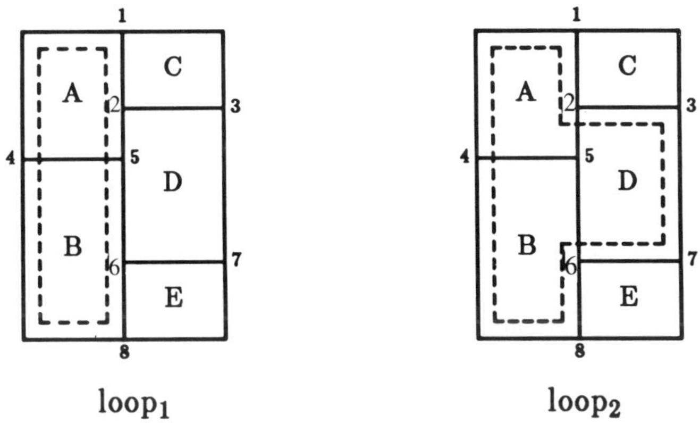

Fig. 7.5 Rule 3 example.

improvement will be achieved as compared to locating the pickup or delivery station of that department at nodes 6 or 2. The distance parts have to travel is increased because of the additional arcs that are added.

7.2.5 Model *SLSLP*

SLSLP (Single Loop Station Location Problem) finds the optimal location of the pickup and delivery stations along a given loop. SLSLP also sets the directions of the arcs that belong to a given loop.

(a) Notation

$$I_{ij} = \begin{cases} 0 & \text{if } D_i \geqslant P_j \\ 1 & \text{otherwise} \end{cases}$$

P_i = location of pickup station department i.
D_i = location of delivery station department i.
U_i = upper bound on the location of station i.
L_i = lower bound on the location of station i.
f_{ij} = flow between departments i, j.
C = single-loop length.

(b) The objective function

The objective function is to minimize the distance unit loads have to travel. The guide path can be viewed as a circle and the stations are positioned along it. If the delivery station is positioned before the pickup station (based on the flow direction of the loop), the distance a part will travel from pickup to delivery is the length of the loop minus the distance between the departments. If the pickup station is positioned before the delivery station, then the distance a part will travel is the distance between the departments. Thus,

$$d_{ij} = \begin{cases} |D_i - P_j| & \text{if } D_i \geqslant P_j, \\ C - |D_i - P_j| & \text{otherwise.} \end{cases}$$

Using the indicator variable I_{ij}, this distance can also be expressed as

$$d_{ij} = (I_{ij}C + D_i - P_j).$$

Thus, the objective function can be written as

$$\text{minimize} \sum_{i=1}^{M} \sum_{j=1}^{M} f_{ij}(I_{ij}C + D_i - P_j).$$

(c) The constraints

There are three types of constraints in SLSLP:

1. Indicator I_{ij} has to be determined by the program. Since SLSLP is a minimization model, the constraint will force the indicator variable to be 1 if the distance between stations is negative and the objective function will force the indicator be 0 otherwise:

$$D_i - P_j + I_{ij}B \geqslant 0 \quad \forall i, j \text{ where } f_{ij} > 0,$$

where B is a very large number. There is a constraint of this type for all pairs of nodes where a flow exists.
2. If the loop is stretched into a line, the locations of the stations along the line are bounded because of the physical location of the department the station belongs to. Thus,

$$L_i \leqslant P_i \leqslant U_i \quad i = 1, \ldots, M,$$
$$L_i \leqslant D_i \leqslant U_i \quad i = 1, \ldots, M.$$

There is a constraint of this type for every pickup and delivery station.
When deciding upon the station boundaries there might be a problem

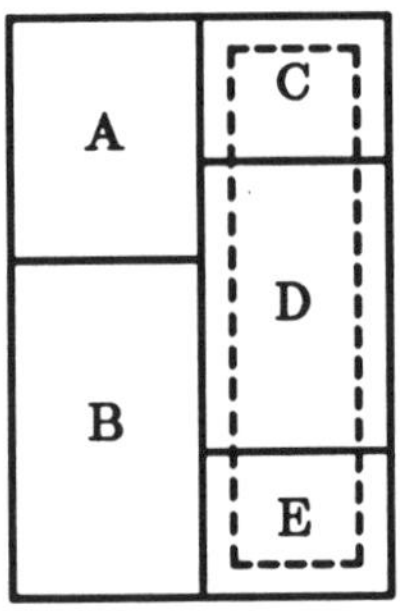

Fig. 7.6 Department D with two split boundaries.

that a department will be divided as shown in Fig. 7.6. In this case, department D is split into two parts, i.e. the space the department takes along the line is not continuous. Instead of the above constraint, another constraint has to be used.

2′. If the bounds of a department are split into H parts, then

$$\begin{gathered} L_i^{(1)} \leqslant P_i \leqslant U_i^{(1)} \\ L_i^{(2)} \leqslant P_i \leqslant U_i^{(2)} \\ \vdots \\ L_i^{(H)} \leqslant P_i \leqslant U_i^{(H)}. \end{gathered}$$

The set of constraints for the pickup stations of every split department will be the following:

$$\begin{gathered} P_i \leqslant U_i^{(1)} + B\delta_1^p \\ P_i \leqslant U_i^{(2)} + B\delta_2^p \\ \vdots \\ P_i \leqslant U_i^{(H)} + B\delta_H^p \\ P_i \geqslant L_i^{(1)} - B\delta_1^p \\ P_i \geqslant L_i^{(2)} - B\delta_2^p \\ \vdots \\ P_i \geqslant L_i^{(H)} - B\delta_H^p \\ \sum_{h=1}^{H} \delta_h^p = H - 1 \end{gathered}$$

A similar set of constraints is needed to define the bounds on the location of the delivery stations where δ_h^d replaces δ_h^p.

3.
$$\begin{aligned} \delta_h^p, \delta_h^d &= 0 \text{ or } 1 \quad \forall\, h \\ I_{ij} &= 0 \text{ or } 1 \quad \forall\, i, j. \end{aligned}$$

Theorem 7.2 The pickup/delivery station will always be located at a node.

Proof Assume that the above statement is not true and that the optimal location of the station is at the middle of an arc. If the incoming flow to that station is greater than the outgoing flow, moving the location of the station up-stream will reduce the distance parts have to travel. If the outgoing flow is greater the same effect will be achieved by moving the station down-stream. The station can be moved as long as the upper or lower bounds of its location are not violated. When the flow between two adjacent stations becomes zero, resulting from both stations sharing the same location, by moving both stations up- or down-stream (based on the incoming and outgoing flow), the solution will be improved until one of the stations reaches its bounds. Bounds are set at nodes, therefore the optimal location of the stations has to be at nodes.

7.2.6 A lower bound calculation procedure

This procedure calculates a lower bound value to the solution of a given loop. A layout that contains a loop has several departments inside the loop and several departments outside the loop. Thus, there exists some overlapping between the departments.

S'_i = set of arcs along the guide path associated with department i
S'_j = set of arcs along the guide path associated with department j
Given $S'_i \in S_i$ and $\mathbf{S}'_j \in \mathbf{S}_j$
If $S'_i \cap S'_j \neq \varnothing$ then departments i and j are overlapping.

Two departments overlap if there is an intersection between the sets of arcs that combine those departments along a given guide path. The procedure calculates only the flow between non-overlapping departments, and gives a zero value to the rest. The non-overlapping relationships is illustrated in Fig. 7.7. The lower bound value between two non-overlapping departments i and j is calculated by the formula:

$$\text{Lower bound value} = (L_i - U_j)f_{ij} + (L_j - U_i)f_{ji}$$

where U_i = the upper bound location of the pickup/delivery station i, and L_i = the lower bound location of the pickup/delivery station i.

In the example shown in Fig. 7.7, the marked departments A and E do not share a common node along the guide path. The lower bound value to the flow between departments A and E is calculated by the following procedure:

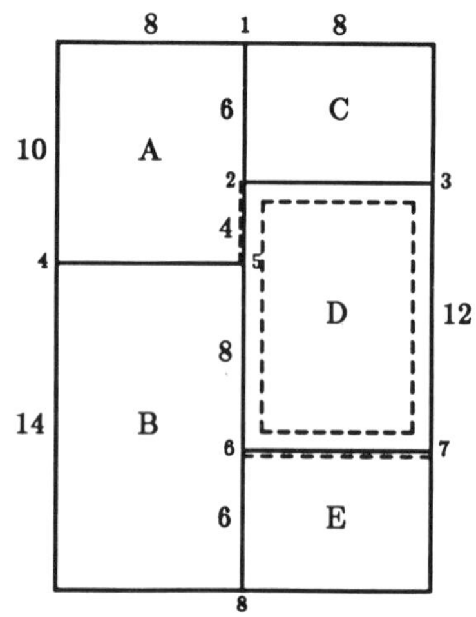

Fig. 7.7 Non-overlapping relationship.

For clockwise flow:

$U_A = \text{node } 2$
$L_A = \text{node } 5$
$U_E = \text{node } 6$
$L_E = \text{node } 7$

Low bound value $= 20f_{AE} + 8f_{EA}$.

For counter-clockwise flow:

$U_A = \text{node } 5$
$L_A = \text{node } 2$
$U_E = \text{node } 7$
$L_E = \text{node } 6$

Low bound value $= 20f_{EA} + 8f_{AE}$.

7.2.7 The OLS (Optimal Single Loop) algorithm

```
begin
 1. find an initial loop (use Model VSLP)
 2. branch from the initial loop and find all the valid loops
 3. eliminate inferior loops by using the 3 rules
 4. calculate the lower bounds of all the remaining loops
 5. arrange all results in {REST} in an ascending order
 6. solve Model SLSLP for the first loop
 7. store the solution in SOLN
 8. check SOLN against results in {REST}
 9. if SOLN is less than {REST_i}
10.     remove all results with an index greater then i
11.     go to 13
    else
12.     go to 13
    end if
13. if {REST} is empty
14.     go to 21
    else
15.     solve the next loop in {REST}
16.     if the solution is greater than the one in SOLN
17.        remove that result from {REST}
18.        go to 13
        else
19.        switch the result in SOLN with the new result
20.        go to 8
        end if
    end if
21. The optimum loop is the one in SOLN.
```

7.3 ILLUSTRATIVE EXAMPLE

To illustrate the solution procedure, the layout in Fig. 7.8 was used as the facility layout. The length of the arcs are shown in the figure. The process plan data and the job mix are given in Table 1.

Table 7.1 Process plan and job mix

Part	*Lot size*	*Job mix*	*Routing*
1	4	0.37	A–D–C–B
2	4	0.185	A–C–D–B–E
3	3	0.11	A–B–D–B
4	4	0.075	A–B–C–D
5	4	0.075	A–B–D–E
6	4	0.075	A–B–E
7	5	0.11	A–C–D

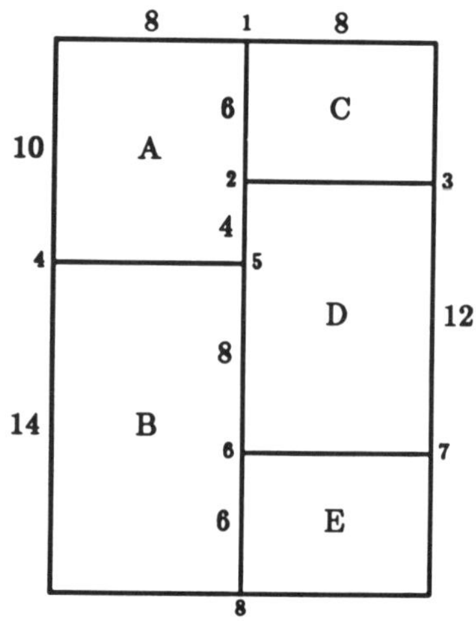

Fig. 7.8 Facility layout.

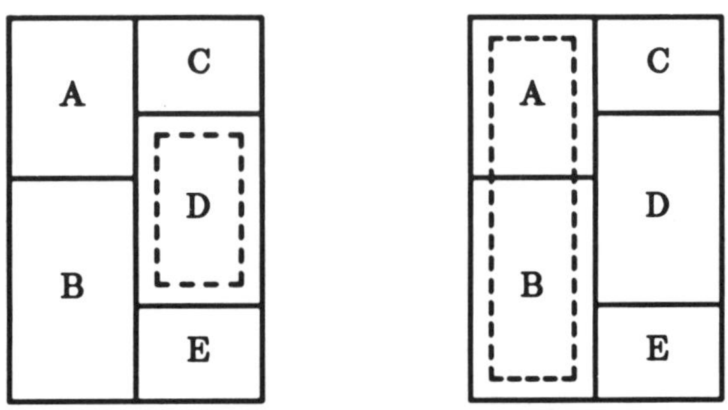

Fig. 7.9 Loops remaining after the elimination procedure.

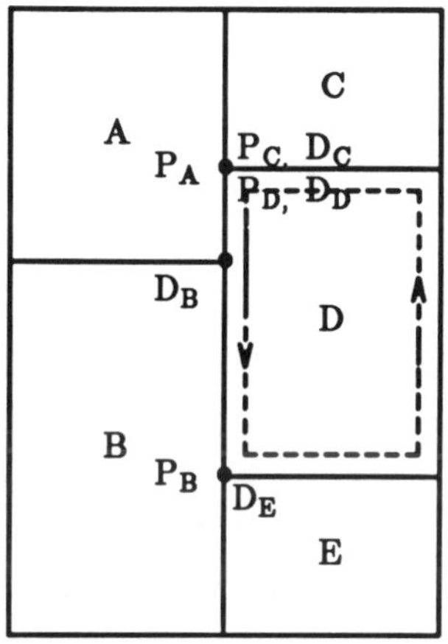

Fig. 7.10 Single-loop guide path as a result of *IP-2*.

The enumeration procedure found 20 valid loops in the layout. The elimination rules eliminated 18 of those loops. *CAN* eliminated four loops, *ANIN* eliminated 10 loops, and *FNIN* eliminated four loops. The two loops left are shown in Fig. 7.9. In this case, because only two loops are left after the elimination procedure, it is not necessary to go through steps (4) to (21) in the **OSL** procedure; it is simpler to solve the SLSLP model for both loops. The optimal single loop guide path found is shown in Fig. 7.10.

7.3.1 Simulation results

To evaluate the performance of the single-loop guide path, the simulation package AGVSim, developed by Egbelu and Tanchoco (1982), was used. The

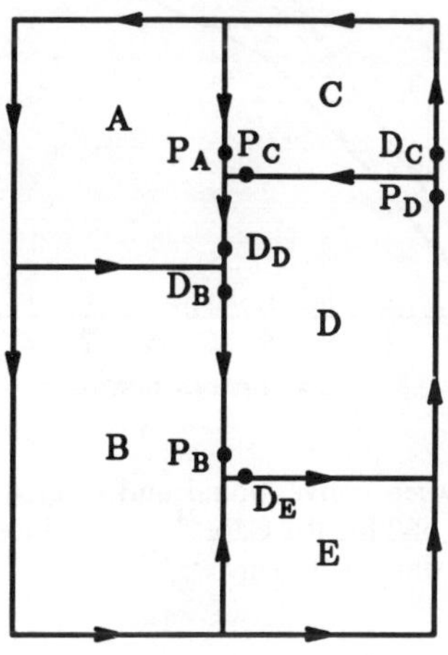

Fig. 7.11 Traditional guide path.

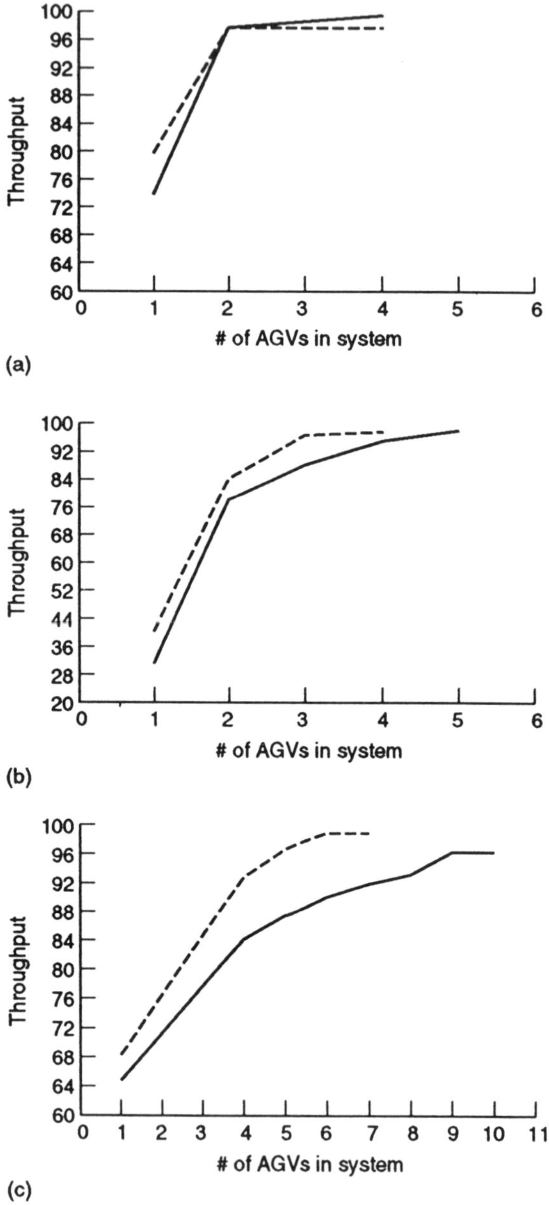

Fig. 7.12 Comparison between conventional and single-loop guide paths. (a) Case 1 – Shop interarrival time of 1 hr; (b) Case 2 – Shop interarrival time of 0.5 hr; (c) Case 3 – Shop interarrival time of 0.25 hr.

performance of the optimal single-loop guide path designed by the **OSL** procedure was compared to a conventional guide path which was designed based on a heuristic approach. The performance measure used was the shop's throughput. Both systems were tested under three different workloads. The workload was changed by increasing the arrival rate of parts to the shop. The light workload was achieved using an interarrival time of 1 hour. The medium workload was achieved using an interarrival time of 0.5 hour, and an interarrival time of 0.25 was used to achieve a heavy workload. The conventional guide path that was simulated is shown in Fig. 7.11. The results shown in the graph in Fig. 7.12 are the average throughput of three runs made for each case.

7.4 EXTENSIONS

Several extensions were later made to the operation and design of single-loop flow path systems.

7.4.1 The impact of the empty vehicle flow

When designing and operating manufacturing systems, it is desirable to achieve a high degree of robustness in performance. This feature reduces the impact changes in operation that the parameters have on the operation and design of such systems. One such operating parameter in any discrete material flow systems is the empty vehicle flow. The dynamic features of these parameter make it very difficult to estimate under different operation policies, which in turn affects the quality of the design outcome. Using the single-loop configuration reduces the impact that the empty vehicle flow has on the system's

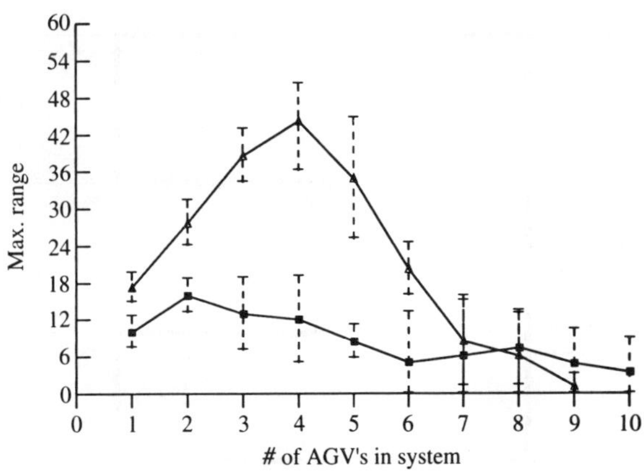

Fig. 7.13 Maximum throughput performance differences. △ = conventional; ■ = single-loop.

performance. This attribute is analysed by Sinriech and Tanchoco (1992a). The study shows the robustness of the single-loop configuration. This feature enhances the accuracy of the design stage, due to the fact that some of the dynamic features and the uncertainties associated with this system are eliminated. It also reduces the importance of choosing a dispatching rule, due to the similar performance across all rules. The maximum throughput differences between the best performing dispatching rule and the worst performing rule for both conventional and single-loop systems are plotted in Fig. 7.13.

7.4.2 Adding within department flow consideration

Most flow path design models presented in the literature overlook the impact the material flow within a department has in determining the location of pickup and delivery stations. Two major factors need to be considered when deciding whether or not to include within department flows in the design model: (1) the cost of moving loads within the department, and (2) the size of the department. Sinriech and Tanchoco (1992b) developed a model which takes into account the flow within a department, as well as the flow between departments, in determining the location of pickup and delivery stations along a prespecified single-loop flow path. The model defines the within department flow as the flows between pickup and delivery stations and the department's centroid, which is used as an aggregate representation of all the activity points in a department. The model also assumes that the cost ratio of the two flows is given. The model shows that by defining special points on the loop flow path called **Centroid Projection Points** (CPP), a trip

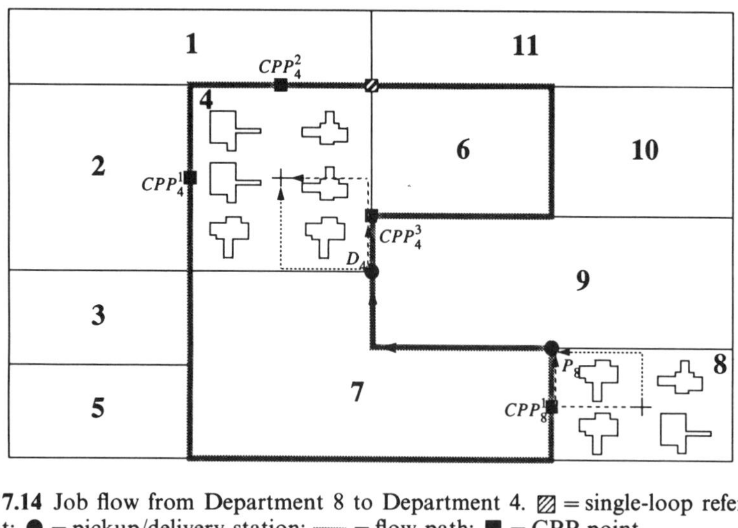

Fig. 7.14 Job flow from Department 8 to Department 4. ▨ = single-loop reference point; ● = pickup/delivery station; ▬ = flow path; ■ = CPP point.

from an origin department to a destination department can be divided into the following segments:

1. Centroid of the origin department to the department's CPP.
2. Origin department's CPP to the pickup station.
3. Pickup station to delivery station along the loop flow path network.
4. Delivery station to the destination department's CPP.
5. Destination department's CPP to the centroid of the department.

All the transportation activities take place along the loop except the first and last moves between the CPP and centroids. However, these distances are fixed and known. Therefore, all elements in the problem can be combined into one optimization model, which minimizes the transportation cost in the system. One such trip is shown in Fig. 7.14.

7.4.3 The Segmented Bidirectional Single-Loop (SBSL)

The use of bidirectional flow paths has been shown to significantly improve system performance. Egbelu and Tanchoco (1986) report an improvement of 40–100% in throughput, depending upon the type of facility layout and the number of carriers used. Kim and Tanchoco (1991) present an efficient routing for carriers operating in a bidirectional system, and show the complexity associated with this problem. Some of this complexity can be reduced through the use of the single-loop flow topology. In such a loop system, a carrier can travel clockwise or counterclockwise, whichever is shorter, to its destination point. Bidirectional single-loop systems pose many operational problems. This is due to the fact that the single-loop flow path contains no shortcuts and no alternative routes. Therefore, a bidirectional flow will probably cause heavy congestion and an eventual gridlock, even if bypass spurs are used. One way in which to avoid the problem is to limit the number of carriers in a loop to one. An alternative approach is to use the Segmented Bidirectional Single-Loop (SBSL) suggested by Sinriech and Tanchoco (1992c).

The term SBSL, denotes a single-loop flow path which is divided into non-overlapping single carrier segments. Transfer buffers are located at both ends of each segment, and serve as input/output buffers. Thus, a carrier can deposit loads which are headed to other segments and pick up loads from the other segments. The carrier has the capability to travel clockwise or counterclockwise on each segment, whichever direction is the shorter to its destination point. The SBSL flow topology reduces the time losses contributed to blocking and congestion due to the mutual exclusive operation mode used, and also shortens the travel distances between origin and destination points due to the bidirectional flows used. These factors contribute to the higher throughput capability of this flow topology compared to the unidirectional single-loop. An illustration of a SBSL flow topology with four segments is shown in Fig. 7.15.

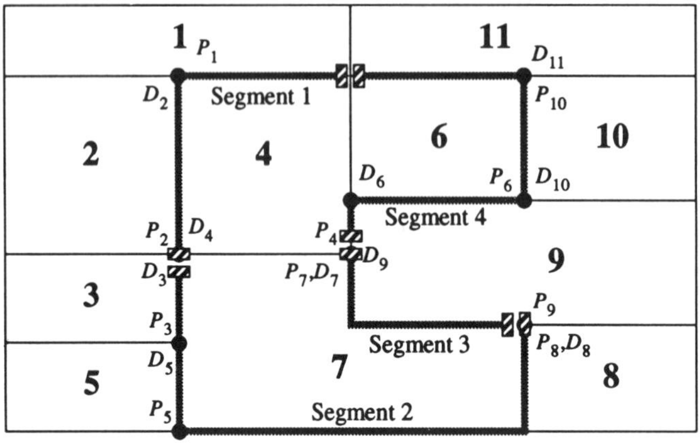

Fig. 7.15 SBSL flow path topology with four segments. ● = pickup/delivery station; ▬ = flow path; □ = segment transfer buffer.

7.5 CONCLUSIONS

The results shown in Fig. 7.12 give a better understanding of the single-loop guidc path capabilitics. Figure 7.12 shows that under light and average shop workload, the single-loop guide path performs as well as the conventional guide path. As the workload goes up the single-loop's performance goes down, and there is a difference of less than 6% in the performance of both systems. The conventional guide path achieves a throughput of 98.5%, while the single-loop guide path achieves a throughput of 92.7%. However, the single-loop needs three more vehicles, i.e. 10 vs. 7. This result is not surprising, since the single-loop guide path needs more vehicle capacity due to the extra distance that unit loads have to travel.

A benefit of using a single-loop guide path is that a simpler controller can be used. The conventional guide path contains four conflict intersections where collisions might occur (an intersection where traffic enters from more than one direction). Controlling traffic through those intersections is a complicated task. The single-loop guide path has no intersections at all.

Further evaluation has to be done on the single-loop guide path configuration's performance, but this study so far points out that there are substantial benefits in using this configuration. The decision maker who looks at implementing this configuration has to make the trade-off between reduced throughput and flexibility vs. gaining simplicity in operational control. The question is one of economics.

The use of multi-load AGVs can give a solution to some of the drawbacks of this configuration. Using multi-load vehicles will solve the need for more vehicle capacity without adding blockage and congestion to the system. This will improve even further the single-loop guide path's performance.

REFERENCES

Bartholdi, J.J., III and Platzman, L.K. (1989) Decentralized control of automated guided vehicles on a simple loop. *IEEE Transactions*, **21**(1), 76–81.

Bozer, Y.A. and Srinivasan, M.M. (1989) Tandem configuration for AGV systems offer simplicity and flexibility. *Industrial Engineering*, **21**(2), 23–7.

Egbelu, P.J. and Tanchoco, J.M.A. (1982) AGVSim User's Manual. *Technical Report No. 8204*. Department of Industrial Engineering and Operations Research, Virginia Tech., Blacksburg, VA.

Egbelu, P.J. and Tanchoco, J.M.A. (1986) Potential for bidirectional guide path for automatic guided vehicles based systems. *International Journal of Production Research*, **24**(5), 1075–99.

Fuller, S.H. (1972) An optimal drum scheduling algorithm. *IIE Transactions on Computers*. **21**(11), 1153–65.

Gaskins, R.J. and Tanchoco, J.M.A. (1987) Flow path design for automated guided vehicle systems. *International Journal of Production Research*, **25**(5), 667–76.

Goetz, W.G. and Egbelu, P.J. (1990) Guide path design and location of load pick-up/drop-off points for an automated guided vehicle system. *International Journal of Production Research*, **28**(5), 927–41.

Haines, C.L. (1985) An algorithm for carrier routing in a flexible material handling system. *IBM Journal of Research of Development*, **29**(4), 356–62.

Kaspi, M. and Tanchoco, J.M.A. (1990) Optimal flow path design of unidirectional AGV systems. *International Journal of Production Research*, **28**(6), 915–26.

Kim, C.W. and Tanchoco, J.M.A. (1991) Conflict-free shortest-time bidirectional AGV routing. *International Journal of Production Research*, **29**(12), 2377–91.

Miller, C.E., Tucker, A.W. and Zemlin, R.A. (1960) Integer programming formulation of Traveling Salesman Problems. *Journal of ACM*, **7**, 326, 329.

Phillips, D.T., Ravindran, A. and Solberg, J.J. (1976) *Operations Research*, John Wiley, New York.

Sinriech, D. (1990) Design and Evaluation of Optimal Single Loop Guide Paths for Automated Guided Vehicle Systems. MSIE Thesis, Purdue University, West Lafayette, IN.

Sinriech, D. and Tanchoco, J.M.A. (1992a) The impact of empty vehicle flow on the performance of single loop guide paths. *International Journal of Production Research*, **30**(10), 2237–52.

Sinriech, D. and Tanchoco, J.M.A. (1992b) The centroid projection method for locating pick-up and delivery stations in single-loop AGVs, *Journal of Manufacturing Systems*, **11**(4), 297–307.

Sinriech, D. and Tanchoco, J.M.A. (1992c) The Segmented Bidirectional Single-Loop for AGV Systems. *Working Paper*, School of Industrial Engineering, Purdue University, West Lafayette, IN.

Stein, D.M. (1978) Scheduling dial-a-ride transportation systems. *Transportation Science*, **12**(3), 232–49.

Taghaboni, F. and Tanchoco, J.M.A. (1988) A LISP-based controller for free-ranging automated guided vehicle systems. *International Journal of Production Research*, **26**(2), 173–88.

Tanchoco, J.M.A. and Sinriech, D. (1992) OSL – Optimal single-loop guide paths for AGVS, *International Journal of Production Research*, **30**(3), pp. 665–81.

Usher, J.A., Evans, G.W. and Wilhelm, M.R. (1988) *AGV Flow Path Design and Load Transfer Point Location*. Proceedings of the 1988 IIE Conference, Orlando, FL, pp. 174–9.

CHAPTER 8

SFT – Segmented Flow Topology

D. Sinriech and J.M.A. Tanchoco

8.1 INTRODUCTION

The material handling system plays a critical role in many manufacturing systems. It is used mainly for transporting and storage of raw material, Work-In-Progress (WIP) and finished goods. The quality of a material handling system is measured in terms of its efficiency in moving the right material to the right place at the right time while maintaining the cost associated with its design and operation as low as possible.

The use of carefully designed facility layouts in general, and efficient material handling flow paths in particular, have a direct impact on the performance of any manufacturing system. The effectiveness of a material handling system depends on a well thought out flow path design. The flow path has a significant effect on the travel time, the operating expenses and the installation costs of the system. Moreover, the flow path configuration has a significant impact on the complexity of the system's control software. The more alternative routes, the more conflicting intersections there are in the flow path, the more complicated it is to control traffic in an effective way.

8.2 MATERIAL FLOW NETWORKS

Several flow network design concepts can be found in the literature. A formal categorization of these concepts is illustrated in Fig. 8.1. The rest of the introduction will describe each of these models.

8.2.1 Bidirectional conventional flow path network

The conventional bidirectional flow path network is the most general system. This means that every carrier can travel to any point on any aisle segment in whichever direction it sees fit within the boundaries of the guidance and

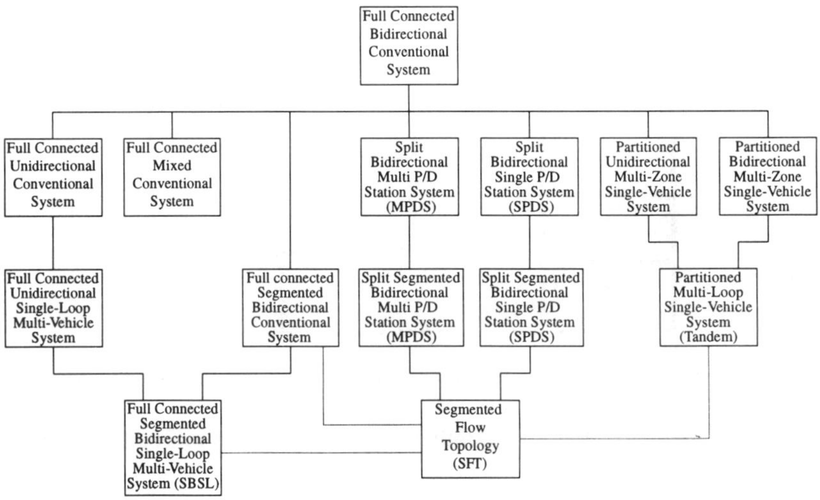

Fig. 8.1 Different system configurations used in the design of material flow systems.

control system. All studies dealing with bidirectional systems conclude that these systems are highly efficient in terms of throughput and performance. Egbelu and Tanchoco (1986) make a distinction between multiple path and switchable path configurations in bidirectional conventional systems, as shown in Figs. 8.2 and 8.3, respectively.

1. *Multiple Paths* – This configuration, shown in Fig. 8.2, is comprised of two parallel unidirectional systems. It is the easiest bidirectional system to control. However, a double set of tracks is needed. This configuration is usually used in conjunction with heavy flow volumes.
2. *Switchable Paths* – In this configuration, shown in Fig. 8.3, a single track is used for flows in both directions. It is the most complicated configuration

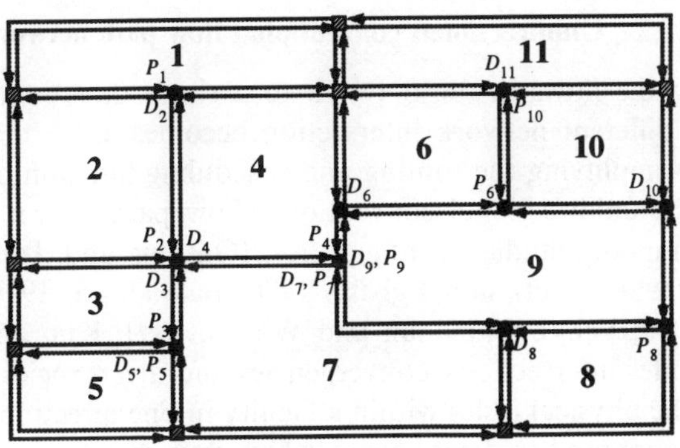

Fig. 8.2 Bidirectional conventional parallel (dual lane) flow path system.

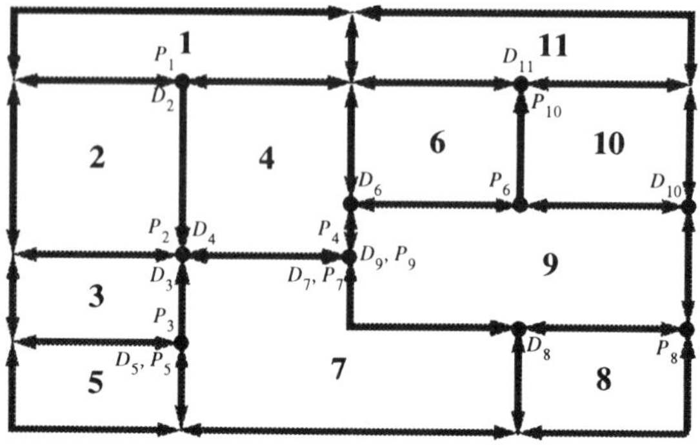

Fig. 8.3 Bidirectional conventional switchable (single lane) flow path system.

to control, due to the flow conflicts, and may not be even efficient when high volume flows are involved.

In general, they point out the potential to create more efficient and cost-effective material handling systems by using bidirectional flows. They report an improvement of 40–100% in the system's throughput while operating in a bidirectional mode, depending upon the type of facility layout and the number of carriers used. However, controlling such a system is a very complex task. A recent paper by Kim and Tanchoco (1991a,b) presents a conflict-free routing procedure for free-path bidirectional carrier systems, and illustrates the complexity of the problem. The study points out that although a free-path bidirectional system promises greater flexibility and efficiency due to the absence of a physical flow path and the bidirectional flow mode, the control system needed to achieve that is much more complicated.

Therefore, there exists an incentive to develop and use simpler flow path configurations. The goal of these studies is to design a configuration which eliminates some of the tasks the system's supervisory controller is in charge of by reducing the number of conflict points in the flow path layout.

8.2.2 Unidirectional conventional flow path network

By restricting the flows in the network to a single direction, some of the traffic at the different network intersection becomes much more simple to control, thus simplifying the routing and scheduling function of the systems controller. The unidirectional conventional flow path shown in Fig. 8.4 is the most commonly studied configuration (Gaskins and Tanchoco, 1987; Usher *et al.*, 1988; Goetz and Egbelu, 1990; Bakkalbasi, 1990; Kaspi and Tanchoco, 1990; Venkataramanan and Wilson, 1991; Kim and Tanchoco, 1991). It includes intersections, convergencies and divergencies, and permits flows on all the physical aisles within a facility in one direction. Most of the above models are used to determine to which flow direction an aisle should be set, and where to position the pickup and delivery stations.

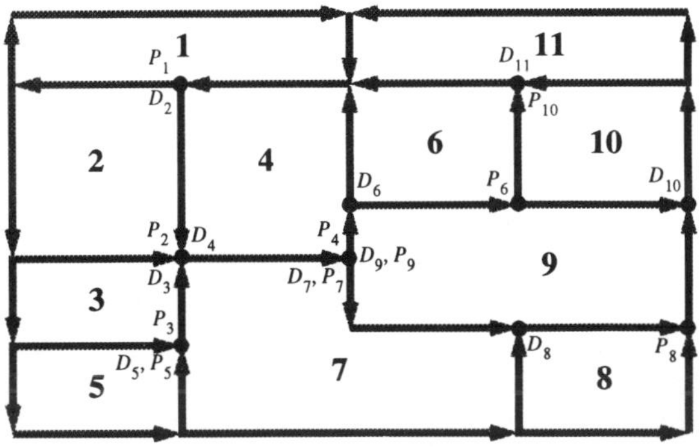

Fig. 8.4 Unidirectional conventional flow path configuration.

Operating a system which is flexible and versatile, like the conventional, unidirectional as well as bidirectional systems, means there is a need for an elaborate and expensive controller. The more sophisticated the physical network becomes, the more alternative decisions there are to explore. In a less complex system, though, the decision making process is much easier, hence controlling and managing such a system is less expensive. Moreover, the use of an inefficient controller in a conventional type system can diminish the efficiency and flexibility of the configuration.

8.2.3 Unidirectional partitioned multi zone flow network

A different approach commonly used in the transportation literature is to simplify the flow network by partitioning the system into non-overlapping zones. In doing so, the flow complexity and size of each zone is reduced, thus reducing the complexity of the traffic control function. The well known dial-a-ride system (Stein, 1978) is such an approach. In this approach, a system is partitioned into zones which are connected by some transfer points.

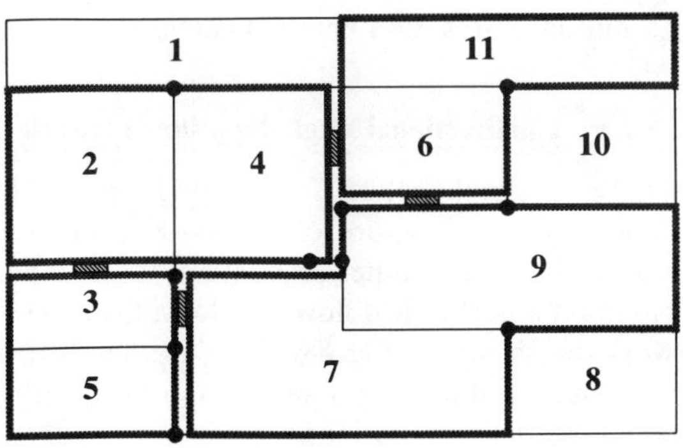

Fig. 8.5 Tandem configuration flow path. ● = pickup/delivery station; ▬ = flow path; ▨ = zone transfer buffer.

A special case of the zoning concept is the tandem configuration suggested by Bozer and Srinivasan (1991a). In the tandem configuration the system is partitioned into non-overlapping zones, where the flow path in each zone is constrained to a single-carrier–single-loop, as shown in Fig. 8.5. By using this approach the congestion and blocking are eliminated provided that there is sufficient capacity at the transfer buffers. Nevertheless, restricting the number of carriers to one per zone means that the workload of a zone cannot exceed that of one carrier. Hence, this restriction forces the way in which the system is going to be partitioned. Therefore, the partitioning procedure needs to balance the workload between the zones as suggested while minimizing the transfers between the zones. This partition also determines the sensitivity of the tandem configuration to carrier failure. Once a carrier fails, an entire part of the manufacturing system is shut down and production is interrupted. Another issue that has to be considered is the Load Routing Problem (LRP), addressed by Lin *et al.* (1992). Although the tandem configuration eliminates the need for routing carriers in the system, it still needs to route loads whenever the load destination resides in a different zone. This routing decision is encountered in each loop which contains more than one transfer buffer.

Simulation runs made by Mahadevan and Narendran (1990) and Bozer and Srinivasan (1991b) suggest that in some cases the tandem configuration will outperform the conventional system. However, it should be noted that no physical aisle network was used in these comparisons. Thus, in each case the flow paths were set up to follow the shortest rectilinear distance between pickup and delivery stations.

8.2.4 Bidirectional partitioned multi zone flow network

A further improvement suggested by Bozer and Srinivasan (1991b) is the bidirectional tandem configuration. In this mode, the carriers can travel clockwise or counterclockwise on each loop, whichever is shorter to the destination point. As expected, this feature improved the system's performance without adding significant complication to the system and its controller, since there is no mutual interaction between carriers.

8.2.5 Unidirectional single-loop flow network

A different flow network simplification is the single-loop. This flow configuration simplifies the physical structure of the system, since it contains no intersections and no alternative routes. In contrast to the tandem configuration, which is based on a partitioned flow topology, the single-loop is a fully connected network, as shown in Fig. 8.6. Jobs are handled by one carrier from origin to destination following a single-loop flow path, which passes through all pickup and delivery stations.

The single-loop flow path problem has previously been addressed by several studies. Afentakis (1989) develops a design procedure for a loop layout

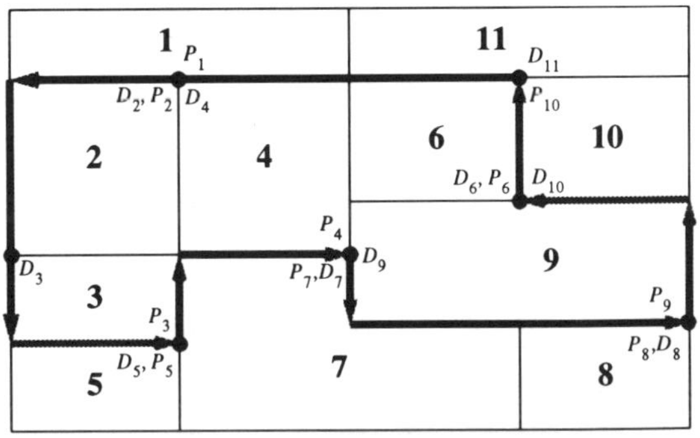

Fig. 8.6 Single-loop flow path configuration.

for Flexible Manufacturing Systems (FMS). Kouvelis and Kim (1992) later denote it as *Unidirectional Loop Network Layout* (ULNL). No explicit facility layout is defined, and the assumption used in these studies is that predetermined sites are located along a loop, and each machine has to be assigned to one of these sites. Rim and Bozer (1991) develop a design procedure for a similar problem. Likewise, in this study, machines have to be assigned to predetermined sites along a given single-loop flow path. However, the unidirectional flow constraint is relaxed and the number of carriers is limited to one. Kiran *et al.* (1992) present a model for locating stations along a unidirectional loop using a balanced flow matrix, assuming that combined P/D stations are used. While this assumption is reasonable when dealing with input/output stations for machines or workcenters, it is too restrictive when dealing with the transportation of parts to and from departments. All of these approaches are more suitable when used to design FMSs, where machines or workcenters need to be arranged in an efficient manner, rather than design a material flow network for a given facility layout where departments have specific geometric shapes, volume and defined boundaries.

The single-loop flow path design problem has also been approached from a different perspective. In a recent paper, Tanchoco and Sinriech (1992) develop the OSL procedure for designing optimal single-loop flow paths when an actual facility layout is given. They point out that to design an optimal single-loop flow path system, both the flow path and P/D station locations have to be determined simultaneously, which is also true using other flow structures. The procedure uses a prespecified facility layout where the shape, area and boundaries of the departments are given. Efficient solution procedures for the mathematical models presented in the OSL procedure study were later developed by Sinriech and Tanchoco (1993).

8.2.6 Bidirectional single-loop (segmented bidirectional single-loop)

There is a big difference between the conventional configuration and the single-loop configuration regarding flow direction. While unidirectional and

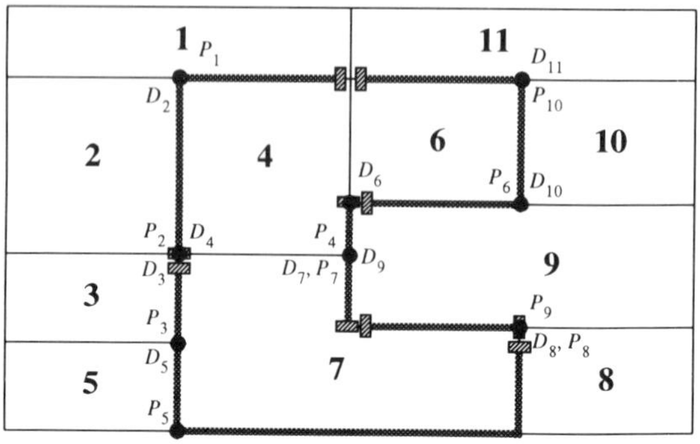

Fig. 8.7 SBSL flow topology with five segment path. ● = pickup/delivery station; ▬ = flow path; ▨ = segment transfer buffer.

bidirectional flows are possible alternatives in conventional systems, the single-loop is associated only with unidirectional flows. This is due to the fact that the single-loop flow path contains no shortcuts and no alternative routes, therefore a bidirectional flow will probably cause heavy congestion and consequently grid-lock, even if spurs are used. However, in cases where the number of carriers is restricted to one per single-loop, it is possible to use a bidirectional flow. The carrier will travel clockwise or counterclockwise, whichever is shorter. To avoid the problems caused by this constraint, Sinriech and Tanchoco (1992a) proposed the Segmented Bidirectional Single-Loop (SBSL), which will be described in the following sections. In this case, all the complexity inherited in the use of bidirectional flows is eliminated due to the fact that no carrier shares the tracks with another carrier.

The term Segmented Bidirectional Single-Loop (SBSL) denotes a single-loop flow path which is divided into non-overlapping single carrier segments. Transfer buffers are located at both ends of each segment and serve as input/output buffers, where a carrier can deposit loads which are headed to other segments and pick up loads from the other segments. The carrier has the capability to travel clockwise or counterclockwise on each segment, whichever direction is shorter to its destination point. An illustration of a SBSL configuration is shown in Fig. 8.7. The SBSL flow topology addresses some of the weak features of the single-loop, hence it can potentially improve the functional efficiency of the single-loop while retaining its simple flow structure. By operating the carriers in a mutually exclusive manner, the congestion, blocking and interferences are eliminated. This results in a more efficient system from both an economical as well as a functional standpoint.

8.3 SEGMENTED FLOW TOPOLOGY (SFT)*

It is clear that some of the goals for simplicity, efficiency and cost-effectiveness were met by using the single-loop. Nevertheless, some design and operation

*Most of this section is derived from Sinriech and Tanchoco (1994a, b).

problems remain. It is with these in mind that the Segmented Flow Topology (SFT) (Sinriech and Tanchoco, 1994a) was developed. This flow topology is a generalization of the Segmented Bidirectional Single-Loop (SBSL) suggested by Sinriech and Tanchoco (1992). The SFT combines all the features discussed earlier, i.e. simple flow structures, shortest paths, bidirectional flow, no carrier interferences, and whenever there is an economic justification for it, split pickup and delivery stations. These, in turn, create an efficient and simple to control material flow system.

8.3.1 General description of a segmented flow topology (SFT) system

The Segmented Flow Topology (SFT) can be used in conjunction with each of the three network types: connected, partitioned and split flow, e.g. the SBSL is a segmented fully connected network. The general SFT is comprised of one or more zones, each of which is separated into non-overlapping segments with each segment serviced by a single material handling device. Transfer buffers are located at both ends of each segment, and they serve as the interface between the segments. These buffers serve as input/output buffers where a material handling device can deposit loads which are headed to other segments and pickup loads from other segments. To eliminate blocking, the buffers have to be able to serve both sides of the segments simultaneously. The material handling device has the capability to travel forwards or backwards on each segment, whichever direction is the shorter, to its destination point. The flow structure in each zone is determined by the logical flow requirements and by an existing aisle network. This means the flow structures can be either points, lines, loops or trees. A SFT type system is shown in Fig. 8.8.

It should be noted that the tandem configuration is a special case of the SFT system. If the shortest bidirectional flow distances in each zone produces a single-loop flow structure, and if one carrier is sufficient to handle the workload around these loops, then the SFT design procedure will result in a tandem configuration. However, in all other cases, the flow structure designed using the SFT will differ from the tandem configuration.

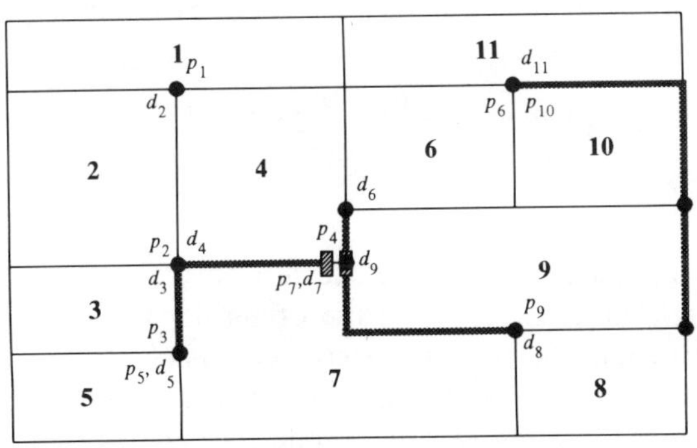

Fig. 8.8 Segmented flow topology (SFT) network. ● = pickup/delivery station; ▬ = flow path; ▨ = segment transfer buffer.

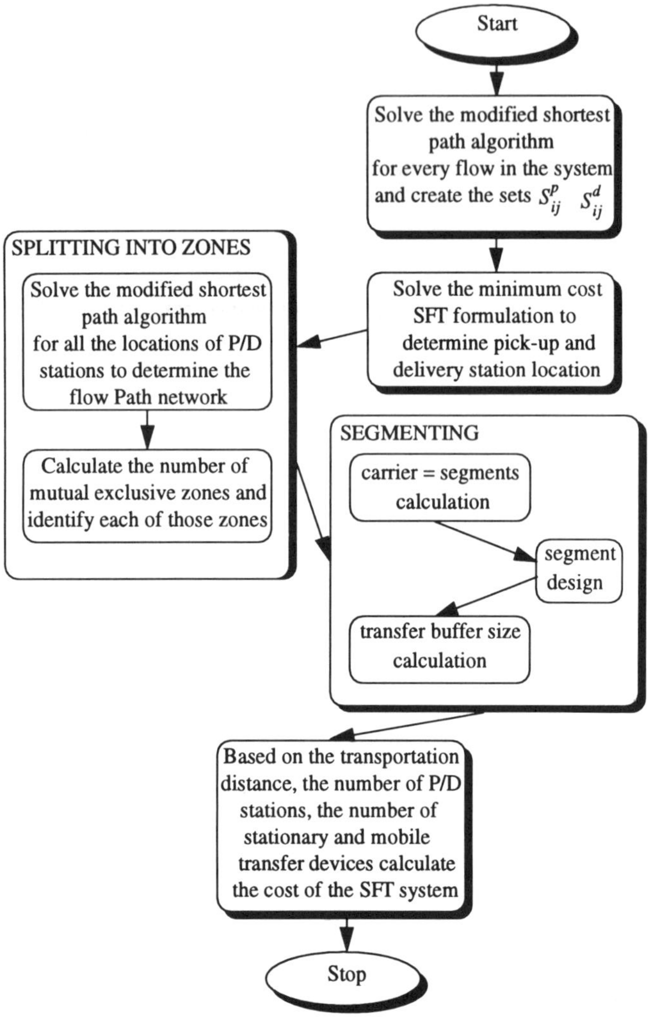

Fig. 8.9 Procedure to design a minimum cost SFT material flow system.

8.3.2 The SFT design procedure

The SFT design procedure comprises of five steps, illustrated in Fig. 8.9 and listed below:

1. The first step determines all the shortest path alternatives for each flow in the system between nodes of the origin departments to nodes of the destination departments. In this step, two sets of nodes are created S^p_i, S^d_j for every department, as shown by equations (8.1) and (8.2). The set S^p_i contains all the alternative origin nodes of department i, and the set S^d_j contains all alternative destination nodes for department j.

2. The second step determines the most cost-effective system design, based on a trade-off between the cost of setting up an additional pickup and/or delivery station compared to the gain in the transportation cost.
3. The third step identifies the flow path network by the use of the pickup and delivery station locations and the shortest path algorithm.
4. The fourth step is the segmentation of the designed flow path based on the number of material handling devices needed in each zone.
5. Based on this information, the fifth and final step is the calculation of the system's cost-effectiveness, which completes the design of the system.

(a) The modified shortest path algorithm (MSPA)

To determine the shortest paths $L_{ij}^{\min}$ between departments for each flow f_{ij} in the system, as dictated by the first step, a modified shortest path algorithm similar to the original shortest path algorithm developed by Dijkstra (1959) is used. However, in this case the origin and destination are sets of nodes which make up the origin and destination departments. For each flow the procedure creates a set of alternative origin nodes S_{ij}^{p}, a set of alternative destination nodes S_{ij}^{d}, and a set of alternative paths S_{ij}^{pd} defined as a set of two tuples (k, k') where k and k' denote the origin and destination nodes of the path, respectively. The number of nodes in each set is determined by the number of alternative minimum length paths from origin to destination. By using equations (8.1) and (8.2), the set of alternative origin and destination nodes S_i^p, S_j^d for each department can be formed:

$$S_i^p = \bigcup_{j=1}^{M} S_{ij}^p \tag{8.1}$$

$$S_j^d = \bigcup_{i=1}^{M} S_{ij}^d \tag{8.2}$$

where M denotes the number of departments in the system.

(b) Designing an optimal SFT system

The design of an optimal SFT system as dictated in step (2) can be achieved by solving a mixed integer programming model as developed by Sinriech and Tanchoco (1994a). Due to the large number of binary variables needed to describe even a relatively small system, it is obvious that solving the mixed integer programming model using commercial Mixed Integer Program solvers is impractical (to reach an optimal solution to a relatively small problem presented by Sinriech and Tanchoco, 1994a, took 1860 CPU minutes). Therefore, a heuristic design procedure is needed to solve the problem. The following sections describe two basic system configurations. The first is the Single Pickup/Delivery Station (SPDS) configuration, and the second is the Multiple Pickup/Delivery Station (MPDS) configuration. These two systems will be the backbone for the SFT heuristic design procedure.

Single pickup/delivery station (SPDS) problem

In the SPDS problem, each department can have only one pickup station and one delivery station, which are not necessarily located at the same node. To design a single pickup and delivery station system, a mixed integer programming model is used. The use of this design model can only be justified in two cases: (a) the use of an SFT-based system, or (b) a multi-lane system. This model uses the origin and destination alternative node sets, as created by the modified shortest path algorithm and equations (8.1) and (8.2).

A similar model was presented by Riopel and Langevin (1991). In this model, they define two types of binary variables, y_i which indicates whether station i is used, and x_{ij} which indicates whether there is a flow between station i and j. The model presented in this chapter will contain only one type of binary variable similar to the y_i variable. The second binary variable is replaced by a real variable and some additional constraints. Therefore, the model presented in this chapter contains a smaller number of binary variables and a larger number of constraints compared to the model presented by Riopel and Langevin (1991).

Notation

Department: $i, j = 1, \ldots, M$.
Pickup/Delivery. Nodes: $k, k' = 1, \ldots, N$.

$$p_{ik} = \begin{cases} 1 & \text{if node } k \text{ serves as a pickup station of department } i \\ 0 & \text{otherwise.} \end{cases}$$

$$d_{jk'} = \begin{cases} 1 & \text{if node } k' \text{ serves as a delivery station of department } j \\ 0 & \text{otherwise.} \end{cases}$$

$$\varsigma_{i_k j_{k'}} = \begin{cases} 1 & \text{if node } k \text{ serves as the pickup station of department } i \text{ and} \\ & \text{node } k' \text{ serves as the delivery station of department } j \\ 0 & \text{otherwise.} \end{cases}$$

$L_{kk'}$ = the shortest path distance from node k to node k'.
C_{stat} = the fixed cost of a pickup or delivery station.
C_{trans} = the variable cost of transporting one unit of material one unit of distance.

Model constraints The model contains six types of constraints. C1 and C2 deal with setting one pickup and one delivery station for each department; constraints C3–C5 are used to linearize the product term in the objective function; and C6 is the integrality constraint.

C1. These constraints ensure that one node from the set of nodes S_i^p will be chosen as a pickup station for each flow:

$$\sum_{k \in S_i^p} p_{ik} = 1 \quad \forall\, i.$$

C2. These constraints ensure that one node from the set of nodes S_j^d will be chosen as a delivery station for each flow:

$$\sum_{k' \in S_j^d} d_{jk'} = 1 \quad \forall j.$$

C3–C5. This set of three constraints is used to linearize the product term in the objective function:

$$p_{ik} + d_{jk'} - \varsigma_{i_k j_{k'}} \leqslant 1 \quad k \in S_i^p, k' \in S_j^d \quad \text{where } f_{ij} > 0$$

$$\varsigma_{i_k j_{k'}} - p_{ik} \leqslant 0 \quad k \in S_i^p, k' \in S_j^d \quad \text{where } f_{ij} > 0$$

$$\varsigma_{i_k j_{k'}} - d_{jk'} \leqslant 0 \quad k \in S_i^p, k' \in S_j^d \quad \text{where } f_{ij} > 0.$$

C6.

$$p_{ik} = 0 \text{ or } 1 \quad \forall\, ik \text{ combination}$$

$$d_{jk'} = 0 \text{ or } 1 \quad \forall\, jk' \text{ combination}$$

$$\varsigma_{i_k j_l} \geqslant 0 \quad \forall\, ijkk' \text{ combination.}$$

Model objective function The first part of the objective function is a constant cost of setting up one pickup and one delivery station for each department in the system. The second part of the objective function is the transportation cost between the pickup stations and delivery stations:

$$\min\left(C_{\text{stat}} \cdot \# Stations + C_{\text{trans}} \cdot \sum_{i=1}^{M} \sum_{j=1}^{M} \sum_{k \in S_i^p} \sum_{k' \in S_j^d} f_{ij} \cdot L_{kk'} \cdot \varsigma_{i_k j_{k'}} \right).$$

Heuristic design procedure for the SPDS problem

The SPDS problem can be solved optimally by a branch-and-bound procedure using the sets S_i^p, S_j^d for each department as created by the modified shortest path algorithm and formulas (8.1) and (8.2). In this case, the branching will be performed on every node in each set which contains several alternatives. For those sets which contain only one origin or one destination node, there is no need to branch. Let us define the two variables Ψ_i^p and Ψ_j^d as follows:

$$\Psi_i^p = \begin{cases} 0 & \text{if } |S_i^p| = 1 \\ n & \text{if } |S_i^p| = n, \end{cases}$$

$$\Psi_j^d = \begin{cases} 0 & \text{if } |S_j^d| = 1 \\ n' & \text{if } |S_j^d| = n'. \end{cases}$$

The total number of nodes that are used in the branching procedure denoted by Ψ^{pd} can be calculated using equation (8.3):

$$\Psi^{pd} = \sum_{i=1}^{M} \Psi_i^p + \sum_{j=1}^{M} \Psi_j^d. \tag{8.3}$$

Therefore, the total number of options in the branch-and-bound tree can be calculated by using equation (8.4). However, as in all branch-and-bound procedures only a small part of these options needs to be explored:

$$Number\ of\ Options = \sum_{i=1}^{\Psi^{pd}} 2^i. \tag{8.4}$$

Solving the model presented in the previous subsection to optimality is possible when dealing with relatively small-to-medium-size problems. However, when dealing with larger size problems, a heuristic procedure will be a better option. Riopel and Langevin (1991) develop a heuristic procedure in conjunction with their model. This procedure calculates a set of penalties which determine the order of choosing a station from among all alternatives. The heuristic design procedure used in this paper is based on solving optimally a subset of the complete problem. As described earlier, the final solution for the MPDS system contains two node sets $\hat{S}_i^p, \hat{S}_j^d$ to be used as pickup and delivery stations for each department. In cases where only one node is assigned as a pickup station and only one node is assigned as a delivery station, both the MPDS system and the SPDS system are identical. However, some departments in the MPDS system may contain multiple pickup and multiple delivery stations. In this case, a branch-and-bound procedure can be used to choose the best nodes to serve as pickup stations and delivery stations. The branching is done using only the nodes in the sets $\hat{S}_i^p, \hat{S}_j^d$. The bounds at each stage can be easily calculated by summing up all the flow × distance values based on the already known minimum path distances and the flows between those nodes.

By replacing the node sets S_i^p, S_j^d with the node sets $\hat{S}_i^p, \hat{S}_j^d$, respectively, and the variables Ψ_i^p, Ψ_j^d with the variables $\hat{\Psi}_i^p, \hat{\Psi}_j^d$, respectively, the total number of options in the new branch-and-bound procedure can be calculated by a similar equation to equation (8.6) using $\hat{\Psi}^{pd}$ instead of Ψ^{pd}.

The order of complexity in both the optimal and heuristic solution procedures is the same, $O(2^m)$, where m denotes the number of the different branching options (either $\hat{\Psi}^{pd}$, or Ψ^{pd}). Nevertheless, it is clear that the heuristic problem is only a fraction of the optimal problem because $\hat{\Psi}^{pd} < \Psi^{pd}$.

Multiple pickup/delivery station (MPDS) problem

In the MPDS problem, each department can have more than one pickup and delivery station. Designing a system based on multiple pickup and delivery stations has the potential to improve the system, both functionally and economically. In the case where the cost ratio $C_{\text{in}}/C_{\text{out}}$ is high, where C_{in} and C_{out} denote the within and between department flow cost, respectively, there is an incentive to position the pickup/delivery stations as closely as possible to each manufacturing activity point in a department. Moreover, by using multiple pickup and delivery stations, flows along aisles can be converted to stationary point-to-point transfers. In these cases, more expensive material handling devices like AGVs and forklifts are replaced by less expensive devices like short conveyors or other stationary transfer devices.

All of these gains have to be compared to the cost of setting up additional pickup/delivery stations.

By summing up the cardinalities of the sets S_i^p, S_j^d over all the pickup and delivery stations, as shown by equation (8.5), the absolute maximum number of pickup and delivery stations Π can be calculated:

$$\Pi = \sum_{i=1}^{M} |S_i^p| + \sum_{j=1}^{M} |S_j^d|. \qquad (8.5)$$

Using this number of pickup and delivery stations ensures that each flow f_{ij} will follow its shortest path $L_{ij}^{\min}$ from origin i to destination j, as dictated by the modified shortest path algorithm. As a result, the total flow distance (flow × distance) is constant, and can be calculated by equation (8.6):

$$\Lambda = \sum_{i=1}^{M} \sum_{j=1}^{M} f_{ij} \cdot L_{ij}^{\min}. \qquad (8.6)$$

Choosing for each flow $f_{ij} > 0$ only one path (a combination of origin and destination nodes as pickup and delivery stations) from among the different alternatives in sets S_{ij}^{pd}, does not effect the total flow distance in the network because each alternative is a shortest path. Nonetheless, each additional splitting up of a pickup or delivery station adds a positive cost of setting up an additional station to the system. Therefore, the objective of the procedure is to choose a subset of shortest paths from the sets S_{ij}^{pd} for each flow $f_{ij} > 0$ in the system, such that the number of pickup and delivery stations required is minimized.

The proposed assignment procedure is an integer programming model which chooses one origin and one destination node for each flow $f_{ij} > 0$ from all of the alternatives available, to serve as pickup stations for department i and delivery stations for department j. The objective of the procedure is to design a minimum cost system which meets the above requirements, meaning determining the minimum number of pickup and delivery stations required.

Notation

Department: $i, j = 1, \ldots, M$.
Pickup/delivery nodes: $k, k' = 1, \ldots, N$.
(k, k') defines a shortest path where k serve as the origin node and k' serves as the destination node.

$$p_{ijk} = \begin{cases} 1 & \text{if node } k \text{ serves as a pickup station for the flow from } i \text{ to } j \\ 0 & \text{otherwise.} \end{cases}$$

$$d_{ijk'} = \begin{cases} 1 & \text{if node } k' \text{ serves as a delivery station for the flow from } i \text{ to } j \\ 0 & \text{otherwise.} \end{cases}$$

$$I_{ik} = \begin{cases} 1 & \text{if node } k \text{ serves as a pickup station for department } i \\ 0 & \text{otherwise.} \end{cases}$$

$$I_{jk'} = \begin{cases} 1 & \text{if node } k' \text{ serves as a delivery station for department } j \\ 0 & \text{otherwise.} \end{cases}$$

S^p_{ij} = the set of nodes from which a pickup station for department i is chosen for the flow to department j.

S^d_{ij} = the set of nodes from which a delivery station for department j is chosen for the flow to department i.

$L^{\min}_{ij}$ = the shortest path distance from department i to department j.

Model constraints The model contains six types of constraints. Constraints C1–C3 deal with setting a pair of nodes given by the modified shortest path algorithm as pickup and delivery stations for each department, Constraints C4 and C5 set the values for the indicator variables, and constraint C6 is the integrality constraint.

C1. These constraints ensure that one node from the set of nodes S^p_{ij} will be chosen as a pickup station for each flow:

$$\sum_{k \in S^p_{ij}} p_{ijk} = 1 \quad \forall i, \forall j \quad \text{where } f_{ij} > 0.$$

C2. These constraints ensure that one node from the set of nodes S^d_{ij} will be chosen as a delivery station for each flow:

$$\sum_{k' \in S^d_{ij}} d_{ijk'} = 1 \quad \forall i, \forall j \quad \text{where } f_{ij} > 0.$$

C3. These constraints ensure that whenever node k is chosen as a pickup station for a flow from department i to department j node k' will be chosen as the delivery station as indicated by (k, k'):

$$p_{ijk} = d_{ijk'} \quad \forall i, \forall j, \forall (k, k').$$

C4. These constraints are used to set the value of the indicator variable I_{ik}. If at least one node k is chosen as a pickup station of department i the indicator is set to 1. Otherwise, it is set to 0 by the minimization objective function:

$$\sum_{j=1}^{M} p_{ijk} - B \cdot I_{ik} \leqslant 0 \quad \forall i, \forall k.$$

C5. These constraints are used to set the value of the indicator variable $I_{jk'}$. If at least one node k' is chosen as a delivery station of department j, the indicator is set to 1. Otherwise, it is set to 0 by the minimization objective function:

$$\sum_{i=1}^{M} d_{ijk'} - B \cdot I_{jk'} \leqslant 0 \quad \forall j, \forall k'.$$

C6.

$$p_{ijk} = 0 \text{ or } 1 \quad \forall ijk \text{ combination}$$

$$d_{ijk'} = 0 \text{ or } 1 \quad \forall ijk' \text{ combination}$$

$$I_{ik}=0 \text{ or } 1 \quad \forall\, ik \text{ combination}$$

$$I_{jk'}=0 \text{ or } 1 \quad \forall\, jk' \text{ combination.}$$

Model objective function The first part of the objective function minimizes the number of pickup and delivery stations in the system. Hence, the minimum split of pickup/delivery stations which still retains the shortest path distance is calculated by the modified shortest path algorithm. The second part of the objective function is the constant cost of the minimum total flow distance, as calculated by equation (8.4):

$$\min\left\{C_{\text{stat}}\cdot\left[\sum_{i=1}^{M}\sum_{k\in S_i^p} I_{ik}+\sum_{j=1}^{M}\sum_{k'\in S_j^d} I_{jk'}\right]+C_{\text{trans}}\cdot\Lambda\right\}.$$

The solution to the MPDS problem contains two types of node sets. The first type of node set, denoted as $\hat{S}_i^p$, contains all the nodes which serve as pickup stations for department i. The second type of node set, denoted by $\hat{S}_j^d$, contains all the nodes which serve as delivery stations for department j, where $\hat{S}_i^p \subseteq S_i^p$ and $\hat{S}_j^d \subseteq S_j^d$.

Heuristic assignment procedure for the MPDS problem

The integer programming model presented in the previous subsection can be used when dealing with small-to-medium-size systems. The computation time may become extensive when dealing with large size systems, therefore an efficient heuristic solution procedure is needed. A heuristic assignment solution procedure is described in the following sections.

The set of origin and destination alternatives for each flow S_{ij}^p, S_{ij}^d, as originally created by the modified shortest path algorithm, can be assigned to two matrices. The first matrix, called the **origin matrix**, contains all the origin node alternatives for each flow from department i to department j, where $f_{ij}>0$. The second matrix, called the **destination matrix**, contains all the destination node alternatives for each flow from department i to department j, where $f_{ij}>0$. Recall that the for every flow $f_{ij}>0$ from department i to department j, the origin node k in the set S_{ij}^p and its corresponding destination node k' in the set S_{ij}^d were determined based on a shortest path distance $L_{ij}^{\min}$ between these departments. Therefore, the objective of selecting a minimum set of nodes to serve as pickup and delivery stations is equivalent to minimizing the objective function of the MPDS problem.

The procedure is based on solving two individual minimum node covering problems. The first on the rows of the origin matrix and the second on the columns of the destination matrix.

Let F_i^p and F_j^d denote the number of flow entries $f_{ij}>0$ in row i of the origin matrix and column j of the destination matrix, respectively. Let O_{ik} denote the number of occurrences of node k in row i of the origin matrix, and let $D_{jk'}$ denote the number of occurrences of node k' in column j of the destination matrix. The numbers for each row and column can be arranged in an descending order such that $O_{ik_1}^{[1]} \geqslant O_{ik_2}^{[2]} \geqslant \cdots \geqslant O_{ik_n}^{[n]}$ and $D_{jk'_1}^{[1]} \geqslant D_{jk'_2}^{[2]} \geqslant \cdots$

$\geqslant D^{[m]}_{jk'_m}$, where $n = |S^p_i|$ and $m = |S^d_j|$. There are two basic cases that need to be distinguished:

1. Assuming that the relationship $O^{[1]}_{ik_1} > O^{[2]}_{ik_2} > \cdots > O^{[n]}_{ik_n}$ holds.
   By choosing the first set of numbers O_{ik} that meet the following requirement $F^p_i \leqslant O^{[1]}_{ik_1} + O^{[2]}_{ik_2} + \cdots$ a minimum node covering for row i of the origin matrix has been found, and nodes that will serve as pickup stations for department i are $k_1, k_2, \ldots$. By choosing these origin nodes the corresponding destination nodes $k'_1, k'_2, \ldots$ as defined by the modified shortest path algorithm, have to be chosen as delivery stations of the j department connected by the flow f_{ij} to department i. This procedure continues until all rows and columns are covered.
2. Assuming that the relationship $O^{[1]}_{ik_1} \geqslant O^{[2]}_{ik_2} \geqslant \cdots \geqslant O^{[n]}_{ik_n}$ holds.
   In cases where the relationships in the ordered lists are not strictly larger, a somewhat different tactic is needed. In these cases, the row node covering proceeds until it is completed, or until the first tie between the O_{ik} numbers is encountered, meaning $O^{[L]}_{ik_L} = O^{[L+1]}_{ik_{L+1}}$. In this case, the following inequality holds:

$$F^p_i > \sum_{l=1}^{L-1} O^{[l]}_{ik_l}. \tag{8.7}$$

This will be only a partial node covering of row i. The procedure continues in the next step to the column node covering phase to check for a tie breaker. That means let the column node covering $k'_1, k'_2, \ldots$ dictate which origin node of those equal will be used. In case node k'_L or k'_{L+1} is used in the node column covering, node k_L or k_{L+1} will be used in the node row covering, respectively. In case the same tie exists in the column covering, or none of these nodes is used in the column covering, the node that will be chosen is that which is already selected and serves as a pickup or delivery station for another department, or the node that is closest to all the rest of the nodes already chosen.

To complete the procedure, a minimum row and column node covering needs to be found for each row and column of the origin and destination matrices, respectively. Only the entries that were not covered in the first pass need to be addressed in the second and third passes. Therefore, the procedure needs to go through $m^2 - p^2$ iterations in the first pass, where m denotes the number of departments in the facility layout and p denotes the number of entries not covered. In the second and third passes, the procedure requires at most p iterations. Bearing in mind that the complexity of the shortest path distance algorithm is $O(n^2)$, where n denotes the number of nodes in the planar graph representing the facility layout, the total complexity of the MPDS heuristic design procedure is $O(m^2 - p + p \cdot n^2)$. For most practical problems, only 4–5 iterations are required in the first pass of the procedure compared to $m^2 - p^2$, and the number of entries p not covered is usually very small.

Improving the current solution In contrast to most heuristic procedures, this procedure enables us to find out relatively easily whether the solution is optimal or not by using a set of propositions and theorems given in Sinriech and Tanchoco (1994a). A nonoptimal solution to the MPDS problem which does not satisfy the user can be improved. According to the proportions given in Sinriech and Tanchoco (1994b), the maximum improvement can be calculated. A simple three step procedure can be used to improve the current solution:

1. Find a row with a non-singular minimal node covering which intersects a column with a non-minimal node covering.
2. Check if a different minimal row covering can actually reduce the δ_j of the column. If yes, the current solution has been improved by 1.
3. If the solution is optimal or satisfies the user then stop; otherwise, go back to step (1).

Heuristic design procedure for the SFT configuration

The final solution to the minimum cost SFT problem can be characterized by the cost of setting up P/D stations, and will be reflected by a set of single and multiple P/D stations for each department, and by the transportation cost reflected by the location of the P/D stations. Two extreme system configurations can be identified. The first includes a minimum expenditure on P/D stations, meaning that only one pickup and delivery station per department is allowed. As a result, the system experiences the maximum expenditure on transportation (if the location of the P/D stations is optimized than this will be a min–max transportation cost). This system configuration was denoted as the SPDS problem. The second extreme configuration includes the minimum possible expenditure on transportation, meaning that each flow follows the shortest possible path in the system. This also means that the maximum expenditure on P/D stations in the system (if the number of splits is kept to a minimum) which is still able to support all the shortest paths in the system will be the min–max P/D fixed cost. This system configuration was denoted as the MPDS problem.

In determining the two extreme system configurations as defined by the SPDS and MPDS problems, the lower and upper bounds on the number of stations needed by a system are set, respectively. Furthermore, the system cost in each case can also be calculated based on the transportation and station setup costs. However, to find the most cost-effective system configuration, more steps are needed. The function that is needed to connect these two extreme points is discrete, due to the fact that specific system configurations are involved. However, a continuous line will be used to represent this function. The function that connects these two points can be either linear or convex, as shown in Fig. 8.10. Due to the fact the function models discrete configurations, any concave function can be replaced by a convex function by just rearranging the points. In case the function is linear, it is obvious that one of the extreme points will be the most cost-effective system configuration, meaning choosing the minimum between {(MPDS), (SPDS)}. However, if the function is convex, there is a new point which corresponds to the most cost-effective system configuration.

The entire heuristic solution procedure is used to replace the mixed integer programming model (step (2) in the solution procedure) of the SFT design procedure. The heuristic procedure is comprised of the following steps:

1. Design the two extreme state system configurations by solving the SPDS and MPDS problems.
2. Calculate all the flow time distance values in the SPDS and MPDS systems.
3. Calculate the maximum benefit, denoted by $\hat{B}_{ij}$, from splitting a pickup or delivery station by subtracting the flow × distance of the MPDS system

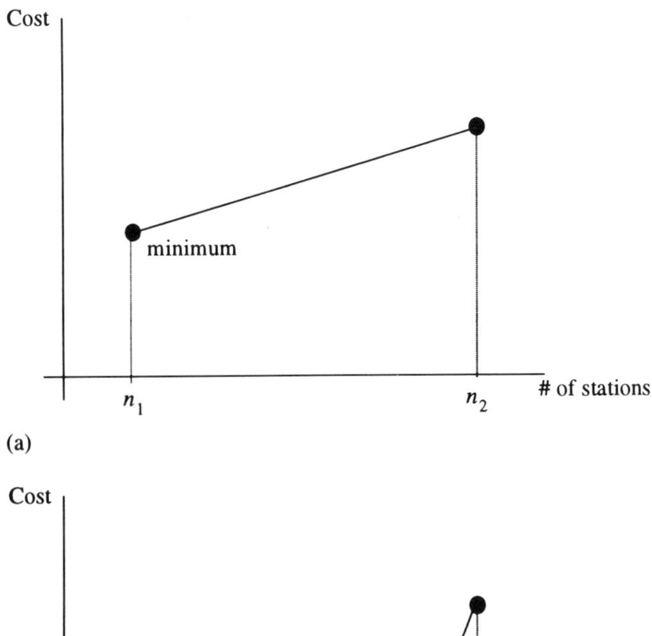

(a)

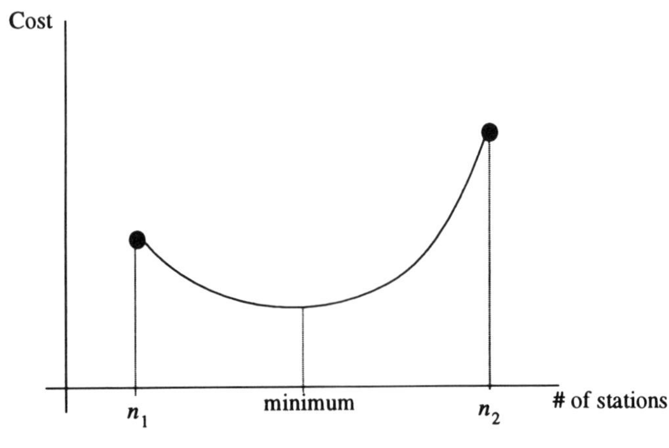

(b)

Fig. 8.10 Different cases of cost functions. (a) linear; (b) concave.

from the flow × distance of the SPDS system for each flow, as follows:

$$\hat{B}_{ij} = \{flow \cdot distance\}_{ij}^{SPDS} - \{flow \cdot distance\}_{ij}^{MPDS} \quad \forall i,j \text{ where } f_{ij} > 0. \quad (8.8)$$

Also, in this step the maximum benefit values $\hat{B}_i^p$, $\hat{B}_j^d$ for each pickup and delivery station have to be calculated from the following equations:

$$\hat{B}_i^p = \sum_{j=1}^{M} \hat{B}_{ij} \quad \forall i. \quad (8.9)$$

$$\hat{B}_j^d = \sum_{i=1}^{M} \hat{B}_{ij} \quad \forall j. \quad (8.10)$$

Choosing $\hat{B} = \max\{\hat{B}_i^p, \hat{B}_j^d\}$, which also satisfies $\hat{B} > (C_{\text{stat}}/C_{\text{trans}})$, identifies the station, pickup or delivery that is most likely to produce the largest savings in the flow × distance value of the system when split, where C_{stat} denotes the cost of setting up a pickup or delivery station and C_{trans} denotes the cost of transporting one job unit one distance unit. The term $C_{\text{stat}}/C_{\text{trans}}$ is the breakeven value which determines if a split in a pickup or delivery station might be cost-effective. In case there is more than one node to choose from in the MPDS system, the splitting will be done one after the other in no particular order. If it turns out that the actual savings in transport cost for a split are smaller than the setup cost of a new station, the split is not performed and the next candidate station is used.

4. After the station is split the procedure is repeated from step (3). In each subsequent pass, the flow × distance of the system designed in the previous pass is used. The procedure continues until $\hat{B} \leqslant (C_{\text{stat}}/C_{\text{trans}})$. This condition will indicate that any additional split of pickup or delivery stations will not be cost-effective.

In case the optimal solution procedures for both the MPDS system and the SPDS system are used, there is no specific order in which the models have to be solved. However, if the heuristic solution procedure to the SPDS problem is used, the set of alternative branching nodes has to be created by first solving the MPDS problem. This also means that the pickup and delivery stations used in the SPDS system are identical to the pickup and delivery stations in the MPDS system. Therefore, the search procedure is simplified somewhat. Once a candidate (pickup or delivery) for splitting is identified based on the rules specified earlier, $\hat{B} = \max\{\hat{B}_i^p, \hat{B}_j^d\}$ and $\hat{B} > (C_{\text{stat}}/C_{\text{trans}})$, the new station location is determined based on the stations location in the MPDS system. The entire heuristic solution procedure is illustrated in Fig. 8.11.

(c) Determining the flow path network

The previous step identified the nodes which are to be used as pickup and delivery stations for each flow in the system. In this step, the flow path network will be determined. By definition, in the SFT system each flow

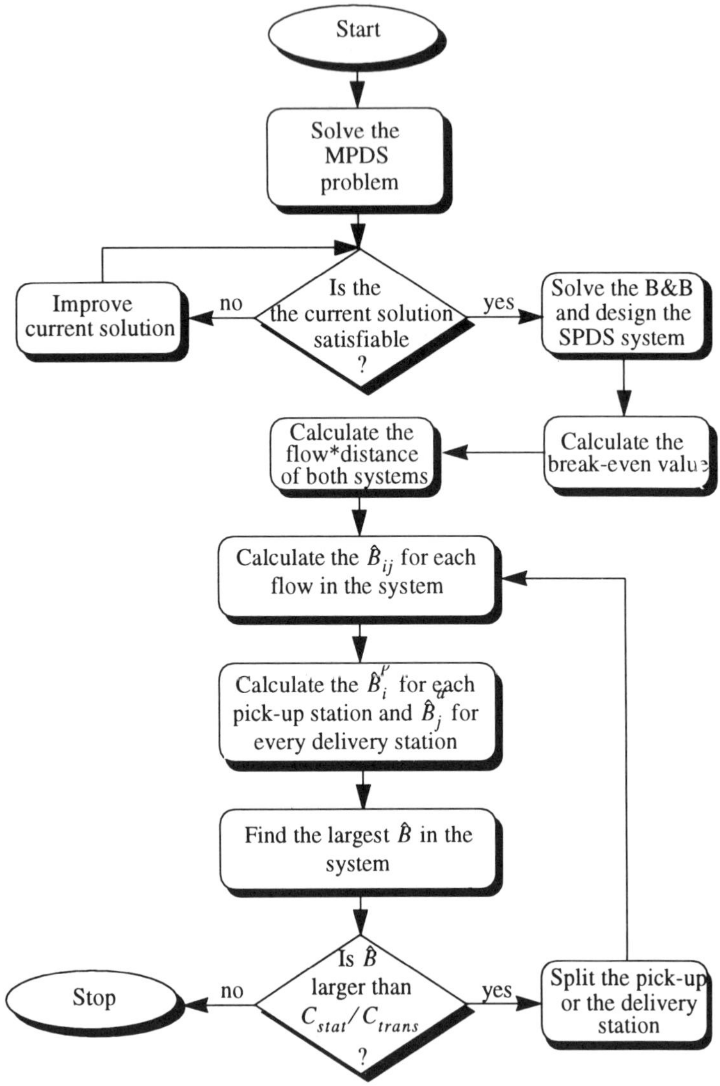

Fig. 8.11 Heuristic solution procedure for the minimum cost SFT formulation.

follows the shortest path from the determined pickup station to the determined delivery station. Consequently, once the layout's active nodes have been identified it is easy to identify the layout's active segments by solving Dijkstra's algorithm. The next part in determining the flow path network is to identify the mutual exclusive zones in the system. To identify these zones, the incidence matrix of the graph representing the network has to be formed. This can be achieved using the information on the active nodes and segments of the layout.

Determining the number of mutual exclusive zones in the flow network
By using a well known procedure from the topology theory literature (Giblin, 1977), the number of components in a graph can be calculated by the following formula:

$$\omega(G_P^*) = \nu(G_P^*) - rank(M(G_P^*)), \tag{8.11}$$

where $\omega(G_P^*)$ denotes the number of components in graph G_P^*, $\nu(G_P^*)$ denotes the number of nodes in graph G_P^*, and $rank(M(G_P^*))$ denotes the column rank of the incidence matrix of graph G_P^*. Graph G_P^*, which is a subgraph of G_P, represents the collection of active nodes and segments, as identified by the previous procedure.

Identifying the mutual exclusive zones
The second step is an allocation procedure used to assign the pickup/delivery station of each department to a specific zone. The proposed procedure uses the adjacency matrix of graph G_P^* in conjunction with a selective elimination algorithm. The algorithm assigns pickup and delivery stations to a zone if their corresponding rows and columns have entries in their common intersection points. These rows and columns are eliminated from the from–to flow matrix, and the procedure starts assigning stations to a new zone until all

1. Set zone counter $k = 0$
2. Set $k = k + 1$
3. If there are any $a_{ij} > 0$ entries in the adjacency matrix A

   find a set of nodes i^* and j^* such that for all i^* in the set and j not in the set, $a_{i^*j} = 0$ and $a_{ji^*} = 0$ and for all j^* in the set and i not in the set, $a_{j^*i} = 0$ and $a_{ij^*} = 0$ and go to step 5.

   else
4. If there are no more unmarked columns

   create a zone for each column and each row and update k.

   stop.
5. Mark all rows i^* and all columns j^* in the adjacency matrix A and erase all $a_{i^*j^*} > 0$.
6. Go to step 2.

Fig. 8.12 Allocation procedure for identifying the mutual exclusive zones.

mutual exclusive sets are detected. The flow chart of the proposed procedure is listed in Fig. 8.12.

(d) Segmenting the flow path network

To achieve good performance from each zone in the system, the workload of the different material handling devices in each segment has to be balanced as much as possible, so that no segment will become the system bottleneck. To segment a zone, the number of segments needed first has to be determined, and then the segmentation can be performed. For the first step, the number of material handling devices in each zone indicates how many segments are required. The second step is to identify each zone and distinguish between the point flow structures which do not need segmentation, and the rest of the flow structures which do need to be segmented. Finally, the sizes of the transfer buffers which are located at the segment boundaries have to be determined.

Calculating the number of carriers

The objective function of the model incorporates the transportation cost and the station setup cost. However, there is an additional substantial cost in the material handling system, which is the material handling device itself. In the SFT system we distinguish between two basic types of material handling devices. The first is a transfer device which deals with stationary material handling between adjacent departments. In this case, the pickup and delivery stations share the same location. This is manifested in the network by a single point flow structure. The second type deals with mobile material handling between departments which do not share common aisles. These devices can be either carriers, conveyors or monorails. There is a large number of studies on the material handling device requirement problem (Maxwell and Muckstadt, 1982; Egbelu, 1987; Tanchoco *et al.*, 1987; Malmborg, 1991). Some of the procedures require very simple calculations by neglecting blocking, congestion and inefficiency. Other methods use very elaborate calculations, which include different dispatching rules, congestion and blocking measures. Using an SFT type system simplifies these calculations because the system does not experience any blocking or congestion, therefore no inefficiency time loss occur. Hence, the carrier requirement calculation includes only the loaded and empty flow time and the load/unload time at pickup/delivery stations and at transfer buffers. This calculation has been suggested by Egbelu (1987), Gaskins *et al.* (1989) and Bakkalbasi (1990).

Segmentation procedure

The segmentation itself can be performed through some workload analysis like queueing theory. However, in this study the segmentation was done with the aid of simulation. An initial rough cut analysis can determine where to locate the segment boundaries, then any refinement of these locations can be done by simulating and analysing the workloads of the material handling devices in each segment for the final result to be balanced as much as possible. The final step is determining the sizes of the transfer buffers which are located at the segment boundaries.

Estimating transfer buffer sizes

The last part of the segmentation procedure is to estimate the transfer buffer capacities. Because these capacities change for different shop workloads, simulation was used to determine the buffer sizes. The simulation runs made to refine the segment boundaries' locations can also indicate the average and maximum levels of WIP in the transfer buffers.

(e) Calculating the total cost of the system

Based on the number of mobile and stationary material handling devices, the total number of pickup and delivery stations in the system and the transportation distance, the total cost of the SFT system can be calculated.

8.4 ILLUSTRATIVE EXAMPLE

To illustrate the solution procedure the 11 department facility layout shown in Fig. 8.13 (indicating the length of the arcs) with the process plan and job mix data given in Table 8.1 are used. Using the process plan in conjunction with a four minute interarrival time, the from–to flow matrix shown in Table 8.2 was formed. The costs used in the design procedure are fixed station cost of $12 000 and a transportation variable cost of $4 per one material unit per one distance unit.

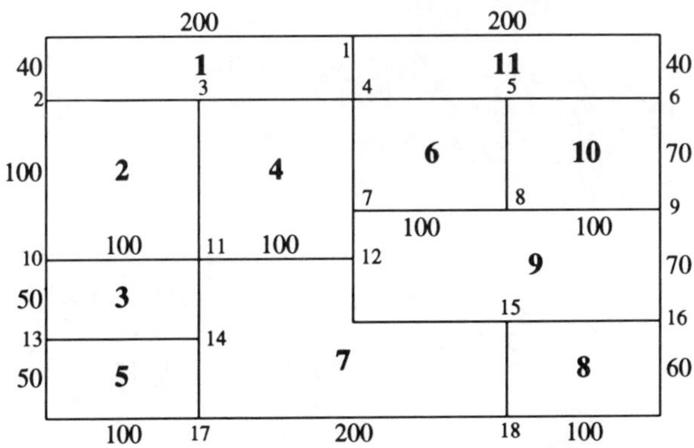

Fig. 8.13 Eleven-department facility layout.

Table 8.1 Process plan and job mix data

Part type	*Job mix*	*Unit load*	*Job routing*
1	0.3125	1	1–2–4–9–8–10–11
2	0.1875	1	1–2–4–7–9–4–6–10–11
3	0.0625	1	1–2–7–9–6–10–11
4	0.125	1	1–2–3–5–9–6–11
5	0.3125	1	1–2–4–8–10–11

Table 8.2 From–to flow matrix

	To										
From	1	2	3	4	5	6	7	8	9	10	11
1	–	120									
2		–	15	97.5			7.5				
3			–		15						
4				–		22.5	22.5	37.5	37.5		
5					–				15		
6						–				30	15
7							–		30		
8								–		75	
9				22.5		22.5		37.5	–		
10										–	105
11											–

Table 8.3 Combined origin and destination matrices of sets S_{ij}^p, S_{ij}^d

	To										
From	1	2	3	4	5	6	7	8	9	10	11
1		2,3 2,3									
2			10,11 10,11	3,11 3,11			11 11				
3					13,14 13,14						
4						4,7 4,7	11,12 11,12	12 15	7,12 7,12		
5									14 12		
6										5,8 5,8	4,5 4,5
7									12,15 12,15		
8										16 9	
9				7,12 7,12		7,8 7,8		16,15 16,15			
10											5,6 5,6
11											

Table 8.4 Node sets used in the minimum cost SFT formulation

Department	*Set* S_i^p	*Set* S_j^d
1	2, 3	–
2	3, 10, 11	2, 3
3	13, 14	10, 11
4	4, 7, 11, 12	3, 7, 11, 12
5	14	13, 14
6	4, 5, 8	4, 7, 8
7	12, 15	11, 12
8	16	15, 16
9	7, 8, 12, 15, 16	7, 12, 15
10	5, 6	5, 8, 9
11	–	4, 5, 6

8.4.1 Optimal and heuristic solution to the MPDS problem

The first step in the procedure is to use the modified shortest path algorithm to create the S_{ij}^p, S_{ij}^d sets and the origin and destination matrix. Table 8.3 superimposes the two matrices. The first line in each matrix entry represents the origin matrix, and the second line represents the destination matrix. These two sets are used to create the S_i^p, S_j^d sets, as shown by equations (8.1) and (8.2). The final result is summarized in Table 8.4. Using the origin and destination matrices shown in Table 8.3, the MPDS problem can be formulated. This model contains 115 binary variables and 220 constraints. The solution to model optimization is shown in Fig. 8.14. This system contains 25 stations, five of which are multiple stations: pickup stations for departments 4 and 9 and delivery stations for departments 4, 7 and 10. The flow × distance value in this case was 12 750 which is the absolute minimum for the list of flows used in the system.

The final solution from the heuristic procedure was identical to the optimal

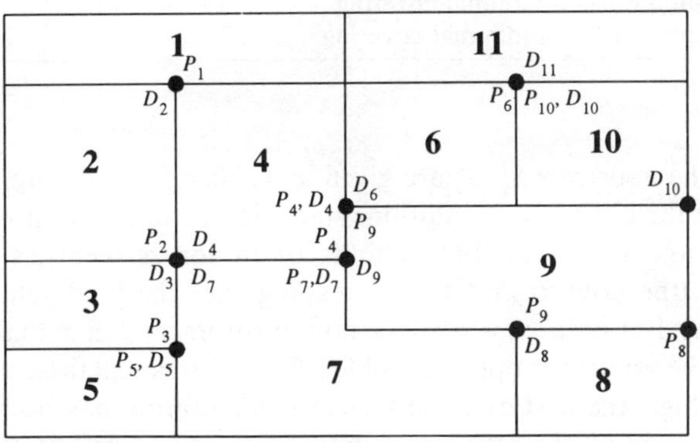

Fig. 8.14 Station location in the MPDS system configuration.

Table 8.5 Solution node sets of the MPDS problem

Department	*Set* $\hat{S}_i^p$	*Set* $\hat{S}_j^d$
1	3	–
2	11	3
3	14	11
4	7, 12	7, 11
5	14	14
6	5	7
7	12	11, 12
8	16	15
9	7, 15	12
10	5	5, 9
11	–	5

Table 8.6 Optimality conditions of the heuristic solution

Entry	*Optimality conditions*
1–2	Column 2 has a minimal covering
2–3	Row 2 has a singular minimal covering and column 3 has a minimal covering
2–4	Row 2 has a singular minimal covering and column 4 has a minimal covering
2–7	Row 2 has a singular minimal covering
3–5	Column 5 has a minimal covering
4–6	Column 6 has a minimal covering
4–7	No conditions can be found
4–8	Column 8 has a minimal covering
4–9	Column 9 has a minimal covering
5–9	Row 5 has a singular minimal covering and column 9 has a minimal covering
6–10	Row 6 has a singular minimal covering and column 10 has a minimal covering
6–11	Row 6 has a singular minimal covering and column 11 has a minimal covering
7–9	Column 9 has a minimal covering
8–10	Row 6 has a singular minimal covering and column 10 has a minimal covering
9–4	Column 4 has a minimal covering
9–6	Column 6 has a minimal covering
9–8	Column 8 has a minimal covering
10–11	Column 11 has a minimal covering

solution. The node sets $\hat{S}_i^p$, $\hat{S}_j^d$ are given in Table 8.5. Analysing the solution reveals that the optimality conditions stated in one of the propositions given in Sinriech and Tanchoco (1992c) holds for all non-zero entries in Table 8.3 except one (the flow from 4 to 7), meaning that the final solution can be improved by 1 at best. Nevertheless, no improvement can actually be made. Therefore, the solution is optimal. Table 8.6 summarizes all these relationships.

At this stage, the first extreme system configuration has been found and the upper bound on the number of stations in the system has been set to 25 by both the optimal and heuristic solution procedures.

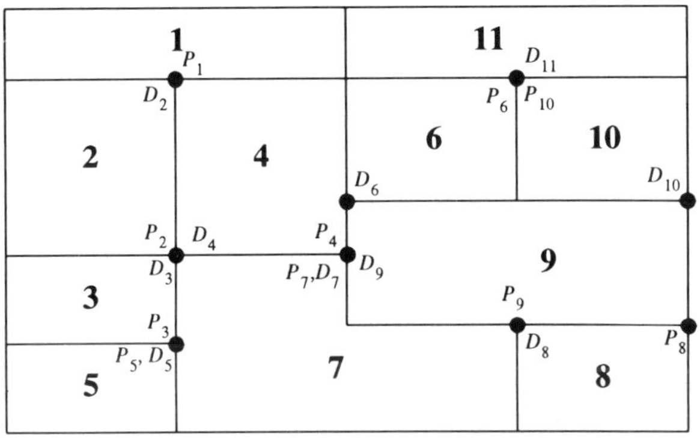

Fig. 8.15 Station location in the SPDS system configuration using the heuristic procedure.

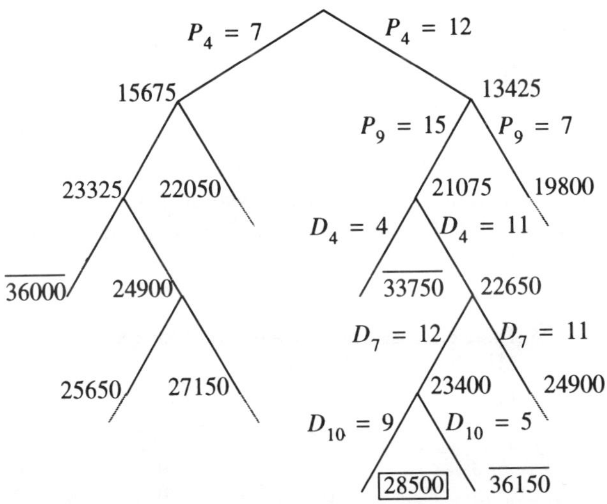

Fig. 8.16 Partial tree of the branch-and-bound procedure.

8.4.2 Optimal and heuristic solutions to the SPDS problem

Using the sets of nodes listed in Table 8.4, the SPDS problem can be formulated. The model contains 51 binary variables, 153 continuous variables and 683 constraints. This system contains the smallest possible number of pickup and delivery stations, 20 (no multiple stations), and a flow × distance value of 26 400.

The branch-and-bound solution procedure ran using the node sets in Table 8.5. In this example, $\hat{\Psi}^{pd} = 5$, i.e. the branch-and-bound tree is five levels deep, and the maximum number of option is $2^1 + \cdots + 2^5$ compared to $\Psi^{pd} = 49$, and as a result, $2^1 + \cdots + 2^{49}$ options in the optimal solution procedure. In all, 38 problems were solved to complete the branch-and-bound procedure, while the best solution was obtained after only solving 17 problems. The solution shown in Fig. 8.15 varied only slightly from the optimal solution. The value of the heuristic solution was 28 500, which is slightly higher as compared to the optimal solution due to a different location of pickup station 6. The partial branch-and-bound tree is shown in Fig. 8.16.

At this stage, the second extreme system configuration has been found and the lower bound on the number of stations in the system has been set to 20 by both the optimal and heuristic solution procedures.

8.4.3 Search procedure and the splitting of the SPDS system

To split the stations in the SPDS system, the flow × distance values of the SPDS and MPDS systems have to be calculated. These values are listed in Table 8.7. Also, the break-even value can easily be calculated as follows: 12 000/4 = 3000. To determine which pickup and/or delivery station to split, a list of candidates has to be formed from all the stations in the system. This list contains only those stations which meet the following requirement: $\hat{B}_i^p$ or $\hat{B}_i^d > C_{\text{stat}}/C_{\text{trans}}$, i.e. the sum of flow × distance in or out exceeds the break-even value of 3000. All the candidate stations are listed in Table 8.8. The procedure will start by splitting pickup station 9 and position the new station at node 7, the same as in the MPDS system. The new system configuration has a total of 21 P/D stations and a flow × distance value of 22 200.

Table 8.7 List of flow × distance for the MPDS and SPDS systems

Flows	*MMPDS system*	*SPDS system*
1–2	0	0
2–3	0	0
2–4	0	0
2–7	0	750
3–5	0	0
4–6	0	675
4–7	0	0
4–8	5250	5250
4–9	0	0
5–9	2250	2250
6–10	0	5100
6–11	0	0
7–9	0	0
8–10	5250	5250
9–4	0	5400
9–6	0	3825
9–8	0	0
10–11	0	0

Table 8.8 List of stations to consider for splitting

Department	*Station*	$\hat{B}_i^p, \hat{B}_j^d$
9	pickup	9225
4	delivery	5400
6	pickup	5100
10	delivery	5100
6	delivery	4500

Table 8.9 List of candidates after the first pass

Department	*Station*	$\hat{B}_i^p, \hat{B}_j^d$
6	pickup	5100
10	delivery	5100
4	delivery	2925
9	pickup	2925

The next step is to recalculate all the $\hat{B}_i^p, \hat{B}_j^d$ in the system. The new list of stations which meet the $\hat{B}_i^p$ or $\hat{B}_i^d > C_{\text{stat}}/C_{\text{trans}}$ requirement is given in Table 8.9. In the next step, both pickup station department 6 or delivery station department 10 can be split. However, in the MPDS system the split station is delivery station 10. Therefore, this will also be the split station in this system; the new delivery station will be position at node 5, as in the MPDS system. The new system has a total 22 P/D stations and a flow × distance of 17 100. At this stage, no more cost-effective splits can be found (the next $\hat{B}$ is equal to 2925, which is less than the break-even value of 3000), and the procedure is done. The final system configuration varies slightly from the optimal SFT system shown in Sinriech and Tanchoco (1994a) (the location of one of the pickup stations, 9, is different). However, both systems have the same flow × distance value of 17 100.

8.4.4 Adding the cost of the mobile and stationary transfer devices to the system

To complete the analysis on the system's cost-effectiveness, the cost of the transport devices has to be added. In the optimal design procedure the cost of the mobile and stationary transport devices was added only to the final system design, because only one configuration was evaluated. While in some cases this does not affect the optimality of the design, in other cases it may change the solution.

Based on the calculations presented in Sinriech and Tanchoco (1994a), the SFT system designed by the optimal design procedure can be split into four zones, one of which requires two mobile carriers, and one which requires one mobile carrier for a total of three \$20 000 mobile carriers. The other two zones require only stationary transfer devices for a total of two \$5000 stationary transfer devices. By summing all these costs, the total cost of the system is \$402 400.

Based on the same calculation, the upper bound system (Fig. 8.14) can be split into five zones, two of which require one mobile carrier for a total of two \$20 000 mobile carriers, and three of which require only stationary transfer devices for a total of three \$5000 stationary transfer devices. By summing all these costs, the total cost of the system is \$406 000. The lower bound system can be split into three zones only (Fig. 8.15), one of which requires two mobile carriers, and one which requires one mobile carrier for a total of three \$20 000 mobile carriers. The last zone requires only one \$5000 stationary transfer device. By summing all these costs, the total cost of the system is \$419 000. These two costs make up the two extreme cases of the system's cost function. Using the same calculation as before, the cost of transport devices can also be added into each intermediate configuration in the search procedure.

After the first split of pickup station 9, no new zone has been added to the system. On the other hand, there was also no reduction in the number of transport (mobile and stationary) devices. Therefore, the total cost of the configuration is \$405 800. However, after the second split of delivery station 10 a new zone has been added to the system. One zone requires two mobile carriers, the other requires one mobile carrier, and the last two require only stationary transfer devices. The system contains 22 P/D stations for a total cost of \$402 400. This is where the optimal design procedure (Sinriech and Tanchoco, 1994a) which does not take the transportation carrier costs into consideration stops. However, an additional split, in this case one which does not meet the $\hat{B}_i^p$ or $\hat{B}_i^d > C_{\text{stat}}/C_{\text{trans}}$ requirement, may result in a more cost-effective system. For example, the next split station is delivery station 4. Although the cost of station setup and transportation goes up (\$332 700

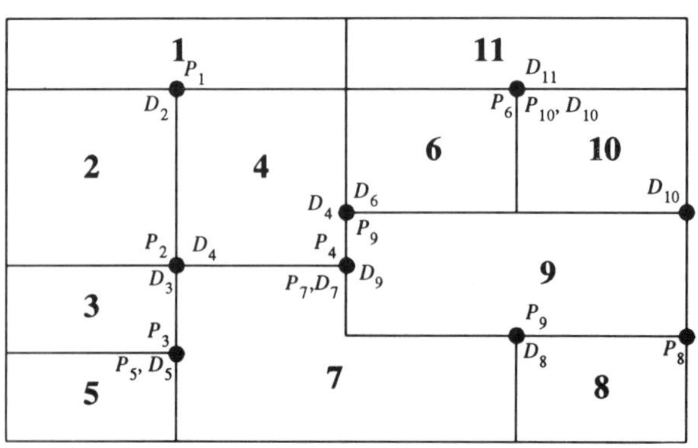

Fig. 8.17 Find station location as a result of the heuristic design procedure.

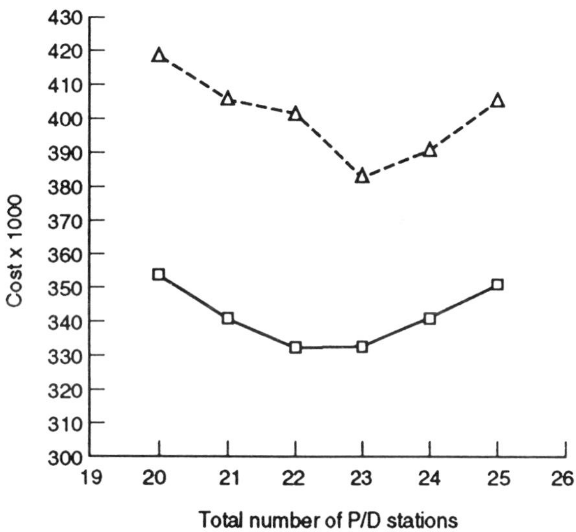

Fig. 8.18 System's total cost function with (△) and without (□) carrier cost.

compared to \$332 400), the total cost including mobile and stationary transport devices actually drops (\$382 700 compared to \$402 400) due to one less mobile transfer device being required). The final station location is illustrated in Fig. 8.17. The entire cost function including the transport devices is illustrated in Fig. 8.18.

8.4.5 Splitting the flow path into mutual exclusive zones

By using the pickup and delivery station locations obtained in the previous step, and the MSPA, the flow path network for the given example can be designed. The final flow path design is shown in Fig. 8.19. The incidence matrix which is used in the splitting procedure, shown in Table 8.10, can be created based on this flow path. The number of nodes $v(G_P^*)$ in the flow path network is 9. By performing column operations, the columns representing nodes 3, 5, 9 and 12 can be set to zero. Therefore, the column rank of the

Table 8.10 Incidence matrix of the flow path network

	Nodes								
Edges	3	5	7	9	11	12	14	15	16
1					1		1		
2					1	1			
3			1			1			
4						1		1	
5				1					1

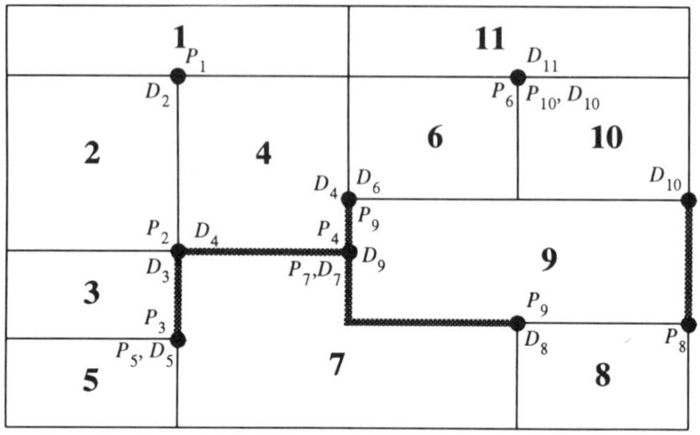

Fig. 8.19 Final SFT flow path network.

incidence matrix is 5. Using equation (8.11), the number of mutual exclusive zones in the graph is 4.

8.4.6 Segmenting the flow path network

In this example, two of the four zones require only stationary transfer devices (point flow structure) and the other two require only one carrier per zone. Therefore, no zone segmentation and transfer buffers were required, and the final flow path network shown in Fig. 8.19 is also the final SFT network.

8.5 FLOW × DISTANCE COMPARISON

The most common criteria measure in the evaluation of material flow systems is the flow × distance of the system. In this case, the flow × distance values of a unidirectional conventional system, a bidirectional conventional system, an optimal single-loop system and an optimal segmented bidirectional single-loop system were compared to those of the MPDS, SPDS and the final SFT systems. The flow × distance of tandem configuration was not included as the tandem configuration does not assume a fixed aisle network.

The models for the unidirectional and bidirectional conventional system assume a predetermined pickup and delivery station location, therefore no optimality can be guaranteed. In this case, the initial station location obtained for the SPDS system was used. The flow × distance of the unidirectional conventional system shown in Fig. 8.1 was obtained by using the model developed by Kaspi and Tanchoco (1990), and was 30 900. The flow × distance of the bidirectional conventional system shown in Figs. 8.2 and 8.3 was obtained using the model developed by Kim and Tanchoco (1991), and was 26 400. The flow × distance of the optimal single loop shown in Fig. 8.4

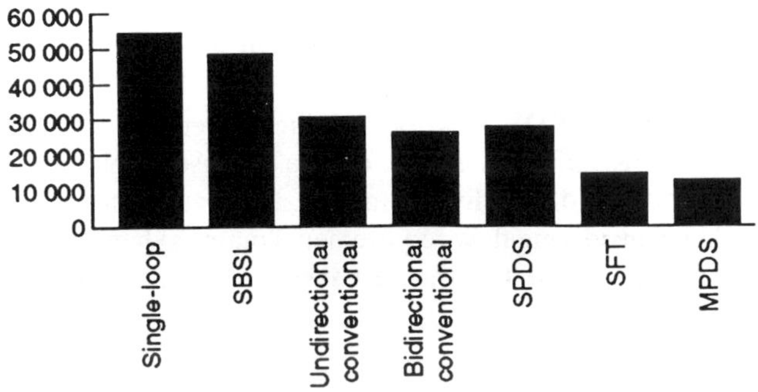

Fig. 8.20 Flow × distance values of the different systems. ■ = flow × distance.

was obtained using the OSL procedure developed by Tanchoco and Sinriech (1992), and was 54 750. The last flow × distance of the optimal segmented bidirectional single-loop shown in Fig. 8.5 was obtained using the model developed by Sinriech and Tanchoco (1992), and was 48 900. These should be compared to a flow × distance of 28 500 and 14 175 for the SPDS and SFT systems, respectively, and 12 750 for the MPDS system, which is the absolute minimum for this network and flow requirements. These results are illustrated in Fig. 8.20.

8.6 CONCLUSIONS

Although the performance measure of flow × distance was calculated for a specific example, some general conclusions can be drawn. The highest value of the flow × distance will always be obtained from the single-loop network. On the other hand, the MPDS system always produces the absolute lowest flow × distance for a specific network and flow requirements, because each flow in the systems follows the shortest path from origin to destination.

Both the unidirectional conventional network and bidirectional single-loop SBSL will produce equal or lower flow × distance results compared to the single-loop; the former due to additional shortcuts in the network, and the latter due the bidirectional flow, which may shorten some or all the transportation distances. However, no specific conclusions can be drawn on the flow × distance relationship of these two networks, because the relationship is problem-dependent. On the other hand, the bidirectional conventional system will produce equal or lower flow × distance results compared to both the unidirectional conventional network and bidirectional single-loop SBSL.

In case both the SPDS network and the bidirectional conventional configuration have the same pickup and delivery station locations, the flow × distance of these systems is identical. In this comparison, the unidirectional conventional configuration used the station location obtained by the SPDS formulation (optimal), while the SPDS network used the station location obtained by the branch-and-bound procedure (heuristic), which explains the difference in flow × distance values.

Both the SFT and MPDS networks produce lower flow × distance results compared to the rest of the systems, due to the use of bidirectional flows and multiple pickup and/or delivery stations. However, in terms of economic efficiency, when considering flow requirements material handling devices and station set up costs, the SFT is the best system that can be obtained.

REFERENCES

Afentakis, P. (1989) A loop layout design problem for flexible manufacturing systems. *International Journal of Flexible Manufacturing Systems*, **1**, 175–96.

Bakkalbasi, O. (1990) *Flow path network and layout configuration for material delivery systems.* PhD Dissertation, Georgia Institute of Technology, Atlanta, GA.

Bozer, Y.A. and Srinivasan, M.M. (1991a) Tandem configuration for automated guided vehicle systems and the analysis of single vehicle loops. *IIE Transactions*, **23**(1), 72–82.

Bozer, Y.A. and Srinivasan, M.M. (1991b) Tandem AGV Systems: A Partitioning Algorithm and Performance Comparison with Conventional AGV Systems. *Working Paper*, Department of Industrial and Operations Engineering, The University of Michigan, Ann Arbor, MI.

Dijkstra, E.W. (1959) A note on two problems in connexion with graphs. *Numerische Mathematik*, **1**, 269–71.

Egbelu, P.J. and Tanchoco, J.M.A. (1986) Potential for bidirectional guide path for automatic guided vehicles based systems. *International Journal of Production Research*, **24**(5), 1075–99.

Egbelu, P.J. (1987) The use of non-simulation approaches in estimating vehicle requirements in an automated guided vehicle based transport system. *Material Flow*, **4**(1 & 2), 17–32.

Gaskins, R.J. and Tanchoco, J.M.A. (1987) Flow path design for automated guided vehicle systems. *International Journal of Production Research*, **25**(5), 667–76.

Gaskins, R.J., Tanchoco, J.M.A. and Taghaboni, F. (1989) Virtual flow paths for free-ranging automated guided vehicles systems. *International Journal of Production Research*, **27**(1), 91–100.

Giblin, P.J. (1977) *Graphs, Surfaces and Homology*, Halsted Press, New York.

Goetz, W.G. and Egbelu, P.J. (1990) Guide path design and location of load pick-up/drop-off points for an automated guided vehicle systems. *International Journal of Production Research*, **28**(5), 927–41.

Kaspi, M. and Tanchoco, J.M.A (1990) Optimal flow path design of unidirectional AGV systems. *International Journal of Production Research*, **28**(6), 1023–30.

Kim, C.W. and Tanchoco, J.M.A. (1991a) Conflict-free shortest-path bidirectional AGV routing. *International Journal of Production Research*, **29**(12), 2377–91.

Kim, C.W. and Tanchoco, J.M.A. (1991b) Operational Control of Bidirectional Automated Guided Vehicle Systems. *Working Paper*, School of Industrial Engineering, Purdue University, West Lafayette, IN.

Kim, K.H. and Tanchoco, J.M.A. (1991) *Flow Path Design of Fixed-Path Material Handling Systems.* Proceedings ASME Winter Annual Meeting, Atlanta, GA.

Kiran, A.S., Unal, A.T. and Karabati, S. (1992) A location problem on a unicyclic network: Balanced case. *European Journal of Operations Research*, **62**(1), 194–202.

Kouvelis, P. and Kim, M. (1992) Unidirectional loop network layout problem in automated manufacturing systems. *Operations Research*, **40**(3), 533–50.

Lin, J.T., Liu, W.-C. and Chang, C.C.K. (1992) *Building a Load Routing Model for a Tandem Configuration AGVS*. Proceedings Second International Conference on Automation Technology, Taipei, Taiwan, **2**, pp. 237–45.

Mahadevan, B. and Narendran, T.T. (1990) Design of an automated guided vehicle based material handling system for flexible manufacturing systems. *International Journal of Production Research*, **28**(9), 1611–22.

Malmborg, C.J. (1991) Tightened analytical bounds on the impact of dispatching rules in automated guided vehicle systems. *Applied Mathematical Modeling*, **15**(6), 305–11.

Maxwell, W.L. and Muckstadt, J.A. (1982) Design of automatic guided vehicle systems. *IIE Transactions*, **14**(2), 114–24.

Rim, S.C. and Bozer, Y.A. (1991) The Bidirectional Circular Layout Problem. *Working Paper*, Industrial Engineering Program, FU-20, University of Washington, Seattle, WA.

Riopel, D. and Langevin, A. (1991) Optimizing the location of material transfer stations within layout analysis. *International Journal of Production Economics*, **22**, 169–76.

Sinriech, D. and Tanchoco, J.M.A. (1992) The Segmented Single-Loop Guide Path for Bidirectional AGV Systems. *Research Memorandum No. 92-18*, School of Industrial Engineering, Purdue University, West Lafayette, IN.

Sinriech, D. and Tanchoco, J.M.A. (1994a) The Segmented Flow Topology (SFT) for Discrete Material Flow Systems. *Working paper*, Technion-Israel Institute of Technology, Haifa 32000, Israel.

Sinriech, D. and Tanchoco, J.M.A. (1994b) Design Procedures and Implementation of the Segmented Flow Topology (SFT) for Discrete Material Flow Systems. *Working paper*, Technion-Israel Institute of Technology, Haifa 32000, Israel.

Sinriech, D. and Tanchoco, J.M.A. (1993) Solution methods for the mathematical models of single-loop AGV systems. *International Journal of Production Research*, **31**(3).

Stein, D.M. (1978) Scheduling dial-a-ride transportation systems. *Transportation Science*, **12**(3), 232–49.

Tanchoco, J.M.A., Egbelu, P.J. and Taghaboni, F. (1987) Determination of the total number of vehicles in an AGV based material transport system. *Material Flow*, **4**(1 & 2), 33–51.

Tanchoco, J.M.A. and Sinriech, D. (1992) OSL – Optimal Single Loop guide paths for AGVs. *International Journal of Production Research*, **30**(3), 665–81.

Usher, J. S., Evans, G. W. and Wilhelm, M.R. (1988) *AGV Flow Path Design and Load Transfer Point Location*. Proceedings IIE Conference, Orlando, FL, pp. 174–9.

Venkataramanan, M.A. and Wilson, K.A. (1991) A branch-and-bound algorithm for flow-path design of automated guided vehicle systems. *Naval Research Logistics Quarterly*, **38**, 431–45.

PART FOUR

Operational Control Issues

CHAPTER 9

Bidirectional Automated Guided Vehicle Systems (AGVS)

C.W. Kim and J.M.A. Tanchoco

9.1 INTRODUCTION

The flow path network of an automated guided vehicle (AGV) system is defined by open aisles between machines, workstations, departments and fixed structures on the shop floor. A lane connecting a pair of nodes in the network can be unidirectional or bidirectional. There can be more than one lane in the same aisle if the traffic requirement is heavy. Therefore, the flow path networks of AGV systems can be classified into a (a) unidirectional model, (b) a multiple-lane model, (c) a bidirectional model, and (d) a mixed model, as depicted in Fig. 9.1.

The unidirectional model is widely used in practice because of its simplicity in operational control. However, the simplicity is gained at the cost of reduced efficiency and network reliability. In Fig. 9.1(a), for example, the only way to get from node B to E is to follow lanes 2 and 6 even though there is a shorter path using lane 5 if lane 5 is bidirectional. Also, if lane 2 becomes unavailable, then there is no way to get from B to E. The multiple-lane model is intended to overcome the deficiencies of the unidirectional model. The same objectives can also be achieved by the bidirectional model if proper control methods are available. The mixed model is an attempt to combine the advantages of the three basic models. It is a generic design; all the basic models are special cases of the mixed model. A more detailed discussion on flow path models can be found in Egbelu and Tanchoco (1986).

The lanes in the bidirectional model are bidirectional in a switchable manner. Although its operation is more complicated than the unidirectional or multiple-lane model, the bidirectional model achieves significant reductions in the total travel distances and space requirements for the flow path network, and may be more economical where there are few vehicles in operation. A bidirectional system can also achieve shorter flow times and higher throughput rates compared to a unidirectional system with the same number of vehicles. For a given production target, a bidirectional system may require fewer

The material in this chapter is obtained from Kim, C.W. and Tanchoco, J.M.A. (1991, 1993).

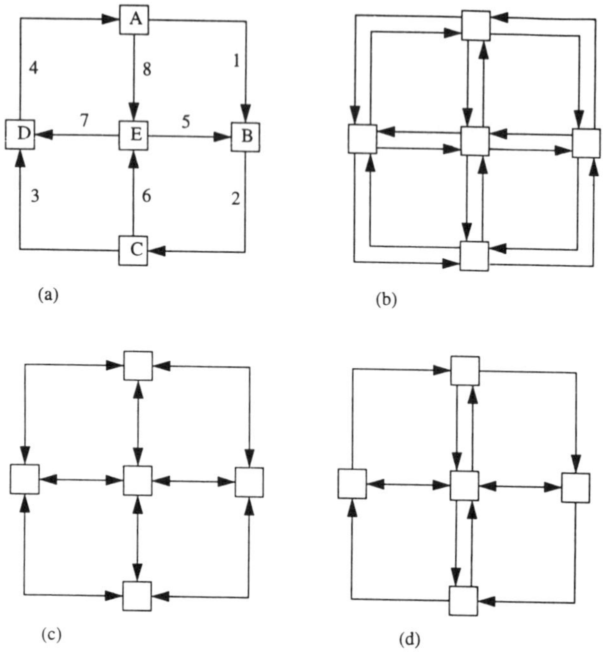

Fig. 9.1 AGV flow path network models: (a) unidirectional; (b) multiple lane; (c) bidirectional; (d) mixed.

vehicles than a unidirectional system. Egbelu and Tanchoco (1986) showed through a simulation study that bidirectional systems can significantly improve the productivity and cut down the fleet size and the makespan. A bidirectional system requires less space for the flow path network compared to an equivalent multiple-lane system. However, bidirectional systems are not generally used in practice, even though many of the AGVs on the market have a bidirectional capability. This seems to be mainly because there has been no proven method for the operational control of such systems, except for a few academic research results reported in the literature.

The purpose of this study is to develop an efficient and simple method for the operational control of bidirectional AGV systems, and to provide a reasonable prediction on the performance behavior of bidirectional AGV systems in comparison to unidirectional systems under a general setting. An efficient algorithm for finding conflict-free shortest-time routes for AGVs moving in a bidirectional flow path network is described in section 9.2. Using this algorithm as a basic tool, a conservative myopic strategy to coordinate the movements of several vehicles in a bidirectional AGV system is described in section 9.3. Then, simulation results to compare the performances of unidirectional and bidirectional systems are presented in section 9.4.

9.2 CONFLICT-FREE SHORTEST-TIME AGV PATH PLANNING

9.2.1 Problem

The *conflict-free shortest-time AGV path planning problem* can be stated informally as follows. There are several vehicles operating in a predefined flow path network, which consists of a set of nodes interconnected by a set of lanes. Nodes represent load transfer stations, intersections, parking lots, battery charging stations and other important locations. Lanes can be unidirectional, bidirectional or mixed, and there may be multiple lanes connecting the same pair of nodes if the traffic is heavy. A dispatch command issued to a vehicle includes a pair of source and destination nodes. Assuming that the vehicle is currently located at the source node, the problem is to find a route starting from the source and arriving at the destination *as early as possible without disrupting other active travel schedules.*

A typical situation is exemplified in Fig. 9.2. The flow path network shown in Fig. 9.2(a) is a portion of a larger network composed entirely of bidirectional lanes. The vehicles can move either forwards or backwards. A safety zone or check zone is a square area defined around a node to protect vehicles from collision (Egbelu and Tanchoco, 1984). Usually, the check zone length is obtained by the following equation:

$$\textit{Check Zone Length} = \textit{Vehicle Length} + 2 \times \textit{Safety Allowance}$$

The safety allowance (minimum headway – vehicle length) is ignored in this example for the sake of simplicity. The response time is the time from the start of algorithm execution until the start of vehicle movement. This includes the computation time of the algorithm and the communications delay needed to transmit the travel schedule to the vehicle. In this example, two vehicles, V1 and V2, have been scheduled to traverse the partial network as shown below:

(Current schedules for V1 and V2)

$$\text{V1:} \rightarrow ([20, 24], 5, 8] \rightarrow ([40, 44], 4, 4) \rightarrow ([60, 64], 1, 1) \rightarrow$$

$$\text{V2:} \rightarrow ([8, 12], 3, 6) \rightarrow [(28, 32], 4, 7) \rightarrow$$

Notation: ([entry time, exit time], node #, next lane #)

Figure 9.2(b) graphically shows the entry and exit times of these vehicles at each node on their routes. In this situation we want to route vehicle V3 from node 2 to node 6. By a careful inspection, it is not difficult to obtain an optimal travel schedule for V3 as shown below:

(Best route for V3)

$$([0, 4], 2, 3) \rightarrow ([10, 14], 1, 4) \rightarrow ([32, 36], 4, 6) \rightarrow$$

$$([44, 48], 4, 8) \rightarrow ([64, 68], 5, 10) \rightarrow ([74, \infty], 6, -)$$

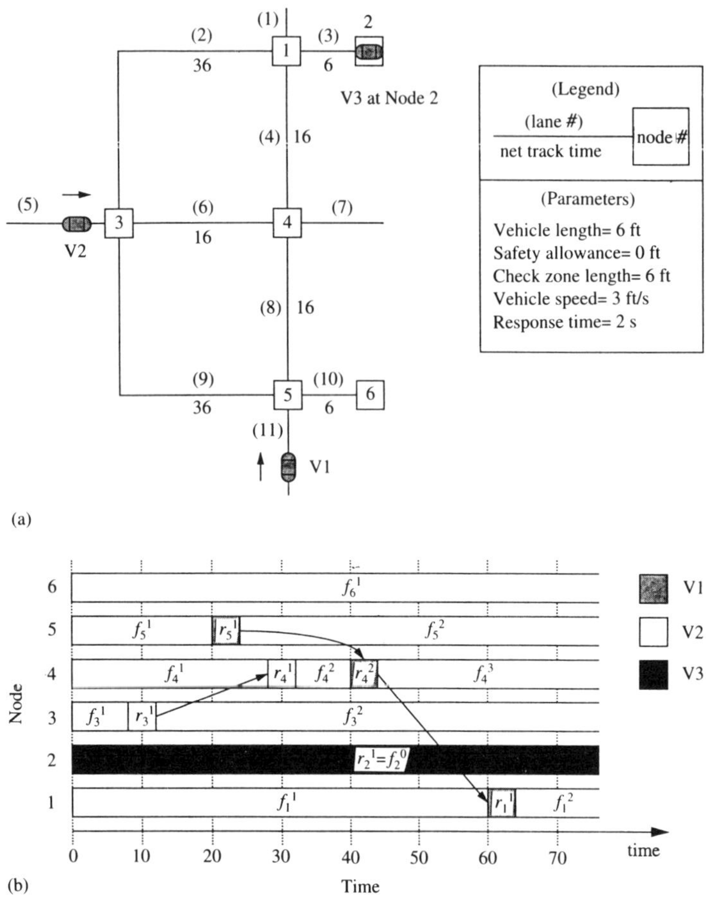

Fig. 9.2 Example situation: (a) flow path network; (b) time windows.

Note that, in the best route for V3, the vehicle stays idle in lane (6) for 8 seconds after crossing node 4 to make way for V1, and then crosses node 4 again. Loops like this, as well as cycles, may give better results in conflict-free shortest-time routing. In the standard terminology of graph theory, a cycle in a graph whose intermediate nodes are all distinct but which starts and ends at the same node, and a loop is a cycle with only one arc. The term 'loop' used here, however, bears an additional meaning; the vehicle has to move in the opposite direction (if it cannot turn around) to re-enter the node if the lane is not a physical loop whose both ends are the same.

9.2.2 Previous work

The concept of conflict-free shortest-time AGV routing was first introduced by Broadbent *et al.* (1985). The procedure employs Dijkstra's shortest path algorithm and generates a matrix describing node occupation times of vehicles. Potential conflicts are detected by comparing node occupation times. Junction and catching-up conflicts are resolved by slowing down the vehicle yet to be scheduled. Head-on conflicts are resolved by finding another shortest path excluding the congested segment. The procedure is suitable for the case of sensor blocking (a minimum headway is maintained between two adjacent vehicles), which results in better utilization of the guide path network. It can be applied to both unidirectional and bidirectional network models. However, it does not guarantee optimal solutions for bidirectional models.

Taghaboni and Tanchoco (1988) introduce the concept of a virtual lane in their description of a real-time supervisory controller for a free-ranging AGV system. The underlying assumption made in the concept of a virtual lane is that bidirectional aisles have multiple unidirectional lanes and single-lane aisles are all unidirectional. Therefore, the possibilities of head-on collisions in lanes are eliminated, hence optimal solutions are guaranteed. The procedure also allows multiple simultaneous crossings at intersections.

The labeling algorithm proposed by Huang *et al.* (1989) assigns labels to free time windows defined for each node, rather than to physical nodes. It guarantees optimal solutions for both unidirectional and bidirectional flows. The algorithm treats the physical arcs of the original guide path network as nodes in the converted network. The journey time on an arc of the original network is converted to the dwell time on the corresponding node of the converted network. Since cycles are allowed in the shortest-time route in the converted network, both cycles and loops are allowed in the shortest-time route in the original network. However, application is restricted to the case where zone blocking is used for traffic control (one vehicle at most is allowed in a node or arc of the guide path network).

Fujii *et al.* (1989) present an A* algorithm in which the travel time of a shortest path from an intermediate node to the destination without considering the time window restrictions is used as the heuristic function. The authors introduce the concepts of modified and shifted time windows for the purpose of conflict detection and resolution. The way in which time windows are defined on arcs allows sensor blocking. It is good for both unidirectional and bidirectional models. Cycles are also allowed. However, since the algorithm does not allow loops in the shortest-time path tree, it may sometimes result in a sub-optimal solution.

9.2.3 Basic procedure

The routing algorithm described here takes into account the current traffic status to calculate conflict-free fastest routes for AGVs in a bidirectional flow path network. The traffic status can be modeled by maintaining a table of scheduled entry and exit times of vehicles at each node. The time interval or time window between a pair of entry and exit times of a vehicle at a node is exclusively reserved by that vehicle, and other vehicles should not be allowed to cross the node during this reserved time window. The free time windows between the reserved time windows are available for scheduling other vehicle crossings. The algorithm routes the vehicle to be scheduled through the free time windows instead of physical nodes. An immediate question is how to determine whether the vehicle can reach from one time window to another. To do this, we will introduce the concept of *a time window graph.* A time window graph is a directed graph in which the node set represents the free time windows and the arc set the reachability between the free time windows.

Let us first introduce some notation. Let the set of all reserved time windows be denoted by:

$$R = \{r_n^k = [c_n^k, d_n^k]\},$$

where r_n^k is the kth reserved time window at node n, c_n^k is the beginning time and d_n^k is the end time of the time window. Likewise, let the set of all free time windows be denoted by:

$$F = \{f_n^k = [a_n^k, b_n^k]\},$$

where f_n^k is the kth free time window at node n, a_n^k is the beginning time and b_n^k is the end time of the time window. By definition, $b_n^k - a_n^k \geqslant t_n$, where t_n is the time needed for vehicle to cross node n. If $b_n^k - a_n^k < t_n$, then it is not considered a free time window. The sets R and F are mutually exclusive. The union of R and F exhausts the set W of all time windows. Furthermore, no two time windows in W overlap each other. Figure 9.2(b) shows the reserved and free time windows corresponding to the example situation. Note that the source time window r_2^1, reserved by vehicle V3 at the source node, is renamed f_2^0 as a free time window, and moved to the set F. The algorithm treats the source time window as a free time window, since it is available for scheduling the vehicle in question even though it is a reserved time window to all the other vehicles.

The current time is denoted by t_{now}. The maximum of the response time of the routing algorithm is denoted by t_{rsp}. The node crossing time at node i, denoted t_i, is normally defined as the amount of time required for a vehicle

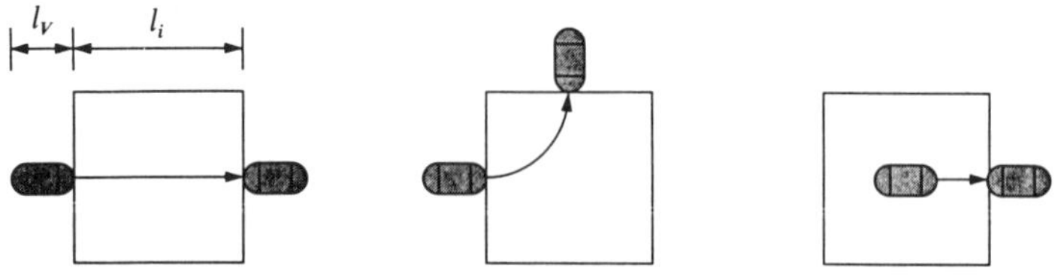

Fig. 9.3 Node crossing times: (a) straight crossing $t_i = (l_v + l_i)/S_1$, where S_1 = speed in straight-line segments; (b) right-angle crossing $t_i = (l_v + \pi \times l_i/2)/S_c$, where S_c = speed in curved segments; (c) at the source node: $t_i = (l_v/2 + l_i/2)/S_1 + t_{\text{rsp}}$, where t_{rsp} = the response time of the controller.

to completely clear the check zone around node i since the entry time (see Fig. 9.3). Depending upon the angle between the in-lane and out-lane at the node, we may assign different values for t_i in practice. If node i is the source node, then $t_i = t_{rsp} + t_i/2$. The net track time in lane (i, j), denoted $t_{(i,j)}$, is defined as the minimum time required for a vehicle to traverse lane (i, j) since leaving node i until arriving node j (see Fig. 9.4). The label of a time window f_i^p, $L(f_i^p)$, represents the earliest time to *enter* node i within f_i^p. The earliest time to *arrive* at node j within f_j^q, given its predecessor time window being f_i^p, is defined by:

$$\alpha(f_j^q | f_i^p) = L(f_i^p) + t_i + t_{(i,j)}.$$

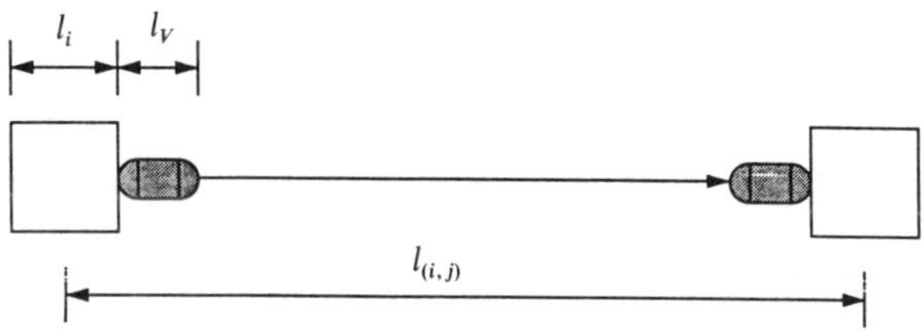

Fig. 9.4 Net track time in a lane: $t_{(i,j)} = (l_{(i,j)} - l_i - l_v)/S$, where S = vehicle speed.

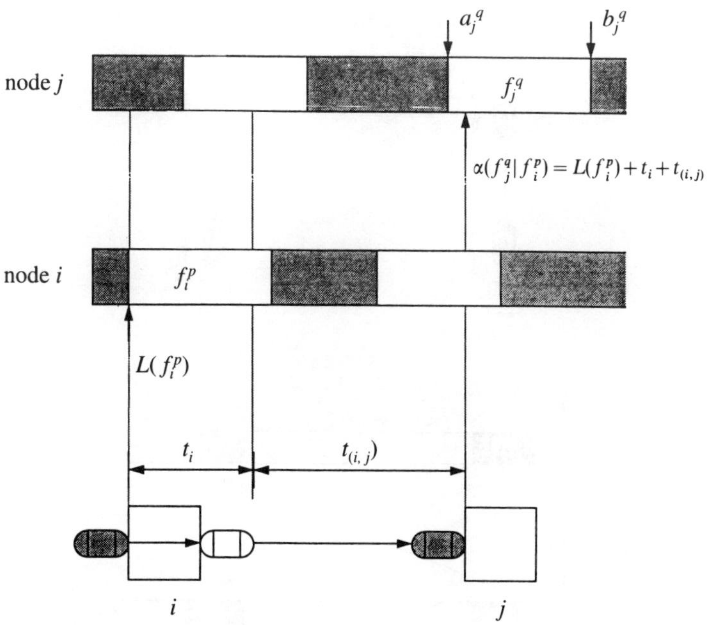

$\tau(f_j^q | f_i^p) = \max\{a_j^q, \alpha(f_j^q | f_i^p\} = \alpha(f_j^q | f_i^p)$

Fig. 9.5 Definitions of $\alpha(f_j^q | f_i^p)$ and $\tau(f_j^q | f_i^p)$.

The earliest time to *enter* node j within time window f_j^q, given its predecessor time window being f_i^p, is defined by:

$$\tau(f_j^q | f_i^p) = \max\{a_j^q, \alpha(f_j^q | f_i^q)\}.$$

The definitions of α and τ are explained graphically in Fig. 9.5. If τ is greater than α, the vehicle can either slow down or stop temporarily at some convenient point in the lane. The lane to be used between time windows f_i^p and f_j^q will be denoted by $\lambda(f_j^q | f_i^p)$. We need to explicitly specify $\lambda(f_j^q | f_i^p)$ since it is not uniquely defined when the vehicle makes a loop to re-enter a node, i.e. when $i = j$. The variables $\tau(f_j^q | f_i^p)$, $\alpha(f_j^q | f_i^p)$ and $\lambda(f_j^q | f_i^p)$ will be abbreviated to τ, α and λ, respectively, in the description of the algorithm if such abbreviations do not cause any confusion.

The reachability from time windows f_i^p to f_j^q can be tested as follows. Say we want to decide if f_j^q is *reachable* from $f_i^p (f_i^p \Rightarrow f_j^q)$ or *blocked* $(f_i^p \Rightarrow | f_j^q)$. Time window f_i^p is called the tail time window and f_j^q the head time window. Likewise, node i is called the tail node and j the head node. We distinguish two cases: (case 1) the tail and head nodes are different ($i \neq j$); and (case 2) the same ($i = $ j).

If $i \neq j$, we first check if $(i, j) \in A$ (the set of lanes), since it is obvious that $f_i^p \Rightarrow | f_j^q$ if $(i, j) \notin A$ (Space-Feasibility). Second, the vehicle should be able to leave node j within time window f_j^q after entering node j. This can be easily tested by comparing the earliest time to enter f_j^q, $\tau(f_j^q | f_i^p)$, and the end time

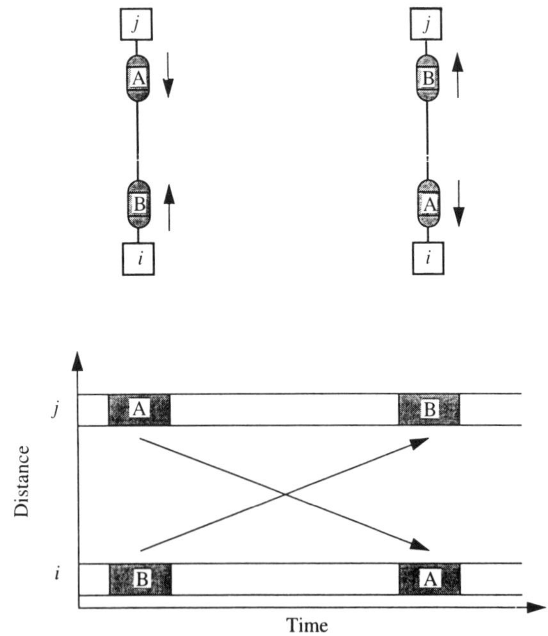

Fig. 9.6 Potential head-on conflict.

of the head time window, b_j^q (Time-Feasibility). Third, there should be no potential conflicts in the lane (Conflict-Free). A conflict in a lane can occur when two vehicles travel in the same lane but at different velocities. Two types of potential conflicts in lanes have been distinguished by Broadbent *et al.* (1985): a head-on conflict (Fig. 9.6) can occur when two vehicles travel in the opposite directions; and a catching-up conflict (Fig. 9.7) can occur when two vehicles travel in the same direction but at different speeds. Unidirectional lanes can have only catching-up conflicts. Potential conflicts in lane (i, j) can be detected by comparing the vehicle crossing schedules at its end nodes i and j. If the orders of the time windows occupied by any pair of vehicles are reversed at their end nodes, then the two vehicles conflict somewhere in the lane. The procedure returns (τ, λ) to the calling procedure. If it is *blocked*, then $\tau = \infty$ and $\lambda = 0$.

If the tail and head nodes are the same $(i = j)$, then a different procedure is applied. To satisfy the space-feasibility requirement, a candidate lane for looping or side-stepping should be bidirectional. It is not necessary to check for time-feasibility. Potential conflicts in lanes can be detected by inspecting the number of available lanes in which the vehicle can stay idle to make way for the vehicles reserving the time windows between the two free time windows f_i^p and f_i^q. Let B be the set of all the bidirectional lanes incident to node i. We identify all the vehicles which reserve the time windows between the two time windows f_i^p and f_i^q, and all the lanes used by these vehicles before and

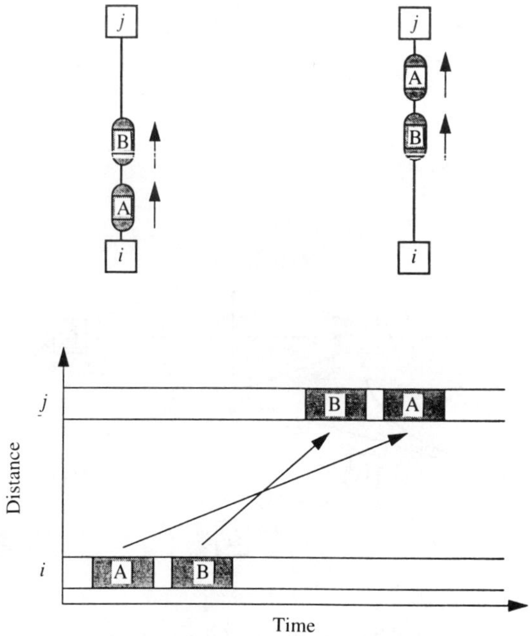

Fig. 9.7 Potential catching-up conflict.

after crossing the node. Let C be the set of these lanes. If $B - C = \varnothing$, then $f_i^p \Rightarrow | f_i^q$. Take, for example, f_4^2 and f_4^3 in Fig. 9.2(b). The time window r_4^2 is reserved by V1, and V2 uses lanes 8 and 4 before and after crossing node 4. Therefore, $B - C = \{4, 6, 7, 8\} - \{4, 8\} = \{6, 7\}$ is the set of lanes available for V3 when it makes a loop to re-enter node 4. A formal description of the procedure for reachability test is shown in Algorithm 1.

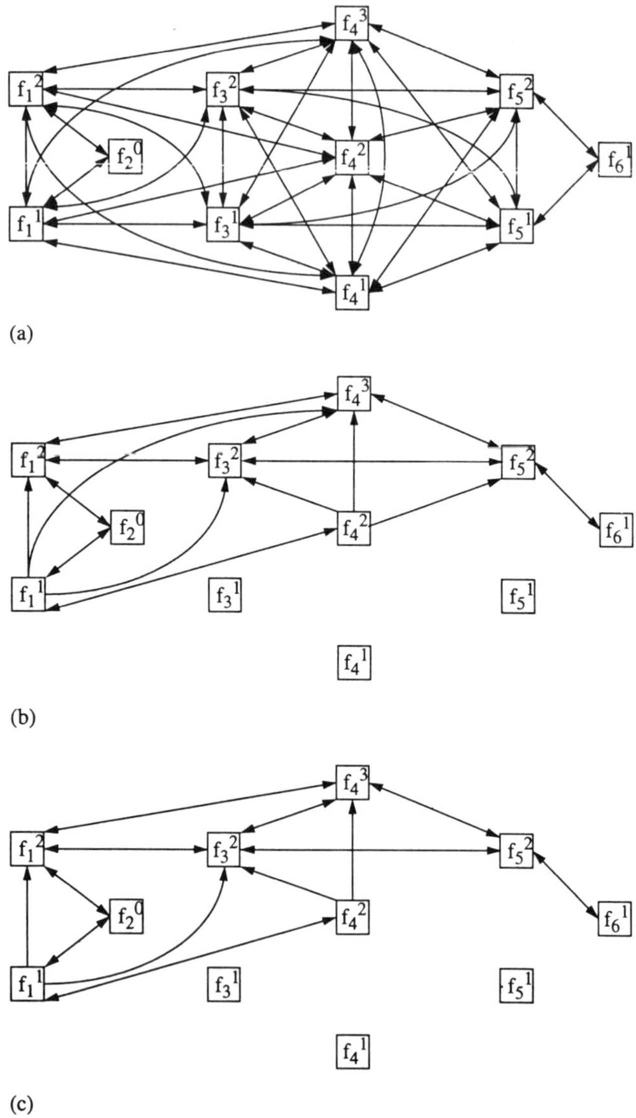

Fig. 9.8 Examples of time window graphs: (a) space-feasible; (b) space and time feasible; (c) space and time feasible conflict-free.

Figure 9.8 shows three different time window graphs corresponding to the example situation. Figure 9.8(a) shows the time window graph when only the space-feasibility tests are applied to set *F*. Figure 9.8(b) results when the time-feasibility tests are added. The space- and time-feasible conflict-free time window graph shown in Fig. 9.8(c) is the result of the whole procedure for the reachability tests. It is uniquely defined only for a specific source node and start time. Because of its dynamic nature, and to avoid unnecessary computations involved in the reachability tests, the time window graphs will not be explicitly constructed but will be implicitly considered during execution of the main procedure. The main procedure is described in Algorithm 2.

The main procedure closely follows the shortest path method by Dijkstra (1959). The set *F* is divided into three mutually exclusive subsets. The first subset, denoted *T*, contains those time windows for which conflict-free shortest-time routes from the source time window are already obtained. The second set, denoted *U*, contains those time windows in $F - T$ which are directly reachable from some time window in *T*. The third set contains the rest of the time windows. If the last free time window defined at the destination node *d* is in *T*, then we have found a solution. If not, we add to the set *T* a time window in the second set *U* which has the minimum label value among the members of *U*. A tie is broken in favor of the one with the fewest hops from the source time window to prevent a vehicle from side-stepping at a node more than once. Its label becomes permanent at this point, and the predecessor time window as well as an available lane between this time window and its predecessor are recorded in the arrays *P* and *Q*, respectively. The procedure keeps improving the labels of the time windows in *U* at each iteration.

Algorithm 1. *test_reachability*

function *test_reachability*( ){
  /* **Case 1** */
  if $i \neq j${
1.   /* Check for Space-Feasibility */
    if $(i,j) \notin A$ **then** Set $\tau = \infty$ and $\lambda = 0$, and **return** (τ, λ);
2.   /* Check for Time-Feasibility */
    Compute $\alpha = L(f_i^p) + t_i + t_{(i,j)}$, and $\tau = \max\{a_j^q, \alpha\}$;
    If $\tau > b_j^q - t_j$ **then** Set $\tau = \infty$ and $\lambda = 0$, and **return** (τ, λ);
3.   /* Check for Catching-up Conflicts */
    for each reserved time window on node i which starts after f_i^p{
3.1   Identify the vehicle which reserves this time window; Call this vehicle v;
3.2   **for** each reserved time window on node j which ends before f_j^q{
      Identify the vehicle that reserves this time window; Call this vehicle u; **if** $v = u$ **then** set $\tau = \infty$ and $\lambda = 0$, and **return** (τ, λ);
    }
  }

4. /* Check for Head-on conflicts */
Repeat step 3 exchanging the roles of f_i^p and f_j^q;
5. Set $\lambda = (i, j)$, and **return** (τ, λ);
}
/* **Case 2** */
else {
6. Let B be the set of all bidirectional lanes incident to node i;
if $B = \varnothing$ **then** Set $\tau = \infty$ and $\lambda = 0$, and **return** (τ, λ);
7. Set $\tau = a_j^q$;
for each reserved time window between the two time windows {
7.1. Identify the vehicle which reserves this time window;
7.2. By searching its travel schedule, check if node i is on its route;
7.3. **if** yes **then** Subtract from B two lanes used before and after crossing node i;
}
8. **if** $B = \varnothing$ **then** Set $\tau = \infty$ and $\lambda = 0$; **else** Let λ be any element of B;
return (τ, λ);
}
}

Algorithm 2. ***cfstp***

function *cfstp*( ){
1. /* Initialization */
Let f_s^0 be the source time window;
/* The vehicle to be scheduled is located at node s occupying time window f_s^0. */
Set $L(f_s^0) = t_{now}$, $L(f_p^i) = \infty$ for all $f_i^p \in F - \{f_s^0\}$;
Set $P(f_i^p) = Q(f_i^p) = 0$ for all $f_i^p \in F$;
Set $T = \{f_s^p\}$ and $U = \varnothing$;
2. /* Labelling */
for each $f_j^q \in F - T${
2.1. **for** each $f_i^p \in T${
Check for the reachability from f_i^p to f_j^q and compute $\tau(f_j^q | f_i^p)$ and $\lambda(f_j^q | f_i^p)$ by calling *test_reachability*( );
}
}
2.2. Find a time window $f_k^r \in T$ which satisfies $\tau(f_j^q | f_k^r) = \min\{\tau(f_j^q | f_i^p)\}$ for all $f_i^p \in T$;
2.3. **if** $L(f_j^q) > \tau(f_j^q | f_k^r)$ **then** Set $L(f_j^q) = \tau(f_j^q | f_k^r)$, $P(f_j^q) = f_k^r$ and $Q(f_j^q) = \lambda(f_j^q | f_k^r)$;
2.4. **if** $L(f_j^q) < \infty$ **then** Add $\{f_j^q\}$ to U;
if $U = \varnothing$ **return** INFEASIBLE;
3. /* Label Setting */
Find a time window $f_i^p \in U$ which satisfies $L(f_i^p) = \min\{L(f_j^q)\}$ for all $f_j^q \in U$; A tie is broken in favor of the one with fewest hops from the source time window; Add $\{f_i^p\}$ to T and subtract it from U;
4. /* Check for Termination */

if The last free time window at the destination node d is in T **then** **goto** step 5; **else goto** step 2;

4. ?: Schedule Generation */

Fix the route by tracing back the arrays P and Q;
Compute the travel schedule by setting the entry time = $L(f_i^p)$ and the exit time = $L(f_i^p) + t_i$ for each time window f_i^p on the route;
At the destination time window f_d^q, set the exit time = b_d^q;
Update the reserved and free time windows, and **return**;

}

9.2.4 Numerical example

The numerical illustration given below is a step-by-step application of the algorithm to the example situation.

1. (Initialization) The set F of free time windows and R of reserved time windows are as shown in Fig. 9.2(b). The vehicle to be scheduled is currently located at node 2 occupying time window $r_2^1 = f_2^0$. Set $L(f_2^0) = 0, L(f_i^q) = \infty$ for all $f_i^q \in F - \{f_2^0\}, P(f_i^q) = Q(f_i^q) = 0$ for all $f_i^q \in F, T = \{f_2^0\}$, and $U = \varnothing$.
2. (Labelling) The labelling step is operated on each member of $F - T$.
   For f_1^1: 2.1. $\alpha(f_1^1|f_2^0) = L(f_2^0) + t_2 + t_{(2,1)} = 10, \tau(f_1^1|f_2^0) = \max\{a_1^1, \alpha(f_1^1|f_2^0)\} = 10, \lambda(f_1^1|f_2^0) = (3)$.
   2.2. $\tau(f_1^1|f_2^0)$ is the trivial minimum.
   2.3. $L(f_1^1) = \tau(f_1^1|f_2^0) = 10, P(f_1^1) = f_2^0, Q(f_1^1) = \lambda(f_1^1|f_2^0)) = (3)$.
   2.4. $U = \{f_1^1\}$.
   for f_1^2: 2.1. $\alpha(f_1^2|f_2^0) = L(f_2^0) + t_2 + t_{(2,1)} = 10, \tau(f_1^2|f_2^0) = \max\{a_1^2, \alpha(f_1^2|f_2^0)\} = 64, \lambda(f_1^2|f_2^0) = (3)$.
   2.2. $\tau(f_1^2|f_2^0)$ is the trivial minimum.
   2.3. $L(f_1^2) = \tau(f_1^2|f_2^0) = 64, P(f_1^2) = f_2^0, Q(f_1^2) = \lambda(f_1^2|f_2^0) = (3)$.
   2.4. $U = \{f_1^1, f_1^2\}$.
   The rest of the members of $F - T$ are labelled ∞. (Fig. 9.2(c).)
3. (Lable Setting) $L(f_1^1) = \min\{L(f_1^1), L(f_1^2)\}$. Set $T = \{f_2^0, f_1^1\}$ and $U = \{f_1^2\}$.
4. (Check for Termination) Go to step (2) since $f_6^1 \notin T$.

By continuing the procedure, we can get the results summarized in Table 9.1. Figure 9.9 shows the time windows after the travel schedule for the vehicle in question is determined. By applying step (5) of the main procedure, we can get the 'best route for V3' as given in section 9.2.1.

9.2.5 Computational complexity

The computational burden of the algorithm is dominated by the labelling step (step 2). At each iteration, $|F - T|$ labelling operations are required, since the labelling step is operated on each member of $F - T$. (The notation $|S|$ represents the size of set S.) After each iteration, $|T|$ is increased by one and $|F - T|$ is decreased by one. Thus, the algorithm may require $(|F| - 1)$

Table 9.1 Completed numerical example

Iteration	U	$L(\cdot)$											T	$P(\cdot)$	$Q(\cdot)$
		f_1^1	f_1^2	f_2^0	f_3^1	f_3^2	f_4^1	f_4^2	f_4^3	f_5^1	f_5^2	f_6^1			
0	$\varnothing$	∞	∞	$\underline{0}$	∞	∞	∞	∞	∞	∞	∞	∞	$\{f_2^0\}$		
1	$\{f_1^1, f_1^2\}$	$\underline{10}$	64	0	∞	∞	∞	∞	∞	∞	∞	∞	$\{f_2^0, f_1^1\}$	$P(f_1^1) = f_2^0$	$Q(f_1^1) = (3)$
2	$\{f_1^2, f_3^2, f_4^2\}$	10	64	0	∞	50	∞	$\underline{32}$	∞	∞	∞	∞	$\{f_2^0, f_1^1, f_4^2\}$	$P(f_4^2) = f_1^1$	$Q(f_4^2) = (4)$
3	$\{f_1^2, f_3^2, f_4^3\}$	10	64	0	∞	50	∞	32	$\underline{44}$	∞	∞	∞	$\{f_2^0, f_1^1, f_4^2, f_4^3\}$	$P(f_4^3) = f_4^2$	$Q(f_4^3) = (6)$
4	$\{f_1^2, f_3^2, f_5^2\}$	10	64	0	∞	$\underline{50}$	∞	32	44	∞	64	∞	$\{f_2^0, f_1^1, f_4^2, f_4^3, f_3^2\}$	$P(f_3^2) = f_1^1$	$Q(f_3^2) = (2)$
5	$\{f_1^2, f_5^2\}$	10	$\underline{64}$	0	∞	50	∞	32	44	∞	64	∞	$\{f_2^0, f_1^1, f_4^2, f_4^3, f_3^2, f_1^2\}$	$P(f_1^2) = f_1^1$	$Q(f_1^2) = (2)$
6	$\{f_5^2\}$	10	64	0	∞	50	∞	32	44	∞	$\underline{64}$	∞	$\{f_2^0, f_1^1, f_4^2, f_4^3, f_3^2, f_1^2, f_5^2\}$	$P(f_5^2) = f_4^3$	$Q(f_5^2) = (8)$
7	$\{f_6^1\}$	10	64	0	∞	50	∞	32	44	∞	64	$\underline{74}$	$\{f_2^0, f_1^1, f_4^2, f_4^3, f_3^2, f_1^2, f_5^2, f_6^1\}$	$P(f_6^1) = f_5^2$	$Q(f_6^1) = (10)$

*Underlined entries represent the minimum label values among the elements of U.

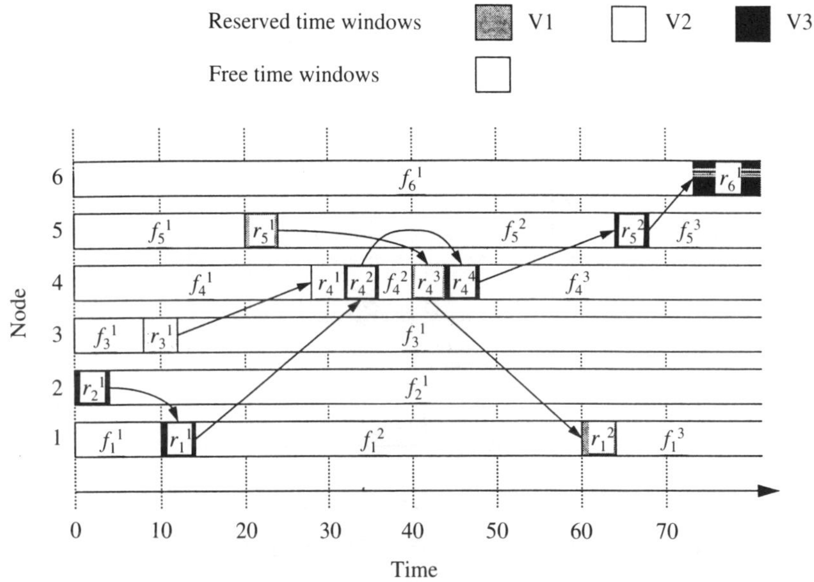

Fig. 9.9 Time windows after Vehicle 3 has been scheduled.

iterations in the worst case. And hence the algorithm may require $o(|F|^2)$ labelling operations in the worst case. Let v be the number of vehicles and n the number of physical nodes in the system. Since there can be at most v free time windows at any node, there are at most vn free time windows in the set F. Therefore, the proposed algorithm requires $o(v^2n^2)$ labelling operations in the worst case. Now let us analyse the labelling step. The computational burden for the labelling step is dominated by the reachability test (step 2.1), which is operated on each member of T. However, by a careful implementation, we can restrict the reachability test operated only on the newest member of T. The reachability test in turn is dominated by the tests for conflicts in lanes. Given a pair of time windows, it requires $o(v^2)$ computations to test for conflicts in lanes, since it operates on a doubly nested loop, each loop being operated on a list of reserved time windows at either head or tail node. Therefore, the whole procedure requires $o(v^4n^2)$ computations in the worst case.

Figure 9.10 shows the computational performance of the proposed algorithm implemented on an 80386-based 25 MHz microcomputer with a math co-processor. The flow path network tested is a bidirectional one with 28 nodes and 30 lanes, as shown in Fig. 9.12. The average performance is almost flat at around 0.015 s per call, regardless of the number of vehicles. The worst case performance shows a polynomial explosion as the number of vehicles increases.

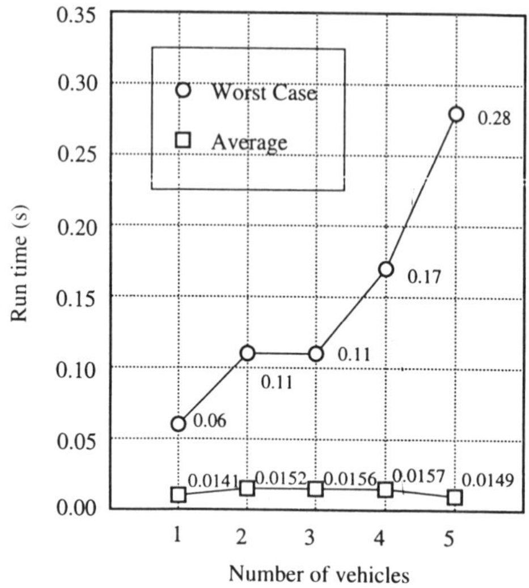

Fig. 9.10 Computational performance of the path planning algorithm.

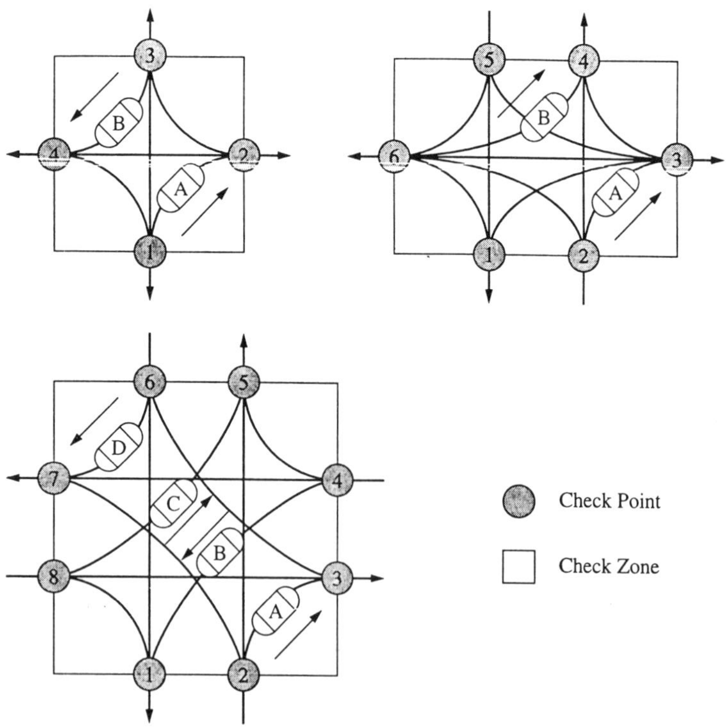

Fig. 9.11 Multiple simultaneous crossings at intersections.

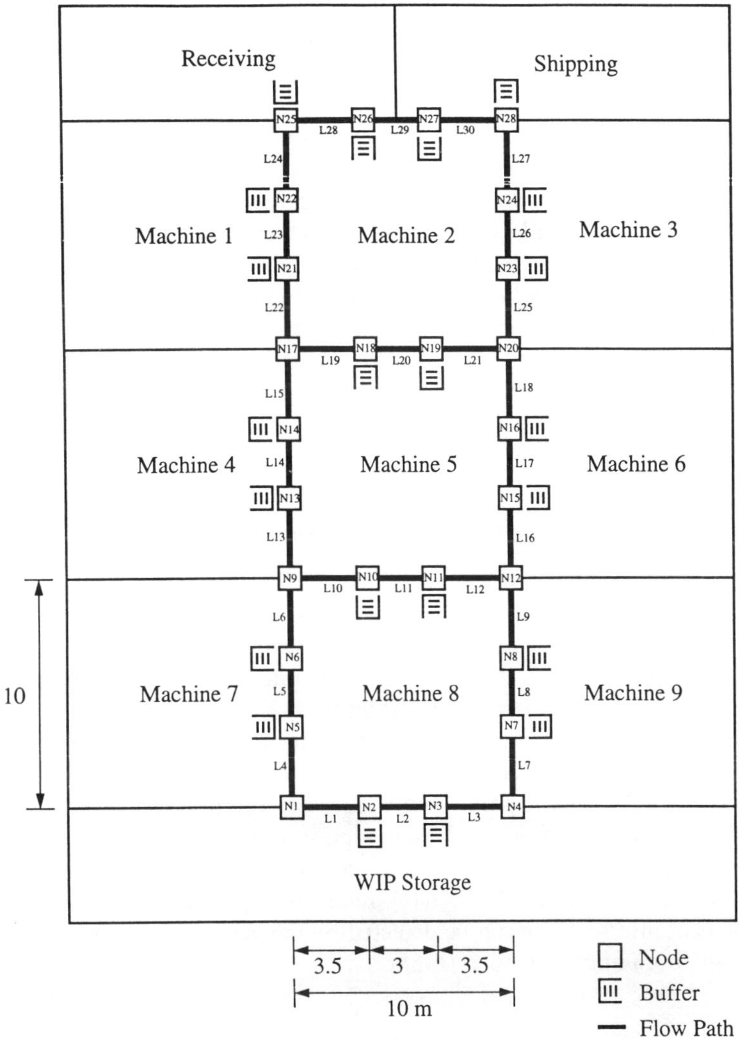

Fig. 9.12 Nine-machine job shop with a bidirectional AGV system.

9.2.6 Extensions, variations and operational considerations

(a) A search*

A graph search algorithm is called an *algorithm A* if the algorithm uses an evaluation function of the form $f(n) = g(n) + h(n)$, where $g(n)$ is an estimate of $g^*(n)$, the minimal cost from the source node to node n, and $h(n)$ is an estimate of $h^*(n)$, the minimal cost from node n to the destination node.

Therefore, $f(n)$ is an estimate of the cost of a minimum-cost path from the source to the destination constrained to go through node n. If $g(n) \geqslant g^*(n)$ and $h(n) \leqslant h^*(n)$, then the algorithm is called an *A* algorithm* and will find an optimal path if one exists. The function $h(n)$ is called the *heuristic function*, since it is decided by heuristic information on the problem domain (Nilsson, 1980, pp. 74–6). Dijkstra's algorithm is an A* algorithm where the function $g(n)$ is the label value of node n and $h(n) = 0$. As noted earlier, the procedure described by Fujii *et al.* (1989) is also an A* algorithm, where $h(n)$ is given by the travel time on a shortest path from node n to the destination without considering the time window and conflict-free restrictions. The same idea can be applied to the label setting step (step 3) of the main procedure. In this case, we will find a time window $f_i^p \in U$ which satisfies $L(f_i^p) + h(i) = \min\{L(f_j^q) + h(j)\}$ for all $f_j^q \in U$. Introducing the new heuristic function h will result in at most the same number of iterations, if not less, since its heuristic power is stronger than 0.

(b) Multiple simultaneous crossing at intersections

Introducing the concept of time windows provides a method for the detection and resolution of potential conflicts at nodes by allowing only one vehicle to reserve a time window at a node. However, there are cases where we want to allow two or more vehicles to cross an intersection at the same time if the space permits. Figure 9.11 illustrates these situations. The procedure can utilize a check point introduced at each entry/exit point of an intersection. If the check zone of a node is too small to allow multiple simultaneous crossing, a single check point is introduced in the middle of the node. The vehicle crossing schedules and time windows are maintained at the check points instead of nodes, and the vehicles are routed through the time windows maintained at the check points. In this model of a flow path network, potential conflicts at intersections can be detected by the geometric method described by Taghaboni and Tanchoco (1988).

(c) Distributed processing

The AGV routing algorithm can be executed either by a supervisory controller if the system is under centralized control, or by the vehicle controllers if the system is under distributed control. In either case, the host of the algorithm should have good information on the travel schedules of all the vehicles. If routing is done by the supervisory controller, then all the necessary information is readily available in the memory of the central computer. The travel schedules are then transmitted to the vehicles. If the routing decision is delegated to the vehicle controllers, then each vehicle has to transmit its travel schedule to all the other vehicles, and the supervisory controller, via a common communications channel. Since the data transmission can take place at any random time, the vehicles should always be alert to receive data from all the other vehicles. This implies that multi-tasking capability is required on the vehicle control computers. The executions of the routing algorithm by the vehicle controllers do not have to be sequenced if the network is all unidirectional. (By sequencing, we mean that at most one vehicle controller executes the routing algorithm at any time.) However, sequencing is necessary if the network includes bidirectional lanes. Parallel execution of the routing algorithm may result in conflicting travel schedules or gridlocks in a bidirectional network.

(d) Rescheduling

The vehicle travel schedules generated by the proposed algorithm are free of conflicts. Therefore, if the vehicles follow the schedules exactly, traffic control is not necessary. However, this is not the case in a real system. Minor changes in schedules can be easily avoided by manipulating the length of the safety allowance between vehicles. If a vehicle is behind its schedule by more than the safety allowance, then the vehicle should be rerouted from the current location to its destination.

(e) Application in unidirectional flow path networks

There are no head-on conflicts if the lanes are all unidirectional. Further, loops are not allowed in unidirectional lanes. The proposed algorithm can be applied to unidirectional networks without any change. However, it will be faster if we remove the step for checking head-on conflicts (step 4) of case 1, and all the steps of case 2 of the reachability test.

(f) Consideration of system size

The computational complexity of the proposed algorithm is more sensitive to the number of vehicles than to the number of physical nodes. Therefore, it will be more suitable for small systems with few vehicles.

9.3 COOPERATIVE PATH PLANNING

9.3.1 Previous work

There can be several different strategies in AGV path planning. Under a conservative strategy, a previous decision on path planning is never reconsidered for further improvement once it is made. Under a myopic strategy, only one vehicle is considered at a time. At the other extreme, we may have a very liberal, hyperopic strategy; whenever a new decision has to be made, we can discard all the future schedules and reoptimize the travel schedules for all the vehicles during a prespecified scheduling horizon. Some researchers suggest a compromised static version of this problem by assuming that the transport demands are known in advance, and they propose exact solution methods or heuristic procedures based on exact formulations. Among them are Maxwell (1981), Maxwell and Muckstadt (1982), Kusiak and Cyrus (1985) and Chen *et al.* (1987). The problems typically considered by these researchers are combined vehicle dispatching and path planning problems. However, these approaches are not appropriate for implementation in real-time control systems because of the assumption of static known demands and the computational complexity; rather, these are more suitable for analysis and planning purposes. A few other researchers suggest adaptive strategies, which are more suitable in dynamic environments.

Egbelu and Tanchoco (1986) describe a method to control AGVs in a bidirectional network in which buffering areas are provided at intersections to resolve vehicular contentions in the use of lanes. In this method, the movements of the vehicles are monitored at discrete points in the network, called **check points**. When there is a contention between vehicles for the use of the same lane, detected by the monitoring system, the control system routes one or more of the contending vehicles to buffer areas to clear the lane. This approach is basically an attempt to solve a control problem by a supplementary design of the flow path network. A drawback of this method is the assumption of infinite capacities of the buffering areas, which can be achieved in practice only by providing fairly large buffering areas.

Taghaboni (1989) suggests a highly adaptive strategy, node-to-node routing, in which each vehicle is routed only to the next node on a path to the destination considering the current status of neighboring nodes and the global network information available. Simultion results on comparison of the node-to-node routing and the source-to-goal routing show that node-to-node routing generally performs poorly because of the difficulties in estimating the actual traffic conditions and the lack of a global look-ahead scheduling capability. This strategy seems to be a viable solution when there are many vehicles in the system and the response time of the route planner is a critical factor in the system performance.

Krishnamurthy (1990) and Krishnamurthy *et al.* (1991) consider the combined problem of assigning vehicles to transport demands and generating conflict-free routes for several vehicles simultaneously in a rolling-horizon framework, and present an integer programming formulation to minimize the latest travel completion time for a given set of transport demands. Because of the NP-completeness of the formulation, they propose a heuristic procedure in which AGVs are first assigned to transport demands by a heuristic rule, and the set of conflict-free routes for the vehicles are evaluated iteratively using a column generation technique, a column being a route for a vehicle.

9.3.2 A conservative myopic strategy

The strategy employed in this research is a myopic one, in which only one vehicle is considered at a time, and a highly conservative one in which all the previous decisions are strictly respected and a subsequent travel schedule is assigned only after the vehicle becomes idle. Yet it is suitable in dynamic environments, since a decision is made whenever it is needed. Under this strategy, it may suffice for the AGV system controller to find a path from the current position of the vehicle to the pickup point and then to the drop-off point by calling the conflict-free shortest-time path planning algorithm described by Kim (1991) and Kim and Tanchoco (1991), named *cfstp*( ), twice after the decision on vehicle dispatching is made.

Algorithm 3 ***make_way***

```
function make_way( ){
   Define the set of preferred nodes. /* Pickup nodes are possible candidates. */
   Find the shortest path from the source to the destination of the blocked
   vehicle.
   Sort the nodes in the order from the destination to the source.
   Identify the blocking vehicles on the shortest path.
   Sort the blocking vehicles in the increasing order of distance from the
   destination.
   for each blocking vehicle {
      /* Send the blocking vehicle to somewhere else */
      Sort the preferred nodes in the increasing order of distance from the
      current location of the blocking vehicle.
      for each preferred node {
         Route the blocking vehicle to the preferred node by calling cfstp( ).
         if cfstp( ) is successful return
      }
      Route the blocking vehicle to other nodes not on the shortest path by
      calling cfstp( ).
      if cfstp( ) is successful return
      for each node on the shortest path of the blocked vehicle {
         Route the blocking vehicle to the node by calling cfstp( ).
         if cfstp( ) is successful return
      }
   }
}
```

In certain cases, however, function *cfstp*( ) fails to find a feasible path. An obvious case is when there is another vehicle occupying the destination node. In checking terminating conditions, function *cfstp*( ) always tries to reserve the last free time window at the destination node. If the vehicle has no further travel schedule, it occupies its final destination eternally. Other cases where several other vehicles are blocking all the feasible paths cannot be detected easily, and function *cfstp*( ) recognizes the failure after evaluating all the possible paths. If one of these situations is detected, the vehicle which blocks the destination node should be routed somewhere else, preferably to a pickup station, but not to a node on the shortest path from the current location of the vehicle to be scheduled to the destination. This procedure is summarized in the function called *make_way*( ), described in Algorithm 3.

Algorithm 4 ***squeeze***
function *squeeze*( ){
  Identify the blocking vehicle and call it *v*;
  Set *block*[*top*] = *v*; Increment *top* by 1;
  Route the blocking vehicle to anywhere it can go;
  if successful {
    Decrement *top* by 1, and **return**;
  }
  Find a vehicle that blocks vehicle *v* but is not in *block*[ ] and call it *u*;
  Move away vehicle *u* by calling *squeeze*( );
  Route vehicle *v* to anywhere it can go; Decrement *top* by 1;
  return;
}

If function *make_way*( ) fails to clear the blocking vehicles out of the shortest path, then a stronger measure is applied. This procedure, called *squeeze*( ), moves the blocked vehicle node by node along the shortest path and pushes the blocking vehicle from behind as it moves until the blocked vehicle reaches the destination. If the blocking vehicle is blocked by another vehicle, then function *squeeze*( ) is called recursively. A formal description of *squeeze*( ) is given in Algorithm 4.

To ensure that a feasible route is obtained for vehicle *veh* to be scheduled, functions *cfstp*( ), *make_way*( ) and *squeeze*( ) are called sequentially. This three-level stratified strategy to find a path from the current location to the destination is implemented in the function called *schedule_travel*( ), described in Algorithm 5.

A transport mission of a vehicle consists of two parts: the empty (deadhead) travel from the current location to the pickup station, and the loaded travel from the pickup station to the drop-off station. Once a vehicle is assigned to a transport demand, the travel schedule from the point where the vehicle becomes idle to the drop-off station of the transport demand can be found by the procedure described in Algorithm 6.

Algorithm 5 ***schedule_travel***
function *schedule_travel*( ) {
  Route vehicle *veh* to the destination by calling *cfstp*( );
  if *cfstp*( ) is successful **return**;
  Clear the vehicles on the shortest path by calling *make_way*( );
  Route vehicle *veh* to the destination by calling *cfstp*( );
  if *cfstp*( ) is successful **return**;
  while vehicle *veh* not on the destination {
    Route vehicle *veh* to the next node on the shortest path by calling *cfstp*( );

```
    if cfstp( ) fails {
      Set block[0] = veh and top = 1;
      Move away the blocking vehicle by calling squeeze( );
      Route vehicle veh to the next node on the shortest path by calling
      cfstp( );
    }
  }
}
```

Algorithm 6 ***schedule_transport***

```
function schedule_transport( ) {
  Compute the travel schedule to the pickup station by calling
  schedule_travel( );
  Compute the travel schedule to the drop-off station by calling
  schedule_travel( );
  return the combined travel schedule;
}
```

9.4 SIMULATION EXPERIMENTS

A simulation study has been performed to evaluate the performance of a bidirectional system. The environment of the simulation experiments is an imaginary nine-machine job shop, as shown in Figs. 9.12 and 9.13. The directions of the lanes of the unidirectional flow path network in Fig. 9.13 were optimized using the method developed by Kaspi and Tanchoco (1990).

Each machine has one input buffer and one output buffer with limited capacity. The receiving area does not have an input buffer, and the shipping area does not have an output buffer. All these buffers are assumed to be operated on the first-in-first-out (FIFO) basis. This assumption seems to be rather restrictive, but simplicity and low investment cost are its advantages. Of course, it is further assumed that it is more economical to have FIFO-type buffers than to have random access buffers for the system under study. If we allow random access to the buffers, to have the capability of sequencing the jobs in the buffers, then additional costs are needed to hire handling workers or to install random access load transfer equipment such as mini AS/RSs or carousels.

Material transport between machines in the shop is done by an AGV system. The flow path network of the AGV system is well defined. Any vehicle can use any segment of the network at any time, unless there is a conflict between vehicles. The input and output buffers of the machines and the receiving and shipping areas are located along the flow path network of the AGV system. An input buffer is a drop-off station and an output buffer is a pickup station for the AGV system.

Orders are received at random, an order containing exactly one job. The

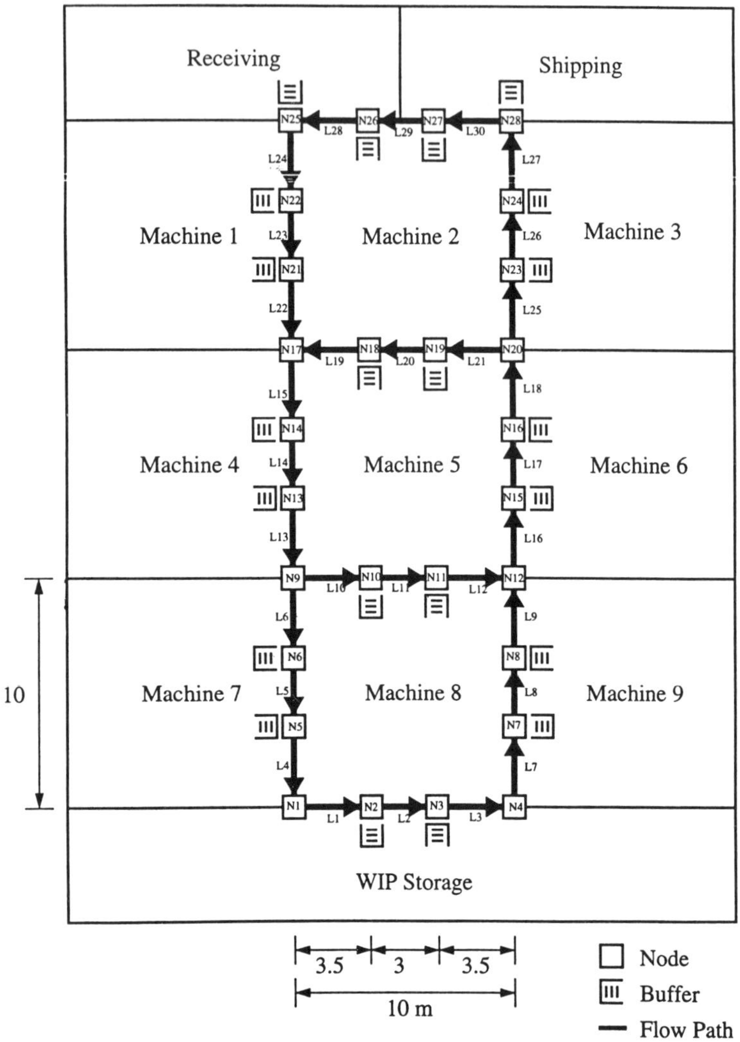

Fig. 9.13 Nine-machine job shop with a unidirectional AGV system.

job arrival is a Poisson process. When an order is received, the raw material is prepared immediately at the receiving area and made ready for release. Released jobs are queued in the output buffer of the receiving area. Each machine can process only one job at a time. Also, each AGV can carry only one job. Therefore, the order unit, the processing unit and the transfer unit area all equal to one job. We also assume that each job is shipped out of the shop as soon as it is dropped off at the shipping area.

The process routing and processing times of each job are those of a job shop, i.e. they are random. Therefore, there may be many job types produced in the system. However, the process routing and processing times are known and fixed once the order is received. We will assume that the setup times are negligible compared to processing times. Therefore, each machine can process different product types in any sequence without losing its efficiency. We further assume that the machines and AGVs are highly reliable.

The first operation of a job is equally likely to be on any machine. A subsequent operation is equally likely to be on any other machine, but a job never revisits any of the previous machines in its process routing. We do not follow the usual assumption on process routings for a pure job shop under which a job may revisit a previous machine if it is not an immediate previous machine (see, for example, Baker, 1984). Under that assumption, a job may visit the same machine several times before completion, which seems to be rare in a real system.

Each job goes through 3–9 machines before completion. The number of operations for a job is chosen from a discrete uniform distribution with the range between 3 and 9, with the mean equal to 6. The distribution for the number of operations is chosen such that the number of operations does not exceed the total number of machines in the shop and such that three times the standard deviation is roughly equal to the mean ($3\sigma \cong \mu$), and under the assumption that only the range is known for the number of operations, and any integer is equally likely within the range.

The processing time on a machine is chosen from a uniform distribution with the range between 0.634 and 2.366. The distribution for the processing times is chosen such that three times the standard deviation is equal to the mean ($3\sigma = \mu$), and under the assumption that only the range is known for the processing times, and any processing time within the range is equally likely. The usual assumption of exponential processing times is not employed here because with exponential processing times it is highly probable to have 0 processing times and very long processing times are also probable, which are not realistic in usual manufacturing applications. Since the mean processing time is equal to 1.5, a job requires in average nine time units of processing before completion. Because there are nine machines in the shop, the maximum production capacity is 1 job per time unit.

The job set consists of 3000 successful arrivals after deleting the first 500 for warm-up. With this job set the 95% confidence intervals for the average machine utilization, job processing time and flow time are within 2% of the theoretical mean. It may be possible to reduce the run length by applying an appropriate variance reduction technique.

The capacities of the input and output buffers of the machines are all equal. The buffer capacity will be varied from 1–5 to see the effects of the buffer sizes. Each buffer attached to the receiving and shipping areas, and the central storage can hold only one job at a time. The receiving area can hold 20 arrivals. With this number, the probability that there will be at least one job which can be started at a given machine is approximately 0.9 when the receiving area is full (see Han and McGinnis, 1989). If a new arrival finds no space in the receiving area, it is discarded. The capacity of the WIP storage is assumed to be infinite. However, the number of jobs allowed in the shop after release from the receiving area is set to the sum of the capacities of the machines and buffers in the shop. Therefore, the actual production rate of the shop may be limited by the capacities of the buffers and the receiving area.

The vehicles travel at the speed of 1 meter per second, regardless of whether they are loaded or empty. The loading/unloading cycle time is 30 s. The ratio of the average processing time per operation to the average transport time per transfer, called the P/T ratio†, is approximately 10 when the time unit is 600 s. Here the processing time is the time from the start of loading until completion of unloading. Similarly, the transport time is the time from the start of pickup until completion of drop-off. To see the effects of the P/T ratio on the system performance, several different levels of P/T ratios will be used in the range from 5–20.

The control of material flow is assumed to be on the first-come-first-served (FCFS) basis, i.e. the job that enters an output buffer or the WIP storage the earliest has the first priority in making dispatching decision. A job can be transferred to the next machine only if there is an empty space in the input buffer of the next machine. If the job with the first priority finds no empty space in the next buffer, a job with the next priority is considered for dispatching. Blocked jobs wait in the current buffer until the machine is blocked, then it is routed to the WIP storage to avoid machine blocking. Decisions on flow control are made based only on the current system status, and no look-ahead feature is provided.

The performance measures chosen are the throughput rate and flow time. The throughput rate is defined to be the number of jobs completed per unit time. The flow time is defined to be the time from release to the output buffer of the receiving area until completion. The throughput performances were tested under five different levels of P/T ratios (5, 6.7, 10, 15 and 20) and five different levels of fleet sizes (1, 2, 3, 4 and 5). The flow time performances were tested under five levels of P/T ratios (5, 6.7, 10, 15 and 20) and six levels

†The P/T ratio is an index that represents the criticality of the transport resource in a manufacturing system. The criticality of a resource can be defined as a measure which is positively correlated to the probability that the resource becomes a bottleneck resource in the system. It may be worth comparing the P/T ratio defined in this chapter with the cycle ratio defined by Tanchoco *et al.* (1987) and the travel time/process time ratio mentioned by Han and McGinnis (1989). The cycle ratio is defined as the ratio of the average time that a part is processed on a workstation to the sum of pure transport times required to move a part from the first workstation to the last workstation. With this definition, the cycle ratio may not provide enough information on the criticality of the transport system. The number of workstations visited by a part and the time required to do pickup or drop-off operations are also major factors that affect the criticality of the transport system. The travel time/process time ratio is the ratio of the pure travel time for one transfer to the pure processing time on a workstation. This also does not consider the pickup or drop-off times. Therefore, we prefer to use the P/T ratio as defined here, because it seems to be a more appropriate measure of the criticality of the transport system.

Table 9.2 Throughput performance of the bidirectional system

P/T ratio (time unit)	*Fleet size*	*Throughput rate*	*Flow time (time units)*	*Vehicle util. (%)*	*% travel*	*Cong. index*
5 (300 s)	1	0.36	33.43	100.0	48.81	1.000
	2	0.69	31.37	99.46	50.37	1.110
	3	0.86	42.14	95.27	51.96	1.261
	4	0.90	42.96	82.92	47.36	1.413
	5	0.91	42.57	73.47	44.43	1.609
6.7 (400 s)	1	0.49	29.68	100.0	48.65	1.000
	2	0.84	43.67	97.43	50.34	1.110
	3	0.90	41.72	77.24	41.86	1.243
	4	0.90	40.01	61.39	34.81	1.412
	5	0.91	38.95	52.07	30.89	1.595
10 (600 s)	1	0.73	31.94	99.72	48.61	1.000
	2	0.91	41.27	74.28	38.35	1.110
	3	0.91	38.07	49.99	26.59	1.241
	4	0.91	37.00	39.00	21.60	1.397
	5	0.91	38.28	34.77	20.64	1.645
15 (900 s)	1	0.89	42.96	91.59	45.77	1.000
	2	0.91	39.37	49.41	25.53	1.116
	3	0.91	38.09	33.71	17.92	1.246
	4	0.91	37.37	26.38	14.63	1.425
	5	0.91	38.53	24.14	14.66	1.776
20 (1200 s)	1	0.90	39.94	70.37	34.95	1.000
	2	0.91	38.02	26.61	18.92	1.124
	3	0.91	37.59	25.14	13.38	1.257
	4	0.91	37.18	20.04	11.28	1.481
	5	0.90	36.36	18.28	11.34	1.911

of arrival rates (0.85, 0.80, 0.75, 0.70, 0.65 and 0.60). The arrival rates determine the throughput targets. The fleet size was fixed at four for the P/T ratio of 5, at three for the P/T ratio of 6.7, and at two for other P/T ratios. The selected fleet sizes are the minima required to achieve the given throughput targets in the unidirectional system.

The results of the simulation study are summarized in Tables 9.2–9.5.† Figure 9.14 is a comparison of the throughput performances of the bidirectional and unidirectional systems at the P/T ratio of 6.7. For this P/T ratio, the

† The results obtained in this study are comparable to those presented by Egbelu and Tanchoco (1986), only at very low P/T ratios. This is mainly because the systems studied by Egbelu and Tanchoco (1986) do not have enough vehicles to satisfy the transport demands, and thus the AGV system is a major critical resource of the system. Examples of such material handling-intensive applications are warehousing and material transport between workstations and WIP storage of large-scale manufacturing facilities. In some recent applications, however, such as flexible manufacturing systems where machining on a workstation may take several hours, the AGV system is hardly a critical resource. It is expected that the applications of AGVs will be more diversified in future. Therefore, it seems to be fair to perform a simulation study for a wide range of the P/T ratios if the study involves the performance of a transport system. This is the approach adopted by Tanchoco *et al.* (1987) and Han and McGinnis (1989).

Table 9.3 Throughput performance of the unidirectional system

P/T ratio (time unit)	*Fleet size*	*Throughput rate*	*Flow time (time units)*	*Vehicle util. (%)*	*% travel*	*Cong. index*
5	1	0.26	41.40	100.0	64.04	1.000
(300 s)	2	0.49	30.81	99.27	64.73	1.072
	3	0.69	31.77	99.18	66.59	1.146
	4	0.81	37.27	98.25	68.80	1.254
	5	0.84	39.48	91.37	65.91	1.450
6.7	1	0.34	34.63	100.0	64.08	1.000
(400 s)	2	0.65	30.63	98.82	64.49	1.068
	3	0.83	41.51	96.05	64.41	1.131
	4	0.89	42.05	86.76	60.76	1.262
	5	0.89	45.46	83.43	61.41	1.488
10	1	0.51	29.07	100.0	64.13	1.000
(600 s)	2	0.87	43.23	94.89	61.81	1.059
	3	0.90	40.42	71.01	47.38	1.141
	4	0.91	38.69	61.46	43.76	1.377
	5	0.89	38.18	58.28	44.48	1.776
15	1	0.76	34.07	99.79	64.02	1.000
(900 s)	2	0.90	39.59	67.00	43.50	1.072
	3	0.91	39.14	49.99	34.15	1.232
	4	0.90	38.46	45.46	33.68	1.621
	5	0.90	37.57	45.38	36.05	2.166
20	1	0.88	43.00	94.75	61.20	1.000
(1200 s)	2	0.89	37.22	49.67	32.52	1.096
	3	0.92	39.88	40.22	28.04	1.342
	4	0.91	37.61	37.69	28.85	1.854
	5	0.91	38.03	40.68	33.57	2.656

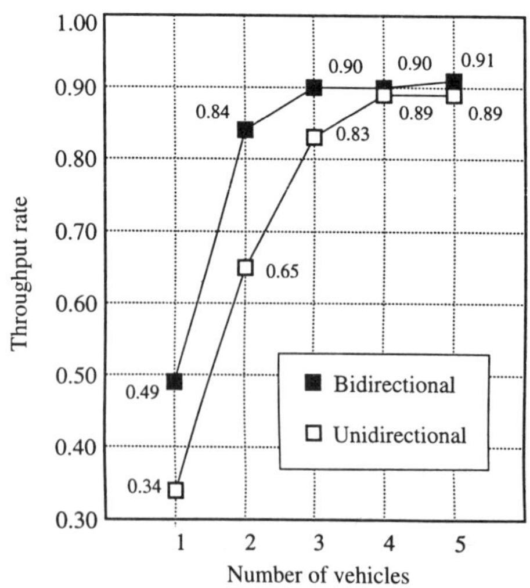

Fig. 9.14 Throughput comparison of bidirectional and unidirectional systems: Job shop with nine machines, buffer size = 3, P/T ratio = 6.7.

Table 9.4 Flow time performance of the bidirectional system

P/T ratio (time unit)	*Fleet size*	*Arrival rate*	*Flow time (time units)*	*Proc. time*	*Move time*	*Time in I. buf.*	*Time in O. buf.*
5	4	0.80	27.46	9.07	2.33	12.17	3.90
(300 s)	4	0.75	24.40	9.03	2.29	10.21	2.87
	4	0.70	22.83	9.03	2.27	9.03	2.50
	4	0.65	20.41	9.06	2.24	7.68	1.43
	4	0.60	18.89	9.04	2.23	6.60	1.02
6.7	3	0.85	31.63	9.05	1.76	14.51	6.31
(400 s)	3	0.80	27.50	9.05	1.71	12.56	4.18
	3	0.75	24.21	9.03	1.68	10.68	2.82
	3	0.70	21.78	9.05	1.67	9.17	1.90
	3	0.65	19.67	9.04	1.65	7.67	1.31
	3	0.60	18.43	9.04	1.64	6.79	0.95
10	2	0.85	30.83	9.08	1.14	14.75	5.86
(600 s)	2	0.80	27.65	9.07	1.13	13.03	4.43
	2	0.75	25.26	9.06	1.12	11.37	3.71
	2	0.70	21.41	9.04	1.09	9.38	1.91
	2	0.65	19.58	9.04	1.08	7.91	1.54
	2	0.60	17.71	9.05	1.07	6.71	0.88
15	2	0.85	29.82	9.09	0.76	15.02	4.95
(900 s)	2	0.80	26.35	9.06	0.75	12.84	3.70
	2	0.75	23.35	9.05	0.73	11.40	2.17
	2	0.70	20.72	9.02	0.72	9.35	1.62
	2	0.65	18.89	9.05	0.72	7.97	1.15
	2	0.60	17.21	9.03	0.72	6.81	0.65
20	2	0.85	29.30	9.07	0.57	14.88	4.78
(1200 s)	2	0.80	26.48	9.08	0.56	13.35	3.48
	2	0.75	22.51	9.05	0.55	11.03	1.89
	2	0.70	20.63	9.03	0.54	9.67	1.39
	2	0.65	18.52	9.04	0.54	8.15	0.80
	2	0.60	16.84	9.05	0.54	6.75	0.51

average processing time for an operation is 10 minutes, and the average processing time for completing a job is 1 hour. With the given system configuration, the maximum achievable throughput rate was at around 0.90. This upper bound is mainly determined by the sizes of the buffers and the capacity of the receiving area. The figure shows that the bidirectional system outperforms the unidirectional system in terms of throughput rate. The bidirectional system achieves the upper bound with three vehicles while the unidirectional system needed four vehicles to achieve a comparable performance. This indicates that the bidirectional system can save a number of vehicles for the same level of throughput target. When there are three or less vehicles, the bidirectional system achieves higher throughput rates than the unidirectional system. However, when there are four or more vehicles, both systems shows similar throughput performances at around 0.90. Tables 9.2 and 9.3 also show that the throughput difference decreases as the P/T ratio increases. This is because the AGV system becomes less important as the P/T ratio increases.

Table 9.5 Flow time performance of the unidirectional system

P/T ratio (time unit)	*Fleet size*	*Arrival rate*	*Flow time (time units)*	*Proc. time*	*Move time*	*Time in I. buf.*	*Time in O. buf.*
5	4	0.80	33.79	9.13	2.85	11.26	10.56
(300 s)	4	0.75	29.30	9.07	2.81	10.02	7.40
	4	0.70	25.35	9.06	2.77	8.48	5.03
	4	0.65	22.61	9.07	2.75	7.49	3.31
	4	0.60	20.58	9.05	2.72	6.36	2.45
6.7	3	0.85	35.46	9.08	2.11	13.55	10.72
(400 s)	3	0.80	31.07	9.05	2.08	12.16	7.78
	3	0.75	26.56	9.05	2.05	10.46	5.01
	3	0.70	23.30	9.06	2.04	8.71	3.48
	3	0.65	21.07	9.04	2.02	7.64	2.36
	3	0.60	19.41	9.05	2.02	6.57	1.76
10	2	0.85	34.65	9.10	1.41	14.31	9.83
(600 s)	2	0.80	29.20	9.07	1.38	12.27	6.47
	2	0.75	26.20	9.06	1.36	10.94	4.84
	2	0.70	22.25	9.05	1.34	9.04	2.81
	2	0.65	20.37	9.05	1.34	7.77	2.21
	2	0.60	18.90	9.05	1.34	6.78	1.73
15	2	0.85	30.70	9.06	0.93	14.73	5.99
(900 s)	2	0.80	26.79	9.07	0.91	12.83	3.98
	2	0.75	23.35	9.06	0.90	10.83	2.56
	2	0.70	21.08	9.05	0.89	9.28	1.86
	2	0.65	19.25	9.05	0.89	8.00	1.31
	2	0.60	17.66	9.05	0.89	6.73	0.98
20	2	0.85	29.17	9.07	0.70	14.52	4.88
(1200 s)	2	0.80	26.10	9.07	0.69	12.82	3.53
	2	0.75	22.74	9.07	0.68	10.89	2.11
	2	0.70	20.57	9.03	0.67	9.47	1.39
	2	0.65	18.86	9.06	0.67	8.07	1.06
	2	0.60	17.30	9.03	0.67	6.74	0.86

Figure 9.15 shows a comparison of the flow time performances of the two systems at the P/T ratio of 6.7, for six different levels of throughput targets, 0.85, 0.80, 0.75, 0.70, 0.65 and 0.60. The fleet size is fixed at three, since both systems can achieve 0.85 throughput rate with three vehicles. The figure shows that the bidirectional system can reduce the flow time by approximately 5–12% for the given P/T ratios and fleet size.

The differences in the throughput and flow time performances of the bidirectional and unidirectional systems are mainly due to the reduction in the travel distance. The columns labelled 'Vehicle Util. (%)' in Tables 9.2 and 9.3 represent the vehicle utilization expressed by the sum of the times the vehicles are used divided by the total available time. The columns labelled '% Travel' represent the proportion of the total travel time relative to the total available time of the vehicles. The results indicate that the vehicles in the unidirectional system had to spend more time in traveling simply because

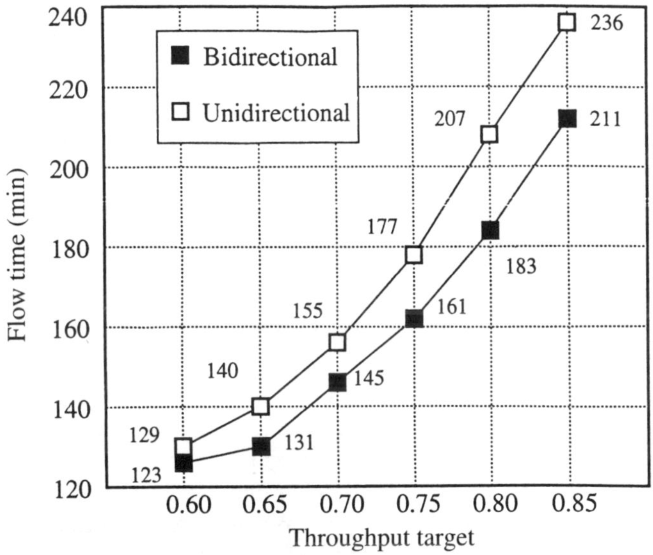

Fig. 9.15 Flow time comparison of bidirectional and unidirectional systems: Job shop with nine machines, buffer size = 3, P/T ratio = 6.7.

the average distance between stations are longer than in the bidirectional system. The effect of increased travel time is significant when the AGV system is a critical resource. The columns labelled 'Time in I. Buf.' and 'Time in O. Buf.' in Tables 9.4 and 9.5 represent the times the jobs spent waiting in the input buffers and output buffers, respectively. The results show that there are significant differences in the times spent in the output buffers waiting for vehicles when the AGV system is a critical resource. This increased waiting time results in increased flow time and reduced throughput rate.

An interesting observation regarding the vehicle performance is the behavior of the percent travel time. When the P/T ratio is rather low, for example at 5 or 6.7, the percent travel time increases as the fleet size increases, because of increased transport demands, then begins to decrease when the throughput rate of the system reaches the upper bound of 0.90. At a higher P/T ratio, however, the percent travel time begins to increase again after the fleet size reaches a certain point. This trend is apparent in the unidirectional system. This behavior can be explained as follows. At a higher P/T ratio, the demands for job transfers are generated less frequently and as a result there are more idle vehicles which block the shortest path of a busy vehicle. The 3-pass cooperative path planning approach to find a feasible path described in the previous section always tries not to disturb other vehicles unless it cannot find a path by *cfstp*( ). Therefore, the answer found by function *schedule_travel*( ) may result in a long path. This indicates more cooperative movements by idle vehicles may be required to enhance the efficiency of the AGV system.

Another indicator, called 'Congestion Index', is listed in the Tables 9.2 and 9.3 to analyse the vehicle performances further. The congestion index is the actual travel time divided by the shortest travel time if there were no congestion. The congestion index quickly grows as the fleet size increases. The growth rate is faster at a higher P/T ratio for the same reason as explained above, and it is even faster in the unidirectional system. This indicates that cooperative path planning is more critical in unidirectional systems to enhance vehicle efficiency.

9.5 DISCUSSION

The benefits of bidirectional AGV systems over unidirectional counterparts are significant in terms of system performances such as throughput rate and flow time when the AGV system is a critical resource in the shop. For the same number of vehicles, a bidirectional system achieves higher throughput rates than its unidirectional counterpart. When there is a certain production target, a bidirectional system may require fewer vehicles than a unidirectional system, and achieves shorter flow times. The reliability of the flow path network also increases, since there are more degrees of freedom in selecting travel paths in the network. A bidirectional system can also save the space requirements for the flow path network when compared to a multi-lanc system, which provides two or more lanes in the same aisle simply to ensure return flows. Therefore, a bidirectional AGV system seems to be a promising alternative to conventional unidirectional systems when the system requires a few vehicles and the traffic requirements do not exceed a certain point where a multi-lane system can be justified.

A prerequisite to realizing bidirectional AGV systems is developing vehicles which have the capability of bidirectional monoeuver and load transfer mechanisms. Otherwise, the path planning algorithm should be modified appropriately such that the vehicles can have proper orientations in main aisles and at load transfer stations. These modifications may result in somewhat degraded system performances and increased complexity in the operational control method. Fortunately, most commercial vehicles on the market have bidirectional monoeuver capabilities, and some unit load carrying vehicles have bidirectional load transfer mechanisms. Appropriate methods for traffic control and collision avoidance should also be developed. The increase in the cost of developing the control software which can handle such complexities also seems to be nontrivial. It is also expected that the first few bidirectional systems may experience some resistance from management and the labor unions because of the fear of complexity and concerns about safety.

The bidirectional AGV system is a controversial research issue. The benefits obtained through improved system efficiency and network reliability may be offset by the costs associated with handling the complexity. Even though it may be possible to obtain reasonably accurate estimates of the benefits by converting the improvements into monetary terms, it may be difficult to estimate the costs of developing and maintaining the control software and managing the system. Further, the magnitude of the benefits of a bidirectional AGV system depends upon the characteristics of the specific system under study. A simulation study seems to be appropriate in estimating the improvements in system performance. Regarding the flow path network model, the mixed design is more likely to be adopted in practice than the pure bidirectional model. The performance of the mixed model should be at least as good as the pure bidirectional model because it is intended to be an improvement of the bidirectional model. However, there still remains the problem of designing the flow path network based on the mixed model.

REFERENCES

Baker, K.R. (1984) Sequencing rules and due-date assignments in a job shop. *Management Science*, **30**(9), 1093–104.

Broadbent, A.J., Besant, C.B., Premi, S.K. and Walker, S.P. (1985) *Free ranging AGV Systems: Promises, Problems and Pathways.* Proceedings 2nd International Conference on Automated Materials Handling, IFS(Publ) Ltd., UK, 221–37. (Reprinted in *Automated Guided Vehicle Systems*, ed R.H. Hollier, IFS(Publ.) Ltd., UK, 1987.)

Chen, C.L., Lee, C.S.G. and McGillem, C.D. (1987) Task assignment and load balancing of autonomous vehicles in a flexible manufacturing system. *IEEE Journal of Robotics and Automation*, **3**(6), 659–71.

Dijkstra, E.W. (1959) A note on two problems in connexion with graphs. *Numerische Mathematik*, **1**, 269–71.

Egbelu, P.J. and Tanchoco, J.M.A. (1984) Characterization of automatic guided vehicle dispatching rules. *International Journal of Production Research*, **22**(3), 359–74.

Egbelu, P.J. and Tanchoco, J.M.A. (1986) Potentials for bi-directional guide-path for automated guided vehicle based systems. *International Journal of Production Research* **24**(5), 1075–97.

Fujii, S., Sandoh, H. and Hohzake, R. (1989) *A Routing Control Method of Automated Guided Vehicles by The Shortest Path with Time-Windows.* 10th International Conference on Production Research.

Han, M.-H. and McGinnis, L.F. (1989) Control of material handling transporter in automated manufacturing. *IIE Transactions*, **21**(2), 184–90.

Huang, J., Palekar, U.S. and Kapoor, S.G. (1989) *A Labeling Algorithm for The Navigation of Automated Guided Vehicles*, Advances in Manufacturing Systems Engineering (Proc. Winter Annual Meeting of The ASME, PED-Vol. 37), eds. M. Anjanappa and D.K. Anand, San Francisco, CA. pp. 181–3.

Kaspi, M. and Tanchoco, J.M.A. (1990) Optimal flow path design of unidirectional AGV systems. *International Journal of Production Research*, **28**(6), 1023–30.

Kim, C.W. (1991) The Operation of an Automated Guided Vehicle System in a Manufacturing Job Shop. PhD Thesis, Purdue University, West Lafayette, IN.

Kim, C.W. and Tanchoco, J.M.A. (1991) Conflict-free shortest-time bidirectional AGV routing. *International Journal of Production Research*, **29**(12), 2377–91.

Kim, C.W. and Tanchoco, J.M.A. (1993) Operational control of a bidirectional automated guided vehicle system. *International Journal of Production Research*, **31**(9), 2123–38.

Krishnamurthy, N.N. (1990) Modeling Blocking in Automated Guided Vehicle Systems. PhD Thesis, State University of New York at Buffalo.

Krishnamurthy, N.N., Batta, R. and Karwan, M.H. (1991) Developing Conflict-Free Routes for Automated Guided Vehicles. *Working Paper*, Department of Industrial Engineering, State University of New York at Buffalo.

Kusiak, A. and Cyrus J.P. (1985) *Routing and Scheduling of Automated Guided Vehicles.* Toward the Factory of the Future (Proc. 8th ICPR), eds. H.-J. Bullinger and H.J. Warnecke, Springer-Verlag, Berlin, pp. 247–51.

Maxwell, W.L. (1981) Solving material handling design problems with OR. *Industrial Engineering*, April, 58–69.

Maxwell, W.L. and Mackstadt, J.A. (1982) Design of automatic guided vehicle systems. *IEE Transactions*, **14**(2), 114–24.

Nilsson, N.J. (1980) *Principles of Artificial Intelligence.* Morgan Kaufmann, Los Altos, CA.

Taghaboni, F. (1989) Scheduling and Control of manufacturing Systems with Critical Material Handling. PhD Thesis, Purdue University, West Lafayette, IN.

Taghaboni, F. and Tanchoco, J.M.A. (1988) A LISP-based controller for free-ranging automated guided vehicle systems. *International Journal of Production Research*, **26**(2), 173–88.

Tanchoco, J.M.A., Egbelu, P.J. and Taghaboni, F. (1987) Determination of the total number of vehicles in an AGV-based material transport system. *Material Flow*, **4**(1–2), 33–51.

CHAPTER 10

Approaches to analysing the load routing problem in tandem AGV systems

J.T. Lin and C.C.K. Chang

10.1 INTRODUCTION

It has been estimated that between 20–50% of the total operating cost within manufacturing can be attributed to material handling. Furthermore, effective facilities design and planning can reduce these costs by at least 10–30% (Tompkins and White, 1984). Even so, material handling is often an overlooked aspect of factory operations.

Automated Guided Vehicle Systems (AGVS) are often used in Flexible Manufacturing Systems (FMSs). They offer a great deal of flexibility that no other conventional material handling systems can match. In a FMS environment, the AGVS provides adaptability to the facilities layout, expandability without production interruption, open floor access, and a high level of automation and integration in computer integrated manufacturing (Vosniakos and Mamalis, 1990). However, an AGV system is expensive to install and maintain. AGVSs can provide an economic solution only if the system is utilized with proper design, operation policies and controls.

10.2 CHARACTERISTICS OF AGVSs

In a review of previous research on AGVSs, three types of guide path design can be identified: conventional network AGVS; optimal single-loop AGVS; and tandem AGVS. The main characteristics of each system are discussed below.

10.2.1 Conventional network AGVS

In conventional AGVSs several vehicles are used, and each vehicle is allowed to serve any station in the system. The guide path orientation can either be unidirectional, bidirectional or mixed (Egbelu and Tanchoco, 1984, 1986). One advantage of the conventional network system lies in its flexibility and reliability. However, implementation of such a system must address the following issues:

1. flow path design;
2. number of vehicles required;
3. location of pickup and drop-off stations;
4. vehicle management and routing;
5. choice of vehicle types; and
6. traffic management.

10.2.2 Optimal single-loop AGVS

The optimal single-loop (OSL) AGVS was suggested by Tanchoco and Sinriech (1992). In this type of layout, there is only one single-loop guide path in the system which serves all the pickup/drop-off stations along the loop. The number of vehicles required is determined by the system workload. The objectives for designing the 'best' single loop are: (1) to minimize the flow in the system based on the material flow between departments; and (2) to minimize the time/distance the AGV has to travel to complete its assignments. The primary benefit of using such a configuration is the simplicity of the traffic control algorithms and the potential elimination of the shop-locking possibility.

10.2.3 Tandem configuration AGVS

The tandem configuration was proposed for AGVSs by Bozer and Srinivasan (1991a,b). It is based on the partitioning of all the stations into non-overlapping, single-vehicle closed loops with an additional transit area provided as an interface between adjacent loops. The vehicle dispatching rule for this configuration is based on First-Encounter-First-Served (FEFS) (Bartholdi and Platzman, 1989). Each transit area has two transit stations which allow loads to be moved bidirectionally between the adjacent loops. The tandem AGVS requires less complicated vehicle dispatching and traffic management. It also removes the problems of vehicle collision and interference. Furthermore, this type of configuration offers distributed control and expansion flexibility.

10.3 PROBLEM STATEMENT

Several new issues are encountered and have to be overcome before the actual implementation of a tandem AGVS. One of the most important issues from a design perspective is how to develop a partitioning algorithm that enables one to group workstations into single vehicle loops with the minimum number of loops while satisfying the transportation requirements. Determining the location of transit stations with an optimal buffer size is yet another issue that has to be addressed. Although vehicle dispatching and routing might be less complicated than in a conventional system, there is one operational issue that was not addressed by Bozer and Srinivasan (1991a,b). This problem, which we call the load routing problem (LRP), is primarily a result of the fact that a load may be handled by several vehicles and moved through several loops before it reaches its destination. The task of scheduling the load to be transferred among loops involves decisions on delivery sequences and quantities. Factors such as the direction of each loop, the throughput capacity of each loop, vehicle speed and routing feasibility can have a significant impact on the overall system performance.

10.3.1 General problem description

A tandem AGVS is comprised of several non-overlapping single vehicle closed loops, where each loop has only one vehicle traveling unidirectionally around it. Vehicle collision, congestion and blocking are completely eliminated in this type of configuration. The complexity of traffic control is therefore reduced. However, in a tandem AGVS with more than three loops, some loads require processing at stations in different loops. Gridlock or deadlock can be avoided by providing sufficient buffer capacity at pickup, drop-off and load transfer stations. Hence, these loads could reach their destinations via two or more routes. The LRP model described in this chapter addresses this problem by determining: (1) which loops to by which travel; (2) which transit stations to use; (3) the total rate of delivery; and (4) the sequence stations to pass, while minimizing total loaded vehicle travel time. It is also important to balance the throughput capacities among all the loops so the bottleneck loops can be avoided.

10.4 MODEL DESCRIPTION

A model is introduced here to demonstrate clearly the concept of the load routing problem in a tandem AGVS. A tandem configuration guide path layout is shown in Fig. 10.1. The facility consists of four areas: (I) conventional machining area; (II) dedicated machining area; (III) heat treatment area; and (IV) inspection and storage area. Each production area constitutes a closed loop that is served by a dedicated vehicle. There are a total of five transit areas, 13 process workstations and three I/O stations in the system. A numerical label is assigned to each station in a clockwise fashion for easy identification. Empty and loaded vehicle speeds are assumed to be 60 m/min. and 54 m/min. respectively, and the direction of each loop is assumed to be clockwise. The distances between two stations in each closed loop are shown in Table 10.1. The vehicle in each closed loop is assumed to travel unidirectionally in a clockwise direction. The delivery sequences and required production rates

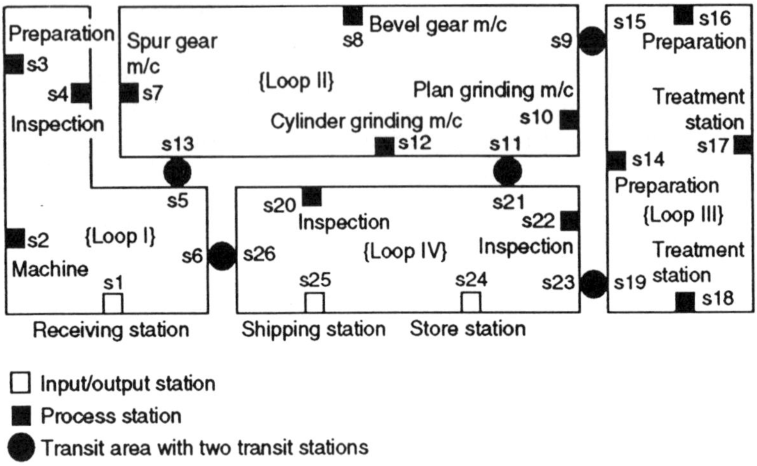

Fig. 10.1 Example of tandem AGVS layout.

for each of the five job types produced in this facility are shown in Table 10.2. Moreover, the pickup and drop-off times for each load are a constant of 0.5 min.

10.5 APPROACHES

Two approaches are presented here for solving the load routing problem. In the first approach, the route of each *across-loop load* is determined by minimizing the total number of loops the loads have to travel. If more than one route is available, the first transit area encountered in the feasible path is selected on the basis of the FEFS rule. The throughput capacity of each loop will be verified through a permutation-based intuitive approach. In the second approach, a linear programming model is formulated to determine the routing decisions. The objective function is to minimize the total loaded vehicle time, which includes the loaded vehicle travel time and the pickup/drop-off times.

10.5.1 Intuitive approach

The intuitive approach is introduced here to forecast the throughput capacity of each loop based on the predetermined route which minimizes the total number of loops to be passed. This approach includes five steps. The first four steps transform a given set of data into a from–to matrix. The final step employs a net-flow concept to determine the number of empty trips, and to estimate the throughput capacity of each loop.

Table 10.1 Travel distance between stations of loops I–IV

Loop I

	s1	s2	s3	s4	s5	s6
s1	0	18	33	54	70	80
s2	78	0	15	36	52	62
s3	63	81	0	21	37	47
s4	42	60	75	0	16	26
s5	26	44	59	80	0	10
s6	16	34	49	70	86	0

Loop II

	s7	s8	s9	s10	s11	s12	s13
s7	0	33	60	69	77	94	114
s8	93	0	27	36	44	61	81
s9	66	99	0	9	17	34	54
s10	57	90	117	0	8	25	45
s11	49	82	109	118	0	17	37
s12	32	65	92	101	109	0	20
s13	12	45	72	81	89	106	0

Loop III

	s14	s15	s16	s17	s18	s19
s14	0	17	27	49	69	78
s15	73	0	10	32	52	61
s16	63	80	0	22	42	51
s17	41	58	68	0	20	29
s18	21	38	48	70	0	9
s19	12	29	39	61	81	0

Loop IV

	s20	s21	s22	s23	s24	s25	s26
s20	0	24	31	39	50	69	81
s21	69	0	7	15	26	45	57
s22	62	86	0	8	19	38	50
s23	54	78	85	0	11	30	42
s24	43	67	74	82	0	19	31
s25	24	48	55	63	74	0	12
s26	12	36	43	51	62	81	0

Table 10.2 Process sequences of the job (loads/hr)

Job (description)	*Process sequences*	*Rate*
A (bevel gear)	s1 → s2 → s4 → s8 → s14 → s18 → s20 → s12 → s22 → s25	$a = 2.5$
B (spur gear)	s1 → s2 → s4 → s7 → s16 → s17 → s12 → s24	$b = 3.3$
C (cyl. prod.)	s1 → s2 → s14 → s18 → s12 → s22 → s24	$c = 1.5$
D (squ. prod.)	s1 → s3 → s4 → s16 → s17 → s10 → s20 → s24	$d = 0.8$
E (input prod.)	s1 → s4 → s22 → s25	$e = 6.0$

(a) Step 1: Creation of the from–to matrix

A 28 × 28 from–to matrix (Table 10.3) can be constructed based on the layout provided in Fig. 10.1. This table includes one title column (row), one quantity sum column (row), and 26 station columns (rows). The table is further partitioned into four divisions. Each division is denoted by one production area (loop) in the system.

(b) Step 2: Generation of across-loop and in-loop deliveries

Based on the information given in Table 10.2, each job process sequence requirements are put into the from–to matrix created in Step 1. For example, the process sequence requirement for job E is s1 → s4 → s22 → s25. The transportation rate, *e* loads/hr, is put into the corresponding cells in Table 10.3. Among them, s1 → s4 and s22 → s25 are called *in-loop load* deliveries in Loops I and IV, respectively, and s4 → s22 is called *across-loop load* delivery of Loop I. The in-loop load, which falls inside the four regions, requires the service of only one vehicle. It has only one route to be done under which the direction of the loop is already decided. The across-loop load, which falls outside the four divisions, needs two or more vehicles before reaching its destination. The transportation requirements for other jobs are inserted into Table 10.3 with their respective transportation rates of *a*, *b*, *c* and *d* load/hr in a similar fashion. The final result is shown in Table 10.4.

(c) Step 3: Determination of travel path for across-loop delivery

All the across-loop loads required in this model are summarized in column 1 of Table 10.5. The travel path of each across-loop load is determined by the following rules:

1. The total number of loops the loads has to travel is minimized.
2. If more than one path with the same number of loops to traverse is available, the first transit area encountered in the feasible path is selected on the basis of the FEFS rule.

For example, the across-loop load s2 → s14 has two feasible paths, I → II → III and I → IV → III, that satisfy the first rule. Since the first transit area encountered by the vehicle in Loop I is adjacent to Loop II, path I → II → III is selected based on the second rule. Table 10.6 shows the final from–to matrix after all the across-loop loads have been regrouped into equivalent in-loop loads.

(d) Step 4: Calculation of arrival and departure rates

To find the empty vehicle travel frequency, arrival and departure rates at each station should be calculated. The calculation for Loop II in Table 10.6 is demonstrated here to illustrate the computation procedure. The transporta-

Table 10.3 Matrix table

	s1	s2	s3	s4	s5	s6	s7	s8	s9	s10	s11	s12	s13	s14	s15	s16	s17	s18	s19	s20	s21	s22	s23	s24	s25	s26	λi
s1																											
s2																											
s3																											
s4																											
s5																											
s6																											
s7																											
s8																											
s9																											
s10																											
s11																											
s12																											
s13																											
s14																											
s15																											
s16																											
s17																											
s18																											
s19																											
s20																											
s21																											
s22																											
s23																											
s24																											
s25																											
s26																											
Λi																											

Table 10.4 From-to chart

	s1	s2	s3	s4	s5	s6	s7	s8	s9	s10	s11	s12	s13	s14	s15	s16	s17	s18	s19	s20	s21	s22	s23	s24	s25	s26	λi
s1		a,b,c	d	e																							a,b,c,d,e
s2				a,b										c													a,b,c
s3				d																							d
s4							b	a								d						e					a,b,d,e
s5																											0
s6																											0
s7																b											b
s8														a													a
s9																											0
s10																				d							d
s11																											0
s12																						a,c		b			a,b,c
s13																											0
s14																		a,c									a,c
s15																											0
s16																	b,d										b,d
s17										d		b															b,d
s18												c								a							a,c
s19																											0
s20												a												d			a,d
s21																											0
s22																								c	a,e		a,c,e
s23																											0
s24																											0
s25																											0
s26																											0
Λi	0	a,b,c	d	a,b,d,e	0	0	b	a	0	d	0	a,b,c	0	a,c	0	b,d	b,d	a,c	0	a,d	0	a,c,e	0	b,c,d	a,c	0	

Table 10.5 Predetermined route of across-loop loads

Across-loop loads	*From–to loop*	*Predetermined path*	*Rate (loads/hr)*
s2→s14	I→III	I→II→III	$c=1.5$
s4→s7	I→II	I→II	$b=3.3$
s4→s8	I→II	I→II	$a=2.5$
s4→s16	I→III	I→II→III	$d=0.8$
s4→s22	I→IV	I→IV	$e=6.0$
s7→s16	II→III	II→III	$b=3.3$
s8→s14	II→III	II→III	$a=2.5$
s10→s20	II→IV	II→IV	$d=0.8$
s12→s22	II→IV	II→IV	$a+c=4.0$
s12→s24	II→IV	II→IV	$b=3.3$
s17→s10	III→II	III→II	$d=0.8$
s17→s12	III→II	III→II	$b=3.3$
s18→s12	III→II	III→II	$c=1.5$
s18→s20	III→IV	III→IV	$a=2.5$
s20→s12	IV→II	IV→II	$a=2.5$

tion rate for the loads in steady state are:

1. s7→s9 (rate = 3.3 loads/hr)
2. s8→s9 (rate = 2.5 loads/hr)
3. s9→s10 (rate = 0.8 loads/hr)
4. s9→s12 (rate = 4.8 loads/hr)
5. s10→s11 (rate = 0.8 loads/hr)
6. s11→s12 (rate = 2.5 loads/hr)
7. s12→s11 (rate = 7.3 loads/hr)
8. s13→s7 (rate = 3.3 loads/hr)
9. s13→s8 (rate = 2.5 loads/hr)
10. s13→s9 (rate = 2.3 loads/hr)

The load departure rates $l_i = \sum_j f_{i,j}$ may be computed for each of the I/O stations s1, s24, s25 and transition stations s5, s6, s9, s11, s13, s15, s19, s21, s23 and s26. The respective values l_i for each I/O station and transit area in the system are:

$l1 = a+b+c+d+e = 14.1$ (loads/hr)
$l5 = 0$ (loads/hr)
$l6 = 0$ (loads/hr)
$l9 = b+c+d = 5.6$ (loads/hr)
$l11 = a = 2.5$ (loads/hr)
$l13 = a+b+c+d = 8.1$ (loads/hr)
$l15 = a+b+c+d = 8.1$ (loads/hr)
$l19 = 0$ (loads/hr)

Table 10.6 From-to chart based on the intuitive approach

	s1	s2	s3	s4	s5	s6	s7	s8	s9	s10	s11	s12	s13	s14	s15	s16	s17	s18	s19	s20	s21	s22	s23	s24	s25	s26	λi
s1		a,b,c	d	e																							a,b,c,d,e
s2				a,b	c																						a,b,c
s3				d																							d
s4					a,b,d	e																					a,b,d,e
s5													a,b,c,d														a,b,c,d
s6																										e	e
s7									b																		b
s8									a																		a
s9										d		b,c			a,b,c,d												a,2b,2c,2d
s10											d																d
s11												a									a,b,c,d						2a,b,c,d
s12											a,b,c																a,b,c
s13							b	a	c,d																		a,b,c,d
s14																		a,c									ac
s15									b,c,d					a,c		b,d											a,2b,2c,2d
s16																	b,d										b,d
s17															b,d												b,d
s18															c				a								a,c
s19																							a				a
s20																					a			d			a,d
s21											a									d		a,c		b			2a,b,c,d
s22																								c	a,e		a,c,e
s23																				a							a
s24																											0
s25																											0
s26																						e					e
Λi	0	a,b,c	d	a,b,d, e	a,b,c, d	e	b	a	a,2b,2c, 2d	d	2a,b,c, d	a,b,c	a,b,c,d	a,c	a,2b,2c, 2d	b,d	b,d	a,c	a	a,d	2a,b,c,d	a,c,e	a	b,c,d	a,e	e	

$l21 = a + b + c + d$	$= 8.1$	(loads/hr)
$l23 = a$	$= 2.5$	(loads/hr)
$l24$	$= 0$	(loads/hr)
$l25$	$= 0$	(loads/hr)
$l26 = e$	$= 6.0$	(loads/hr)

Likewise, the respective load arrival rates $L_j = \sum_i f_{i,j}$ for each I/O station and transit area j in the system are:

$L1$	$= 0$	(loads/hr)
$L5 = a + b + c + d$	$= 8.1$	(loads/hr)
$L6 = e$	$= 6.0$	(loads/hr)
$L9 = a + b + c + d$	$= 8.1$	(loads/hr)
$L11 = a + b + c + d$	$= 8.1$	(loads/hr)
$L13$	$= 0$	(loads/hr)
$L15 = b + c + d$	$= 5.6$	(loads/hr)
$L19 = a$	$= 2.5$	(loads/hr)
$L21 = a$	$= 2.5$	(loads/hr)
$L23$	$= 0$	(loads/hr)
$L24 = b + c + d$	$= 5.6$	(loads/hr)
$L25 = a + e$	$= 8.5$	(loads/hr)
$L26$	$= 0$	(loads/hr)

l_i is assumed here to be equal to L_i at every processing station i, i.e. the flow is conserved at each processing station.

(e) Step 5: Throughput capacity verification

The throughput capacity needs to be verified to make sure that a single vehicle assigned to a loop can handle all the arrival and departure loads on each closed loop. There are three components that can affect the throughput capacity of a loop. The first is the total loaded vehicle travel time. This can be expressed mathematically by the formula:

$$\sum_i \sum_j [f_{i,j} \times \tau_{i,j}] = \sum_i \sum_j \left[\frac{f_{i,j} \times d_{i,j}}{54} \right],$$

where $f_{i,j}$ is the delivery frequency from station i to station j; $\tau_{i,j}$ is the loaded vehicle travel time from station i to station j; $d_{i,j}$ is the distance from station i to station j along the path; and 54 m/min. is the loaded vehicle speed.

The second component has to do with the time taken to pickup and drop-off a load at stations. The mathematical formula associated with this factor can be written as follows:

$$\sum_i \sum_j [2 \times \mu \times f_{i,j}],$$

where $f_{i,j}$ is the delivery frequency from station i to station j, and μ is the pickup and drop-off time.

The last component is the total empty vehicle travel time. It can be expressed mathematically as:

$$\sum_i \sum_j [f'_{i,j} \times \sigma_{ij}] = \sum_i \sum_j \left[\frac{f'_{i,j} \times d_{i,j}}{60} \right],$$

where $f'_{i,j}$ is the frequency of load to be transferred from station i to station j by an empty vehicle; $\sigma_{i,j}$ is the travel time by an empty vehicle from station i to j; and 60 m/min. is the empty vehicle speed.

The first two components can be directly calculated by using the information provided in Table 10.6. The third component, however, depends on the empty vehicle dispatching strategy, and cannot be derived directly. To estimate the empty vehicle travel time for each loop, a vehicle net-flow concept is utilized. The net flow for station i can be shown as:

$$E_i = \sum_j f_{j,i} - \sum_j f_{i,j}.$$

The vehicle is assumed to be made available at the station where it makes the last delivery. This idea was first proposed by Maxwell and Muckstadt (1982) for determining the minimum number of vehicles required in an AGVS. The objective is to allocate the empty vehicle available from the 'surplus' stations ($E_i > 0$) to the 'demand' stations ($E_i < 0$), while the total empty vehicle travel time is minimized. A simple linear program was formulated to determine the number of empty vehicle trips between station i and j:

Minimize: $\sum_i \sum_j [\sigma_{i,j} \times f'_{i,j}]$

Subject to: $\sum_j f'_{i,j} = E_i$ if the net flow for station i is non-negative

$\sum_i f'_{i,j} = -E_j$ if the net flow into station j is negative

$f'_{i,j} \geqslant 0$

Loop IV is used here to illustrate this idea. The from–to matrix for Loop IV is shown in Table 10.7(a) and the net flow for each station within the loop is provided in Table 10.7(b). Given the travel time between two stations: $\sigma_{24,21} = 1.117$ min., $\sigma_{24,23} = 1.367$ min., $\sigma_{24,26} = 0.517$ min., $\sigma_{25,21} = 0.8$ min., $\sigma_{25,23} = 1.05$ min. and $\sigma_{25,26} = 0.2$ min., the linear programming model can be constructed as follows:

$$\text{Minimize:} \quad Z = 1.117 f'_{24,21} + 1.367 f'_{24,23} + 0.517 f'_{24,26} + 0.8 f'_{25,21} + 1.05 f'_{25,23} + 0.2 f'_{25,26}$$

$$\begin{aligned} \text{Subject to:} \quad & f'_{24,21} + f'_{24,23} + f'_{24,26} = 5.6 \\ & f'_{25,21} + f'_{25,23} + f'_{25,26} = 8.5 \\ & f'_{24,21} + f'_{25,21} = 5.6 \\ & f'_{24,23} + f'_{25,23} = 2.5 \\ & f'_{24,26} + f'_{25,26} = 6.0. \end{aligned}$$

Table 10.7 From–to chart of loop IV (a); net-flow of loop IV (b); transportation problem table (c); result of transportation problem table

(*a*) *Unit = loads/hr*

	s20	s21	s22	s23	s24	s25	s26	From
s20		2.5			0.8			3.3
s21	0.8		4.0		3.3			8.1
s22					1.5	8.5		10
s23	2.5							2.5
s24								0
s25								0
s26			6.0					6.0
to	3.3	2.5	10	0	5.6	8.5	0	$\sum = 29.9$

(*b*)

	s20	s21	s22	s23	s24	s25	s26
Total TO	3.3	2.5	10	0	5.6	8.5	0
Total FROM	3.3	8.1	10	2.5	0	0	6.0
New flow	0	−5.6	0	−2.5	5.6	8.5	−6.0

(*c*)

	s21		s23		s26		Supply
s24	**1.117**		**1.367**		**0.517**		5.6
		f'_{2421}		f'_{2423}		f'_{2426}	
s25	**0.800**		**1.050**		**0.200**		8.5
		f'_{2521}		f'_{2523}		f'_{2526}	
Demand		5.6		2.5		6.0	14.1

(*d*)

	s21		s23		s26		Supply
s24	**1.117**		**1.367**		**0.517**		
		5.6		0		0	5.6
s25	**0.800**		**1.050**		**0.200**		
		0		2.5		6.0	8.5
Demand		5.6		2.5		6.0	14.1

The optimal solution to this problem is: $f'_{24,21} = 5.6, f'_{24,23} = 0, f'_{24,26} = 0, f'_{25,21} = 0, f'_{25,23} = 2.5$ and $f'_{25,26} = 6.0$. The throughput capacity of Loop IV can now be calculated as follows:

1. The total loaded vehicle travel time:

$$= \sum_i \sum_j [f_{i,j} \times \tau_{i,j}] = \sum_i \sum_j \left[\frac{f_{i,j} \times d_{i,j}}{54} \right]$$

$$= [2.5]*[0.444] + [0.8]*[0.926] + [0.8]*[1.278] + [4.0]*[0.13] + [3.3]*[0.481] + [1.5]*[0.352] + [8.5]*[0.704] + [2.5]*[1.0] + [6.0]*[0.796]$$

$$= 18.7685 \text{ min/hr.}$$

2. The total pickup and drop-off time:

$$= \sum_i \sum_j [2 \times \mu \times f_{ij}]$$

$$= [2.5 + 0.8 + 0.8 + 4.0 + 3.3 + 1.5 + 8.5 + 2.5 + 6]$$

$$= 29.9\ \text{min/hr}.$$

3. The total empty vehicle travel time:

$$= \sum_i \sum_j [\sigma_{i,j} \times f'_{i,j}] = \sum_i \sum_j \left[\frac{f'_{i,j} \times d_{i,j}}{60}\right]$$

$$= [5.6] * [1.117] + [2.5] * [1.05] + [6.0] * [0.2]$$

$$= 10.0802\ \text{min/hr}.$$

The overall total vehicle travel time in Loop IV is 58.7487 min/hr. The throughput capacity of Loop I (59.2929 min/hr), II (65.4047 min/hr) and III (44.3316 min/hr) can be calculated in a similar fashion. The transportation rate for the vehicle in Loop II has exceeded the maximum capacity.

The predetermined load routing must be altered, since the transportation capacity in Loop II has been exceeded. The procedure presented above is repeated until a 'feasible' load routing is found.

10.5.2 Analytical approach: Linear programming model

In the intuitive approach, we try to minimize the number of loops that the across-loop loads have to traverse. The load routing for each across-loop load has to be predetermined. A designer has to evaluate all the possible combinations of feasible routes. The delivery rates among those feasible routes cannot be easily derived. Furthermore, the final outcome is not guaranteed to be the optimal result.

An analytical approach using a linear programming model is proposed here, which is based on the findings of Bozer and Srinivasan (1991a). The solution derived using the LP model will specify the required routings of each job and its associated delivery rate to transport loads between workstations.

This approach consists of six steps. The first four steps explain how to transform a given set of data into a from–to matrix and decision variables of all the feasible routes in the LP model. The fifth step shows the construction of the LRP in the form of a Linear Program. For the sixth step, the LRP is solved using the LINDO software on a PC.

(a) Step 1: Creation of the from–to matrix

This step is identical to steps (1) and (2) in the intuitive approach. The from–to matrix chart will be generated based on the layout given and the required process sequences.

(b) Step 2: Generation of feasible paths for across-loop load

The tandem AGVS layout can be represented as a graph $G=(N, A)$, where N is the set of nodes representing loops and A is the set of arcs representing transit areas between two adjacent loops. The node–arc network is a undirected graph because each transit area can transport loads bidirectionally between two adjacent loops.

All the across-loop loads are summarized in the first column of Table 10.8. The feasible routes for across-loop load deliveries can be derived using the node–arc network of tandem AGVSs. The delivery path can then be translated into feasible routes in terms of stations. The delivery sequence for each job type can be separated into several delivery routes. However, by the principle of conservation, the sum of the separated delivery rates needs to be equal to the required delivery rate in a steady state. For example, the initial loop for across-loop load s2→s14 is I and its destination loop is III. The four possible paths are I→II→III, I→IV→III, I→II→IV→III and I→IV→II→III. The respective decision variables in term of transportation rates for these feasible paths can be denoted as $R_{02,14,1}$, $R_{02,14,2}$, $R_{02,14,3}$ and $R_{02,14,4}$. The summation of these variables is equal to 1.5 load/hr, i.e. $R_{02,14,1}+R_{02,14,2}+R_{02,14,3}+R_{02,14,4}=1.50$. Other delivery paths can be derived in the same way. All the feasible paths for across-loop loads and their respective decision variables and transportation rates for the example model are shown in the second and fourth columns of Table 10.8.

(c) Step 3: Feasible routes for across-loop loads

$R_{(m,n)}$ denotes the delivery rate from station m (entrance station) to station n (exit station) for the *across-loop load*, which is from station i to station j, in the rth routing, i.e. $R_{(m,n)}=R_{i,j,r}$. Both stations m and n have to be in the same loop. Since each loop in the feasible path is connected by a transit area, the exit station of the first loop must be the exit station of the second loop in each adjacent loop. For example, one of the feasible paths for across-loop load s2→s14 is I→II→III. The delivery rate for this feasible across-loop load can be denoted by $R_{(2,5)}$ in the initial Loop I, $R_{(13,9)}$ in the transit Loop II, and $R_{(15,14)}$ in the final destination Loop III. Other across-loop loads can be acquired in a similar fashion, and the final outcome is shown in the third column of Table 10.8.

(d) Step 4: From–to matrix modification

The across-loop loads are appended to the initial from–to matrix for all of the decision variables on the feasible routes in this step. For example, the first feasible route for across-loop load s_2→s14 consists of s2→s5, s5→s13, s13→s9, s9→s15 and s15→s14. The decision variable $R_{02,14,1}$ is appended to the corresponding cells in the from–to matrix. We repeat this procedure

Table 10.8 Feasible routes of across-loop loads

across-loop loads	feasible paths	feasible routes	variable of route	rate (loads/hr)
s2→s14	I→II→III	R(2,5), R(13,9), R(15,14)	R02141	c = 1.5
	I→IV→III	R(2,6), R(26,23), R(19,14)	R02142	
	I→II→IV→III	R(2,5), R(13,11), R(21,23), R(19,14)	R02143	
	I→IV→II→III	R(2,6), R(26,21), R(11,9), R(15,14)	R02144	
s4→s7	I→II	R(4,5), R(13,7)	R04071	b = 3.3
	I→IV→II	R(4,6), R(26,21), R(11,7)	R04072	
	I→IV→III→II	R(4,6), R(26,23), R(19,15), R(9,7)	R04073	
s4→s8	I→II	R(4,5), R(13,8)	R04081	a = 2.5
	I→IV→II	R(4,6), R(26,21), R(11,8)	R04082	
	I→IV→III→II	R(4,6), R(26,23), R(19,15), R(9,8)	R04083	
s4→s16	I→II→III	R(4,5), R(13,9), R(15,16)	R04161	d = 0.8
	I→IV→III	R(4,6), R(26,23), R(19,16)	R04162	
	I→II→IV→III	R(4,5), R(13,11), R(21,23), R(19,16)	R04163	
	I→IV→II→III	R(4,6), R(26,21), R(11,9), R(15,16)	R04164	
s4→s22	I→IV	R(4,6), R(26,22)	R04221	e = 6.0
	I→II→IV	R(4,5), R(13,11), R(21,22)	R04222	
	I→II→III→IV	R(4,5), R(13,9), R(15,19), R(23,22)	R04223	
s7→s16	II→III	R(7,9), R(15,16)	R07161	b = 3.3
	II→IV→III	R(7,11), R(21,23), R(19,16)	R07162	
	II→I→IV→III	R(7,13), R(5,6), R(26,123), R(19,16)	R07163	
s8→s14	II→III	R(8,9), R(15,14)	R08141	a = 2.5
	II→IV→III	R(8,11), R(21,23), R(19,14)	R08142	
	II→I→IV→III	R(8,13), R(5,6), R(26,23), R(19,14)	R08143	
s10→s20	II→IV	R(10,11), R(21,20)	R10201	d = 0.8
	II→I→IV	R(10,13), R(5,6), R(26,20)	R10202	
	II→III→IV	R(10,9), R(15,19), R(23,20)	R10203	
s12→s22	II→IV	R(12,11), R(21,22)	R12221	a + c = 4.0
	II→I→IV	R(12,13), R(5,6), R(26,22)	R12222	
	II→III→IV	R(12,9), R(15,19), R(23,22)	R12223	
s12→s24	II→IV	R(12,11), R(21,24)	R12241	b = 3.3
	II→I→IV	R(12,13), R(5,6), R(26,24)	R12242	
	II→III→IV	R(12,9), R(15,19), R(23,24)	R12243	
s17→s10	III→II	R(17,15), R(9,10)	R17101	d = 0.8
	III→IV→II	R(17,19), R(23,21), R(11,10)	R17102	
	III→IV→I→II	R(17,19), R(23,26), R(6,5), R(13,10)	R17103	
s17→s12	III→II	R(17,15), R(9,12)	R17121	b = 3.3
	III→IV→II	R(17,19), R(23,21), R(11,12)	R17122	
	III→IV→I→II	R(17,19), R(23,26), R(6,5), R(13,12)	R17123	
s18→s12	III→II	R(18,15), R(9,12)	R18121	c = 1.5
	III→IV→II	R(18,19), R(23,21), R(11,12)	R18122	
	III→IV→I→II	R(18,19), R(23,26), R(6,5), R(13,12)	R18123	
s18→s20	III→IV	R(18,19), R(23,20)	R18201	a = 2.5
	III→II→IV	R(18,15), R(9,11), R(21,20)	R18202	
	III→II→I→IV	R(18,15), R(9,13), R(5,6), R(26,20)	R18203	
s20→s12	IV→II	R(20,11), R(11,12)	R20121	a = 2.5
	IV→I→II	R(20,26), R(6,5), R(13,12)	R20122	
	IV→III→II	R(20,23), R(19,15), R(9,12)	R20123	

Table 10.9 From-to chart of all feasible routes

	s1	s2	s3	s4	s5	s6	s7	s8	s9	s10	s11	s12	s13	s14	s15	s16	s17	s18	s19	s20	s21	s22	s23	s24	s25	s26	λj
s1		a,b,c	d	e																							
s2				a,b	$R_{02141}R_{02143}$	$R_{02142}R_{02144}$																					
s3																											
s4					$R_{04071}R_{04081}$ $R_{04161}R_{04163}$ $R_{04222}R_{04223}$	$R_{04072}R_{04073}$ $R_{04082}R_{04083}$ $R_{04162}R_{04164}$ R_{04221}																					
s5						$R_{07163}R_{08143}$ $R_{10202}R_{12222}$ $R_{12242}R_{18203}$																					
s6					$R_{17103}R_{17123}$ $R_{18123}R_{20122}$																						
s7									R_{07161}		R_{07162}		R_{07163}														
s8									R_{08141}		R_{08142}		R_{08143}														
s9							R_{04073}	R_{04083}		R_{17101}	R_{18202}	$R_{17121}R_{18121}$ R_{20123}	R_{18203}														
s10									R_{10203}		R_{10201}		R_{10202}														
s11							R_{04072}	R_{04082}	$R_{02144}R_{04164}$	R_{17102}		$R_{17122}R_{18122}$ R_{20122}															
s12									$R_{12223}R_{12243}$		R_{12221} R_{12241}		R_{12222} R_{12242}														
s13							R_{04071}	R_{04081}	$R_{02141}R_{04161}$ R_{04223}	R_{17103}	$R_{02143}R_{04163}$ R_{04222}	$R_{17123}R_{18123}$ R_{20122}															
s14																		a,b									
s15														$R_{02141}R_{02144}$ R_{08141}		$R_{04161}R_{04164}$ R_{07161}			$R_{04223}R_{10203}$ $R_{12223}R_{12243}$								
s16																	b,c										
s17															$R_{17101}R_{17121}$				$R_{17102}R_{17103}$ $R_{17122}R_{17123}$								
s18															$R_{18121}R_{18202}$ R_{18203}				$R_{18122}R_{18123}$ R_{18201}								
s19														$R_{02142}R_{02143}$ $R_{08148}R_{08143}$	$R_{04073}R_{04083}$ R_{20123}	$R_{04162}R_{04163}$ $R_{07162}R_{07163}$											
s20																					R_{20121}		R_{20123}	d		R_{20122}	
s21																				R_{10201} R_{18202}		R_{04222} R_{12221}	$R_{02143}R_{04163}$ $R_{07162}R_{08142}$	R_{12241}			
s22																								c	a,e		
s23																				R_{10203} R_{18201}	$R_{17102}R_{17122}$ R_{18122}	R_{04223} R_{12273}		R_{12243}		$R_{17103}R_{17123}$ R_{18123}	
s24																											
s25																											
s26																				R_{10202} R_{18203}	$R_{02144}R_{04072}$ $R_{04082}R_{04164}$	R_{04221} R_{12222}	$R_{02142}R_{04073}$ $R_{04083}R_{04162}$ $R_{07163}R_{08143}$	R_{12242}			
λi																											

until all the across-loop loads are taken into account. Table 10.9 shows the modified from–to chart that includes all the in-loop and across-loop loads.

(e) Step 5: LRP model construction

After all the feasible routes have been identified, the linear programming model presented below can be used to determine optimal delivery routes and their associated rates for across-loop load deliveries:

Minimize: $Z =$ total loaded vehicle travel time in the system,

Subject to: total across-loop loads in each job = total delivery rates of all the feasible routes; (10.1)

vehicle capacity of the loop $\geqslant$ transportation requirement of the loop, for all loops in the system. (10.2)

The objective function includes the total sum of vehicle pickup/drop-off time and loaded vehicle travel time, and it can be written as:

$$\sum_i \sum_j [f_{i,j} \times \tau_{i,j} + 2 \times \mu \times f_{i,j}].$$

Equality (10.1) ensures that the delivery rates of all routes for the across-loop load from station i to j are equal to the required delivery rate. Inequality (10.2) represents the throughput capacity constraint of the loop. It ensures that the vehicle inspection rate in each station (W_i) is greater than the total transportation rate at station i.

An analytical model has been proposed by Bozer and Srinivasan (1991a) to estimate the throughput capacity of a single vehicle operating in a closed loop. It can be used as the basic 'building block' for tandem AGV systems, since the overall throughput performance of the system depends upon the performance of the individual loops. The vehicle dispatching rule is based on FEFS. Under this dispatching policy, an empty vehicle will travel around the loop and check each station along its path until a load is found in an output buffer.

The vehicle inspection rate (W_i) depends upon the direction of the loop, the total track length of the loop, the vehicle's speed and the relative position between the initial station and the destination of each load to be transferred. The vehicle inspection rate (W_i) must be greater than the total arrival rate li at any station i, as the total jobs to be transferred within the work period must be completed. The vehicle inspection rate (W_i) and its capacity constraints can be written as follows:

$$W_1 = \frac{1}{\chi}\left[1 - \alpha - 2\mu Q - \sum_{j=2}^{M} (L_j - l_j)\sigma_{j,1}\right] + \lambda_1$$

$$W_i = W_1 + \sum_{j=2}^{i} (L_j - l_{j-1})$$

$$W_i \geqslant l_i$$

where M is the total number of stations in each loop; α is the total loaded vehicle travel time, which can be expressed mathematically as:

$$\alpha = \sum_i \sum_j [f_{i,j} \times \tau_{i,j}] + \sum_i \sum_j \sum_r [R_{i,j,r} \times \tau_{i,j}],$$

Q is the total frequency of the loads to be transferred, and can also be expressed mathematically as:

$$Q = \sum_i \sum_j [f_{i,j}] + \left[\sum_i \sum_j \sum_r R_{i,j,r} \right],$$

and W_1 is the vehicle inspection rate at the output buffer of first station.

The capacity constraints for Loop I are formulated here for illustration purposes; the completed LRP linear programming model for the example can be found in Appendix 10.A:

$$\begin{aligned} Q:\ & R_{02,14,1} - R_{02,14,3} - R_{02,14,2} - R_{02,14,4} - R_{04,07,1} - R_{04,08,1} \\ & - R_{04,16,1} - R_{04,16,3} - R_{04,22,2} - R_{04,22,3} - R_{04,07,2} - R_{04,07,3} \\ & - R_{04,08,2} - R_{04,08,3} - R_{04,16,2} - R_{04,16,4} - R_{04,22,1} - R_{07,16,3} \\ & - R_{08,14,3} - R_{12,24,2} - R_{10,20,2} - R_{12,22,2} - R_{18,20,3} - R_{20,12,2} \\ & - R_{17,10,3} - R_{17,12,3} - R_{18,12,3} - 2a - 2b - c - 2d - e = 0 \end{aligned}$$

$$\begin{aligned} \alpha:\ & 0.963R_{02,14,1} - 0.963R_{02,14,3} - 1.148R_{02,14,2} - 1.148R_{02,14,4} \\ & - 0.296R_{04,07,1} - 0.296R_{04,08,1} - 0.296R_{04,16,1} - 0.296R_{04,16,3} \\ & - 0.296R_{04,22,2} - 0.296R_{04,22,3} - 0.481R_{04,07,2} - 0.481R_{04,07,3} \\ & - 0.481R_{04,08,2} - 0.481R_{04,08,3} - 0.481R_{04,16,2} - 0.481R_{04,16,4} \\ & - 0.481R_{04,22,1} - 0.185R_{07,16,3} - 0.185R_{08,14,3} - 0.185R_{12,24,2} \\ & - 0.185R_{10,20,2} - 0.185R_{12,22,2} - 0.185R_{18,20,3} - 1.592R_{20,12,2} \\ & - 1.592R_{17,10,3} - 1.592R_{17,12,3} - 1.592R_{18,12,3} - 0.333a - 0.333b \\ & - 0.333c - 0.611d - e - 0.667a - 0.667b - 0.389d = 0 \end{aligned}$$

$$l_{01} - a - b - c - d - e = 0$$

$$l_{02} - R_{02,14,1} - R_{02,14,3} - R_{02,14,2} - R_{02,14,4} - a - b = 0$$

$$l_{03} - d = 0$$

$$\begin{aligned} l_{04} & - R_{04,07,1} - R_{04,08,1} - R_{04,16,1} - R_{04,16,3} - R_{04,22,2} - R_{04,22,3} \\ & - R_{04,07,2} - R_{04,07,3} - R_{04,08,2} - R_{04,08,3} - R_{04,16,2} - R_{04,16,4} \\ & - R_{04,22,1} = 0 \end{aligned}$$

$$l_{05} - R_{07,16,3} - R_{08,14,3} - R_{12,24,2} - R_{10,20,2} - R_{12,22,2} - R_{18,20,3} = 0$$

$$l_{06} - R_{20,12,2} - R_{17,10,3} - R_{17,12,3} - R_{18,12,3} = 0$$

$$L_{01} = 0$$

$$L_{02} - a - b - c = 0$$

$$L_{03} - d = 0$$

$$L_{04} - a - b - d - e = 0$$

$$\begin{aligned} L_{05} & - R_{02,14,1} - R_{02,14,3} - R_{04,07,1} - R_{04,08,1} - R_{04,16,1} - R_{04,16,3} \\ & - R_{04,22,2} - R_{04,22,3} - R_{20,12,2} - R_{17,10,3} - R_{17,12,3} - R_{18,12,3} = 0 \end{aligned}$$

$$L_{06} - R_{02,14,2} - R_{02,14,4} - R_{04,07,2} - R_{04,07,3} - R_{04,08,2} - R_{04,08,3} - R_{04,16,2} - R_{04,16,4} - R_{04,22,1} - R_{07,16,3} - R_{08,14,3} - R_{12,24,2} - R_{10,20,2} - R_{12,22,2} - R_{18,20,3} = 0$$

$$1.601W_{01} + \alpha + Q + 1.301L_{02} + 1.051L_{03} + 0.701L_{04} + 0.434L_{05} + 0.267L_{06} - 1.301l_{02} - 1.051l_{03} - 0.701l_{04} - 0.434l_{05} - 0.267l_{06} - 1.601l_{01} = 60$$

$$W_{02} - W_{01} - L_{02} + l_{01} = 0$$

$$W_{03} - W_{01} - L_{02} + l_{01} - L_{03} + l_{02} = 0$$

$$W_{04} - W_{01} - L_{02} + l_{01} - L_{03} + l_{02} - L_{04} + l_{03} = 0$$

$$W_{05} - W_{01} - L_{02} + l_{01} - L_{03} + l_{02} - L_{04} + l_{03} - L_{05} + l_{04} = 0$$

$$W_{06} - W_{01} - L_{02} + l_{01} - L_{03} + l_{02} - L_{04} + l_{03} - L_{05} + l_{04} - L_{06} + l_{05} = 0$$

$$W_{01} - l_{01} \geqslant 0$$

$$W_{02} - l_{02} \geqslant 0$$

$$W_{03} - l_{03} \geqslant 0$$

$$W_{04} - l_{04} \geqslant 0$$

$$W_{05} - l_{05} \geqslant 0$$

$$W_{06} - l_{06} \geqslant 0$$

(f) Step 6: Solution to the LRP

The linear programming model for the LRP is solved using the LINDO software on a PC. The optimal load routing for the across-loop deliveries for the example model is given in Table 10.10. The total vehicle travel time in the system is 203.92 min.

10.6 CONCLUSIONS

The load routing problem (LRP) is a significant issue in designing a tandem AGV system as it balances the transport capacity among loops. Two approaches have been presented here for solving the LRP. The first, an intuitive approach, is based on a predetermined load routing. The delivery rates among those feasible routes cannot be easily derived. Furthermore, the final outcome is not guaranteed to be the optimal result. The analytical approach uses a linear programming model to solve the LRP. The optimal delivery sequence for the across-loop deliveries can be found by solving the LP. However, to avoid bottleneck loops in the system, the direction of each loop must be carefully examined. It is basically an enumeration method to test all the combinations of loop direction, and then finding the lowest value of the objective function.

Both approaches are easy to implement and straightforward to use. The solutions from either approach will specify the required routings of each job and their associated delivery rate to transport loads between stations. It should be noted, however, that the model is based on the assumption of steady state delivery.

Table 10.10 An optimal load routing for the example problem

across-loop loads	feasible paths	feasible routes	variable of route	rate (loads/hr)
s2→s14	I→II→III	R(2,5), R(13,9), R(15,14)	R02141	0
	I→IV→III	R(2,6), R(26,23), R(19,14)	R02142	1.5
	I→II→IV→III	R(2,5), R(13,11), R(21,23), R(19,14)	R02143	0
	I→IV→II→III	R(2,6), R(26,21), R(11,9), R(15,14)	R02144	0
s4→s7	I→II	R(4,5), R(13,7)	R04071	3.3
	I→IV→II	R(4,6), R(26,21), R(11,7)	R04072	0
	I→IV→III→II	R(4,6), R(26,23), R(19,15), R(9,7)	R04073	0
s4→s8	I→II	R(4,5), R(13,8)	R04081	2.5
	I→IV→II	R(4,6), R(26,21), R(11,8)	R04082	0
	I→IV→III→II	R(4,6), R(26,23), R(19,15), R(9,8)	R04083	0
s4→s16	I→II→III	R(4,5), R(13,9), R(15,16)	R04161	0.8
	I→IV→III	R(4,6), R(26,23), R(19,16)	R04162	0
	I→II→IV→III	R(4,5), R(13,11), R(21,23), R(19,16)	R04163	0
	I→IV→II→III	R(4,6), R(26,21), R(11,9), R(15,16)	R04164	0
s4→s22	I→IV	R(4,6), R(26,22)	R04221	6.0
	I→II→IV	R(4,5), R(13,11), R(21,22)	R04222	0
	I→II→III→IV	R(4,5), R(13,9), R(15,19), R(23,22)	R04223	0
s7→s16	II→III	R(7,9), R(15,16)	R07161	3.3
	II→IV→III	R(7,11), R(21,23), R(19,16)	R07162	0
	II→I→IV→III	R(7,13), R(5,6), R(26,23), R(19,16)	R07163	0
s8→s14	II→III	R(8,9), R(15,14)	R08141	2.5
	II→IV→III	R(8,11), R(21,23), R(19,14)	R08142	0
	II→I→IV→III	R(8,13), R(5,6), R(26,23), R(19,14)	R08143	0
s10→s20	II→IV	R(10,11), R(21,20)	R10201	.553
	II→I→IV	R(10,13), R(5,6), R(26,20)	R10202	.247
	II→III→IV	R(10,9), R(15,19), R(23,20)	R10203	0
s12→s22	II→IV	R(12,11), R(21,22)	R12221	4.0
	II→I→IV	R(12,13), R(5,6), R(26,22)	R12222	0
	II→III→IV	R(12,9), R(15,19), R(23,22)	R12223	0
s12→s24	II→IV	R(12,11), R(21,24)	R12241	3.05
	II→I→IV	R(12,13), R(5,6), R(26,24)	R12242	0.25
	II→III→IV	R(12,9), R(15,19), R(23,24)	R12243	0
s17→s10	III→II	R(17,15), R(9,10)	R17101	0.8
	III→IV→II	R(17,19), R(23,21), R(11,10)	R17102	0
	III→IV→I→II	R(17,19), R(23,26), R(6,5), R(13,10)	R17103	0
s17→s12	III→II	R(17,15), R(9,12)	R17121	3.3
	III→IV→II	R(17,19), R(23,21), R(11,12)	R17122	0
	III→IV→I→II	R(17,19), R(23,26), R(6,5), R(13,12)	R17123	0
s18→s12	III→II	R(18,15), R(9,12)	R18121	1.5
	III→IV→II	R(18,19), R(23,21), R(11,12)	R18122	0
	III→IV→I→II	R(18,19), R(23,26), R(6,5), R(13,12)	R18123	0
s18→s20	III→IV	R(18,19), R(23,20)	R18201	2.5
	III→II→IV	R(18,15), R(9,11), R(21,20)	R18202	0
	III→II→I→IV	R(18,15), R(9,13), R(5,6), R(26,20)	R18203	0
s20→s12	IV→II	R(20,11), R(11,12)	R20121	2.5
	IV→I→II	R(20,26), R(6,5), R(13,12)	R20122	0
	IV→III→II	R(20,23), R(19,15), R(9,12)	R20123	0

APPENDIX 10.A LRP LINEAR PROGRAMMING MODEL

Min
1) AA + AB + AC + AD + QA + QB + QC + QD
SUBJECT TO
2) JOBA = 2.5
3) JOBB = 3.3
4) JOBC = 1.5
5) JOBD = 0.8
6) JOBE = 6.0
10) R02141 + R02142 + R02143 + R02144 – JOBC = 0
11) R04071 + R04072 + R04073 – JOBB = 0
12) R04081 + R04082 + R04083 – JOBA = 0
13) R04161 + R04162 + R04163 + R04164 – JOBD = 0
14) R04221 + R04222 + R04223 – JOBE = 0
15) R07161 + R07162 + R07163 – JOBB = 0
16) R08141 + R08142 + R08143 – JOBA = 0
17) R10201 + R10202 + R10203 – JOBD = 0
18) R12221 + R12222 + R12223 – JOBA – JOBC = 0
19) R12241 + R12242 + R12243 – JOBB = 0
20) R17101 + R17102 + R17103 – JOBD = 0
21) R17121 + R17122 + R17123 – JOBB = 0
22) R18121 + R18122 + R18123 – JOBC = 0
23) R18201 + R18202 + R18203 – JOBA = 0
24) R20121 + R20122 + R20123 – JOBA = 0
40) QA – R02141 – R02143 – R02142 – R02144 – R04071 – R04081 – R04161 – R04163 – R04222 – R04223 – R04072 – R04073 – R04082 – R04083 – R04162 – R04164 – R04221 – R07163 – R08143 – R12242 – R10202 – R12222 – R18203 – R20122 – R17103 – R17123 – R18123 – 2 JOBA – 2 JOBB – JOBC – 2 JOBD – JOBE = 0
41) QB – R07161 – R07162 – R07163 – R08141 – R08142 – R08143 – R04073 – R04083 – R17101 – R18202 – R17121 – R18121 – R20123 – R18203 – R10203 – R10201 – R10202 – R04072 – R04082 – R02144 – R04164 – R17102 – R17122 – R18122 – R20121 – R12223 – R12243 – R12221 – R12241 – R12222 – R12242 – R04071 – R04081 – R02141 – R04161 – R04223 – R17103 – R02143 – R04163 – R04222 – R20122 – R17123 – R18123 = 0
42) QC – R02141 – R02144 – R08141 – R04161 – R04164 – R07161 – R04223 – R10203 – R12223 – R12243 – R17101 – R17121 – R17102 – R17103 – R17122 – R17123 – R18121 – R18202 – R18203 – R18122 – R18123 – R18201 – R02142 – R02143 – R08142 – R08143 – R04073 – R04083 – R20123 – R04162 – R04163 – R07162 – R07163 – JOBA – JOBB – JOBC – JOBD = 0

43) QD – R20121 – R20123 – R20122 – R10201 – R18202 – R04222
– R12221 – R02143 – R04163 – R07162 – R08142 – R12241
– R10203 – R18201 – R17102 – R17122 – R18122 – R04223
– R12223 – R12243 – R17103 – R17123 – R18123 – R10202
– R18203 – R02144 – R04072 – R04082 – R04164 – R04221
– R12222 – R02142 – R04073 – R04083 – R04162 – R07163
– R08143 – R12242 – JOBA – JOBC – JOBD – JOBE = 0

50) AA – 0.963 R02141 – 0.963 R02143 – 1.148 R02142 – 1.148 R02144
– 0.296 R04071 – 0.296 R04081 – 0.296 R04161 – 0.296 R04163
– 0.296 R04222 – 0.296 R04223 – 0.481 R04072 – 0.481 R04073
– 0.481 R04082 – 0.481 R04083 – 0.481 R04162 – 0.481 R04164
– 0.481 R04221 – 0.185 R07163 – 0.185 R08143 – 0.185 R12242
– 0.185 R10202 – 0.185 R12222 – 0.185 R18203 – 1.592 R20122
– 1.592 R17103 – 1.592 R17123 – 1.592 R18123 – 0.333 JOBA
– 0.333 JOBB – 0.333 JOBC – 0.611 JOBD – JOBE – 0.667 JOBA
– 0.667 JOBB – 0.389 JOBD = 0

51) AB – 1.111 R07161 – 1.426 R07162 – 2.111 R07163 – 0.500 R08141
– 0.815 R08142 – 1.500 R08143 – 1.222 R04073 – 1.833 R04083
– 0.167 R17101 – 0.315 R18202 – 0.630 R17121 – 0.630 R18121
– 0.630 R20123 – R18203 – 2.166 R10203 – 0.148 R10201
– 0.833 R10202 – 0.907 R04072 – 1.518 R04082 – 2.018 R02144
– 2.018 R04164 – 2.185 R17102 – 0.315 R17122 – 0.315 R18122
– 0.315 R20121 – 1.703 R12223 – 1.703 R12243 – 2.018 R12221
– 2.018 R12241 – 0.370 R12222 – 0.370 R12242 – 0.222 R04071
– 0.833 R04081 – 1.333 R02141 – 1.333 R04161 – 1.333 R04223
– 1.500 R17103 – 1.648 R02143 – 1.648 R04163 – 1.648 R04222
– 1.963 R20122 – 1.963 R17123 – 1.963 R18123 = 0

52) AC – 1.351 R02141 – 1.351 R02144 – 1.351 R08141 – 0.185 R04161
– 0.185 R04164 – 0.185 R07161 – 1.129 R04223 – 1.129 R10203
– 1.129 R12223 – 1.129 R12243 – 1.074 R17101 – 1.074 R17121
– 0.537 R17102 – 0.537 R17103 – 0.537 R17122 – 0.537 R17123
– 0.704 R18121 – 0.704 R18202 – 0.704 R18203 – 0.167 R18122
– 0.167 R18123 – 0.167 R18201 – 0.222 R02142 – 0.222 R02143
– 0.222 R08142 – 0.222 R08143 – 0.537 R04073 – 0.537 R04083
– 0.537 R20123 – 0.722 R04162 – 0.722 R04163 – 0.722 R07162
– 0.722 R07163 – 1.277 JOBA – 1.277 JOBC – 0.407 JOBB
– 0.407 JOBD = 0

53) AD – 0.444 R20121 – 0.722 R20123 – 1.500 R20122 – 1.278 R10201
– 1.278 R18202 – 0.130 R04222 – 0.130 R12221 – 0.278 R02143
– 0.278 R04163 – 0.278 R07162 – 0.278 R08142 – 0.482 R12241
– R10203 – R18201 – 1.444 R17102 – 1.444 R17122 – 1.444 R18122
– 1.574 R04223 – 1.574 R12223 – 0.204 R12243 – 0.778 R17103
– 0.778 R17123 – 0.778 R18123 – 0.222 R10202 – 0.222 R18203
– 0.666 R02144 – 0.666 R04072 – 0.666 R04082 – 0.666 R04164
– 0.796 R04221 – 0.796 R12222 – 0.944 R02142 – 0.944 R04073

– 0.944 R04083 – 0.944 R04162 – 0.944 R07163 – 0.944 R08143
– 1.148 R12242 – 0.704 JOBA – 0.704 JOBE – 0.352 JOBC
– 0.926 JOBD = 0

60) O01 – JOBA – JOBB – JOBC – JOBD – JOBE = 0
61) O02 – R02141 – R02143 – R02142 – R02144 – JOBA – JOBB = 0
62) O03 – JOBD = 0
63) O04 – R04071 – R04081 – R04161 – R04163 – R04222 – R04223
– R04072 – R04073 – R04082 – R04083 – R04162 – R04164
– R04221 = 0
64) O05 – R07163 – R08143 – R12242 – R10202 – R12222 – R18203 = 0
65) O06 – R20122 – R17103 – R17123 – R18123 = 0
70) I01 = 0
71) I02 – JOBA – JOBB – JOBC = 0
72) I03 – JOBD = 0
73) I04 – JOBA – JOBB – JOBD – JOBE = 0
74) I05 – R02141 – R02143 – R04071 – R04081 – R04161 – R04163
– R04222 – R04223 – R20122 – R17103 – R17123 – R18123 = 0
75) I06 – R02142 – R02144 – R04072 – R04073 – R04082 – R04083
– R04162 – R04164 – R04221 – R07163 – R08143 – R12242
– R10202 – R12222 – R18203 = 0
80) O07 – R07161 – R07162 – R07163 = 0
81) O08 – R08141 – R08142 – R08143 = 0
82) O09 – R04073 – R04083 – R17101 – R18202 – R17121 – R18121
– R20123 – R18203 = 0
83) O10 – R10203 – R10201 – R10202 = 0
84) O11 – R04072 – R04082 – R02144 – R04164 – R17102 – R17122
– R18122 – R20121 = 0
85) O12 – R12223 – R12243 – R12221 – R12241 – R12222 – R12242 = 0
86) O13 – R04071 – R04081 – R02141 – R04161 – R04223 – R17103
– R02143 – R04163 – R04222 – R20122 – R17123 – R18123 = 0
90) I07 – R04073 – R04072 – R04071 = 0
91) I08 – R04083 – R04082 – R04081 = 0
92) I09 – R07161 – R08141 – R10203 – R02144 – R04164 – R12223
– R12243 – R02141 – R04161 – R04223 = 0
93) I10 – R17101 – R17102 – R17103 = 0
94) I11 – R07162 – R08142 – R18202 – R10201 – R12221 – R12241
– R02143 – R04163 – R04222 = 0
95) I12 – R17121 – R18121 – R20123 – R17122 – R18122 – R20121
– R20122 – R17123 – R18123 = 0
96) I13 – R07163 – R08143 – R18203 – R10202 – R12222 – R12242 = 0
100) O14 – JOBA – JOBC = 0
101) O15 – R02141 – R02144 – R08141 – R04161 – R04164 – R07161
– R04223 – R10203 – R12223 – R12243 = 0
102) O16 – JOBB – JOBD = 0

103) O17 − R17101 − R17121 − R17102 − R17103 − R17122 − R17123 = 0
104) O18 − R18121 − R18202 − R18203 − R18122 − R18123 − R18201 = 0
105) O19 − R02142 − R02143 − R08142 − R08143 − R04073 − R04083 − R20123 − R04162 − R04163 − R07162 − R07163 = 0
110) I14 − R02141 − R02144 − R08141 − R02142 − R02143 − R08142 − R08143 = 0
111) I15 − R17101 − R17121 − R18121 − R18202 − R18203 − R04073 − R04083 − R20123 = 0
112) I16 − R04161 − R04164 − R07161 − R04162 − R04163 − R07162 − R07163 = 0
113) I17 − JOBB − JOBD = 0
114) I18 − JOBA − JOBC = 0
115) I19 − R04223 − R10203 − R12223 − R12243 − R17102 − R17103 − R17122 − R17123 − R18122 − R18123 − R18201 = 0
120) O20 − R20121 − R20123 − R20122 − JOBD = 0
121) O21 − R10201 − R18202 − R04222 − R12221 − R02143 − R04163 − R07162 − R08142 − R12241 = 0
122) O22 − JOBA − JOBC − JOBE = 0
123) O23 − R10203 − R18201 − R17102 − R17122 − R18122 − R04223 − R12223 − R12243 − R17103 − R17123 − R18123 = 0
124) O24 = 0
125) O25 = 0
126) O26 − R10202 − R18203 − R02144 − R04072 − R04082 − R04164 − R04221 − R12222 − R02142 − R04073 − R04083 − R04162 − R07163 − R08143 − R12242 = 0
130) I20 − R10201 − R18202 − R10203 − R18201 − R10202 − R18203 = 0
131) I21 − R20121 − R17102 − R17122 − R18122 − R02144 − R04072 − R04082 − R04164 = 0
132) I22 − R04222 − R12221 − R04223 − R12223 − R04221 − R12222 = 0
133) I23 − R20123 − R02143 − R04163 − R07162 − R08142 − R02142 − R04073 − R04083 − R04162 − R07163 − R08143 = 0
134) I24 − R12241 − R12243 − R12242 − JOBC − JOBD = 0
135) I25 − JOBA − JOBE = 0
136) I26 − R20122 − R17103 − R17123 − R18123 = 0
150) 1.601 W01 + AA + QA + 1.301 I02 + 1.051 I03 + 0.701 I04 + 0.434 I05 + 0.267 I06 − 1.301 O02 − 1.051 O03 − 0.701 O04 − 0.434 O05 − 0.267 O06 − 1.601 O01 = 60
151) 2.099 W07 + AB + QB + 1.549 I08 + 1.099 I09 + 0.949 I10 + 0.816 I11 + 0.533 I12 + 0.200 I13 − 1.549 O08 − 1.099 O09 − 0.949 O10 − 0.816 O11 − 0.533 O12 − 0.200 O13 − 2.099 O07 = 60
153) 1.500 W14 + AC + QC + 1.217 I15 + 1.050 I16 + 0.683 I17 + 0.350 I18 + 0.200 I19 − 1.217 O15 − 1.050 O16 − 0.683 O17 − 0.350 O18 − 0.200 O19 − 1.500 O14 = 60

154) $1.550\,W20 + AD + QD$
$+ 1.150\,I21 + 1.033\,I22 + 0.900\,I23 + 0.717\,I24 + 0.400\,I25 + 0.200\,I26$
$- 1.150\,O21 - 1.033\,O22 - 0.900\,O23 - 0.717\,O24 - 0.400\,O25$
$- 0.200\,O26 - 1.550\,O20 = 60$
160) $W02 - W01 - I02 + O01 = 0$
161) $W03 - W01 - I02 + O01 - I03 + O02 = 0$
162) $W04 - W01 - I02 + O01 - I03 + O02 - I04 + O03 = 0$
163) $W05 - W01 - I02 + O01 - I03 + O02 - I04 + O03 - I05 + O04 = 0$
164) $W06 - W01 - I02 + O01 - I03 + O02 - I04 + O03 - I05 + O04$
$- I06 + O05 = 0$
170) $W08 - W07 - I08 + O07 = 0$
171) $W09 - W07 - I08 + O07 - I09 + O08 = 0$
172) $W10 - W07 - I08 + O07 - I09 + O08 - I10 + O09 = 0$
173) $W11 - W07 - I08 + O07 - I09 + O08 - I10 + O09 - I11 + O10 = 0$
174) $W12 - W07 - I08 + O07 - I09 + O08 - I10 + O09 - I11 + O10$
$- I12 + O11 = 0$
175) $W13 - W07 - I08 + O07 - I09 + O08 - I10 + O09 - I11 + O10$
$- I12 + O11 - I13 + O12 = 0$
180) $W15 - W14 - I15 + O14 = 0$
181) $W16 - W14 - I15 + O14 - I16 + O15 = 0$
182) $W17 - W14 - I15 + O14 - I16 + O15 - I17 + O16 = 0$
183) $W18 \quad W14 - I15 + O14 - I16 + O15 - I17 + O16 - I18 + O17 = 0$
184) $W19 - W14 - I15 + O14 - I16 + O15 - I17 + O16 - I18 + O17$
$- I19 + O18 = 0$
190) $W21 - W20 - I21 + O20 = 0$
191) $W22 - W20 - I21 + O20 - I22 + O21 = 0$
192) $W23 - W20 - I21 + O20 - I22 + O21 - I23 + O22 = 0$
193) $W24 - W20 - I21 + O20 - I22 + O21 - I23 + O22 - I24 + O23 = 0$
194) $W25 - W20 - I21 + O20 - I22 + O21 - I23 + O22 - I24 + O23$
$- I25 + O24 = 0$
195) $W26 - W20 - I21 + O20 - I22 + O21 - I23 + O22 - I24 + O23$
$- I25 + O24 - I26 + O25 = 0$
210) $W01 - O01 >= 0$
211) $W02 - O02 >= 0$
212) $W03 - O03 >= 0$
213) $W04 - O04 >= 0$
214) $W05 - O05 >= 0$
215) $W06 - O06 >= 0$
220) $W07 - O07 >= 0$
221) $W08 - O08 >= 0$
222) $W09 - O09 >= 0$
223) $W10 - O10 >= 0$
224) $W11 - O11 >= 0$
225) $W12 - O12 >= 0$
226) $W13 - O13 >= 0$

230) W14 – O14 > = 0
231) W15 – O15 > = 0
232) W16 – O16 > = 0
233) W17 – O17 > = 0
234) W18 – O18 > = 0
235) W19 – O19 > = 0
240) W20 – O20 > = 0
241) W21 – O21 > = 0
242) W22 – O22 > = 0
243) W23 – O23 > = 0
244) W24 – O24 > = 0
245) W25 – O25 > = 0
246) W26 – O26 > = 0
END

REFERENCES

Bartholdi, J.J. and Platzman, L.K. (1989) Decentralized control of automated guided vehicles on a simple loop. *IIE Transactions*, **21**(1), 76–81.

Bozer, Y.A. and Srinivasan, M.M. (1991a) Tandem configurations for automated guided vehicle systems and the analysis of single vehicle loops. *IIE Transactions*, **23**(1), 72–82.

Bozer, Y.A. and Srinivasan, M.M. (1991b) Tandem AGV Systems: A Partitioning Algorithm and Performance Comparison with Conventional AGV Systems. *Working Paper*, University of Michigan, Ann Arbor, MI.

Egbelu, P.J. and Tanchoco, J.M.A. (1984) Characterization of Automatic Guided Vehicle Dispathing Rules. *International Journal of Production Research*, **22**(3), 259–74.

Egbelu, P.J. and Tanchoco, J.M.A. (1986) Potential for bidirectional guide path of automated guided vehicle based systems. *International Journal of Production Research*, **24**(5), 1075–97.

Lin, J.T. and Dgen, P.K. (1992) An Algorithm for Routing Control of a Tandem Automated Guided Vehicle System. *Working Paper*, Industrial Engineering Department, National Tsing Hua University, Hsinchu, Taiwan, ROC.

Lin, J.T., Chang, C.C.K. and Liu, W.-C. (1994) A load routeing problem in a tandem configuration AGVS. *International Journal of Production Research*, **32**(2), 411–27.

Maxwell, W.L. and Muckstadt, J.A. (1982) Design of Automatic Guided Vehicle Systems. *IIE Transactions*, **14**(2), 114–24.

Tanchoco, J.M.A. and Sinriech, D. (1992) OSL – Optimal Single-Loop guide paths for AGVS. *International Journal of Production Research*, **30**(3), 665–81.

Tompkin, J.A. and White, J.A. (1984) *Facilities Planning*, Wiley, New York.

Vosniakos, G.C. and Mamalis, A.G. (1990) Automated guided vehicle system design for FMS applications. *International Journal of Machine Tools and Manufacture*, **30**(1), 85–97.

CHAPTER 11

Real-time control strategies for multiple-load AGVs

J.M.A. Tanchoco and C.G. Co

11.1 INTRODUCTION

One of the main issues in any manufacturing facility is the management of material movement between work areas. With the current trend towards fully automated and integrated manufacturing environments, an efficient material handling system becomes even more important. Productivity and flexibility, which are the primary benefits of automation technology, could be significantly enhanced by a carefully designed material handling system.

The variety of product types processed in a job shop results in fluctuating transport requirements between workstations. It is thus imperative that the transport system is capable of responding readily to changes in the load transport patterns in the shop. Automated guided vehicle systems (AGVS) can provide this flexibility. Another advantage of AGVSs is that they can be integrated into a fully automated intelligent control system. These features make AGV systems a viable alternative to space consuming and labor-intensive material handling methods.

This chapter focuses on the operation of AGVSs in a job shop environment, where jobs that enter the shop are transported in batches called **unit loads**. At the output queue of each workstation, parts wait until the desired number of parts is accumulated. The processed parts are then transported in containers to the next workstation. The vehicle type more commonly used in such systems is the unit load transporter. These vehicles are programmed to pick up a unit load and deliver it to its destination. Usually, the vehicles are designed to carry one unit load at a time.

The advantages of decreasing the size of the transfer loads have been postulated in previous studies. Reducing the unit load size can improve machine utilization, because parts belonging to the same job can be processed simultaneously at different workstations. The parts can also be transported to the next wokstation sooner, since a smaller number of parts has to be collected at the output queues to warrant a move request. Thus, flowtime and work-in-progress inventory could be minimized (Tanchoco *et al.*, 1979; Solberg and McGillem, 1987; Egbelu, 1987; Steudel and Moon, 1987).

However, reducing unit load sizes has a negative effect on the level of transportation demand in the system. Smaller transfer loads imply increased number of trips between workstations. To satisfy the trip requirements, one possible solution is to employ a larger vehicle fleet. Adding more vehicles to the system aggravates traffic congestion in the network and increases problems in communication and control of the larger fleet. An alternative to augmenting the fleet size is to dispatch carriers that have the capability to transport multiple loads simultaneously.

Aside from alleviating communication and congestion problems, the use of multiple load carriers can potentially reduce the deadhead or unproductive time of vehicles. The cycle time a vehicle requires to perform a load pickup and delivery task can be broken down into empty travel time to the pickup station, loading time, loaded travel time to the drop-off station and unloading time. With multiple load carriers, it would be possible to pick up additional loads while the vehicle is *en route* to accomplish a previously assigned task, thus reducing the total distance that vehicles travel.

Unlike conveyors, mobile transport equipment such as an AGV may not continuously be available at the right time and place to satisfy move requests immediately. For this reason, control policies that determine the assignment of vehicles to tasks should be carefully formulated to ensure that materials are brought to the proper workstations in a timely manner. A number of studies have been devoted to topics related to dispatching of single load AGVs (Egbelu and Tanchoco 1982, 1984, 1986). Multiple load AGVs add another dimension to this problem, in that dispatching decisions affect the utilization of the extra capacity on the vehicle. Different logic considerations could be involved in dispatching empty and partially loaded vehicles. Partially loaded vehicles could be rerouted to pick up more loads at the expense of delaying the delivery of loads already on the vehicles. With intelligent vehicle management policies, the overall system performance should improve.

See Co and Tanchoco (1990) for a comprehensive review of AGV vehicle management strategies.

11.2 MODEL DESCRIPTION

This section describes the model of the manufacturing environment considered in this chapter. The description includes assumptions on the job shop structure, the vehicle configuration, network layout and the control system framework. The emphasis of this discussion is on the logical aspects of the model. Nevertheless, physical requirements relevant to the problem are discussed.

11.2.1 The job shop environment

The following assumptions are made in this model:

1. The job shop consists of several machine centers, each having a group of identical machines.
2. Jobs enter the system randomly through a receiving center according to a specified interarrival distribution.
3. There is a finite set of job classes processed in the shop. The probability distribution of the classification of arriving jobs is constant.
4. The process route, i.e. the sequence of tasks required for each job and the corresponding machine centers in which the tasks should be performed, is prescribed for each job class. Alternative process routes are not considered.
5. Each job consists of one or more identical parts. Jobs are divided into **unit loads** upon arrival at the shop. Parts belonging to a unit load are always transported and processed as a single entity. Thus, although the model allows backtracking to previously visited machine centers, there is an implicit assumption that process routes are serial because unit load splitting is not possible.
6. The parts in a unit load are processed on the same machine at every machine center. Processing of a unit load on a machine cannot be interrupted, and the machine is not released until all the parts in the unit load have been completed.
7. Handling of unit loads for transportation between machine centers is facilitated through the use of containers, e.g. tote boxes or pallets.
8. All jobs leave the shop through the shipping center. Unit loads belonging to the same job are accumulated at a shipping center until all unit loads in the job have been completely processed.

11.2.2 Vehicles with multiple load carrying capacities

The carrying capacities of the multiple load AGVs can be expressed in terms of the number of load positions on board a vehicle. There are several possible vertical and/or horizontal configurations for the load positions. The suitable design is dependent on the type of loads transported and the mode of loading and unloading the vehicles.

A single level arrangement as in Fig. 11.1 is assumed. Therefore, details on load adjustment and elevation are not included in the model. Furthermore, each section on the carrier deck can accommodate only one load container at any time. The vehicles are equipped with automatic bilateral transfer mechanisms that allow loading and unloading to and from either side of the vehicle. With current technology, AGVs with shuttle tops or roller decks provide this capability.

The load positions are independently operated, i.e. the load positions can be randomly loaded or unloaded. The vehicles have the capability to index and align any of the load positions for a transfer activity. Neither the pickup

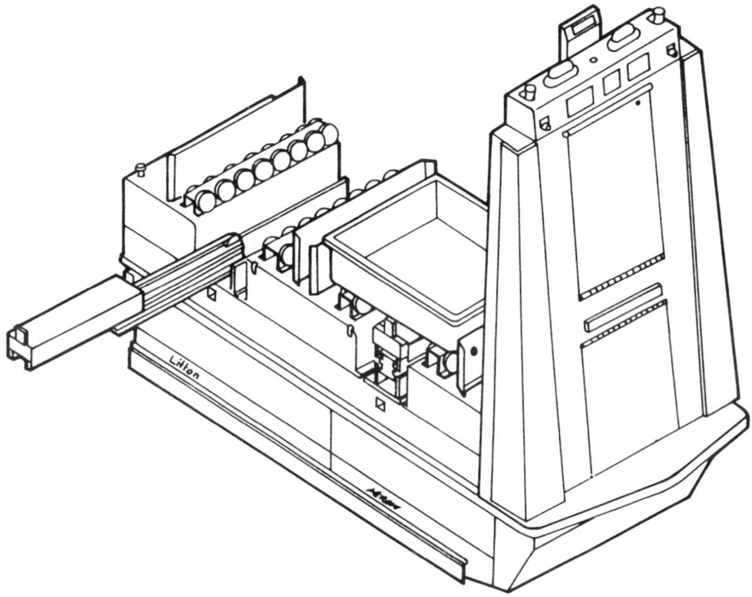

Fig. 11.1 AGV with two-load positions.

nor delivery stations of loads assigned to a vehicle need to be identical. Load sensors and bar code scanners can be installed to identify the load on each section of the vehicle deck.

11.2.3 Network and station layout

A general layout representation scheme using nodes and arcs is adapted in this model (Egbelu and Tanchoco, 1982, 1984; Tanghaboni and Tanchoco, 1988). The arcs indicate the direction of vehicle movement allowed between two nodes. In situations where bidirectional travel between two nodes occur, the assumption is that vehicles traveling in opposite directions will be moving along parallel tracks. Thus, bidirectional segments of the guide path are represented by two arcs oriented appropriately. Figure 11.2 shows a sample guide path network with one-way and two-way aisles.

The nodes represent aisle intersections, battery charging stations and vehicle parking areas. Also represented as nodes are the machine center transfer stations. Transfer stations can function as pickup points, delivery points or both. A machine center has two transfer stations if the load pickup and delivery points are distinct, otherwise only one dual function transfer station is specified for the machine center. All stations, regardless of function, are assumed to have a single transfer dock such that only one load at a time can be transferred between the station and a vehicle.

Load input and output queues at machine centers are independently

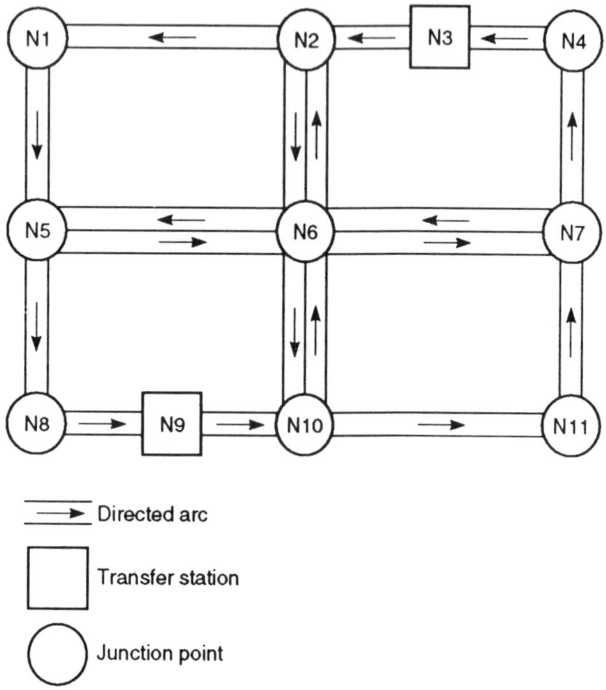

Fig. 11.2 Guide path network representation.

maintained with specified capacities. A vehicle cannot drop off a load if the input queue storage facility for a machine center is full. Likewise, even though operations on a unit load have been completed at a machine, the machine is not released if there is no space in the output queue for the unit load. Loads are accessed from the input and output queues sequentially on a first-come-first-serve (FCFS) basis. The sensor mechanisms at the stations

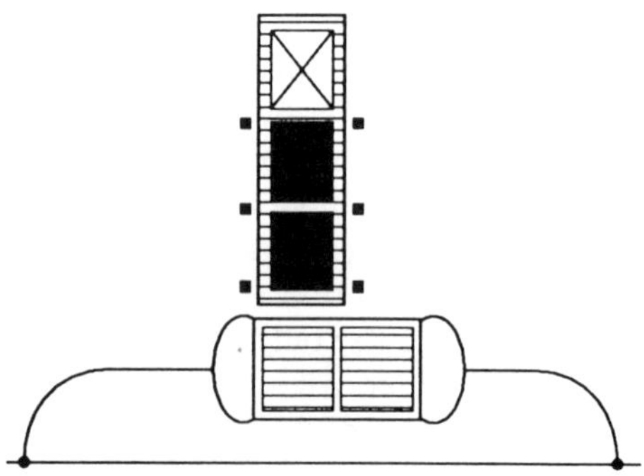

Fig. 11.3 Load/unload station layout.

should be designed to have the following functionalities:

1. Indicate that the output and/or input queue facility is full.
2. Detect and identify the load at the head of the output queue.
3. Activate warning when an outstanding transport request has been unattended beyond a specified time limit.
4. Receive and transmit signals from a properly positioned vehicle to commence automatic load transfer.

Loading and unloading of vehicles will be performed at spurs or along parallel tracks to reduce potential congestion along the main aisles. Vehicles waiting to load or unload at a transfer station behind another vehicle are also parked at the spurs. An illustration of a station layout is provided in Fig. 11.3.

11.2.4 The vehicle controller

(a) Functions of the vehicle controller

The following tasks related to AGV fleet management are performed by a central vehicle controller:

1. receives, prioritizes and tracks outstanding transport orders;
3. assigns transport orders to specific vehicles;
3. builds the itinerary for each AGV based on the list of transport orders assigned to the vehicle;
4. checks the status of machine center input and output queues; and
5. maintains database on status and location of each vehicle.

In dispatching AGVs, the vehicle controller communicates with machine center area controllers and controllers on board the vehicles. Machine centers transmit transport orders with specific origin and destination of 'ready' loads. Transport requests are prioritized on the basis of available capacity at both the origin and destination points. Vehicles with at least one load position that is neither occupied nor assigned to pick up a load can be allocated by the vehicle controller to outstanding transport orders. After assignments are determined, the vehicle controller then revises and sends detailed route itineraries to the affected AGVs. Aside from the node sequence that should be traversed to get to their destinations, the route itineraries also include information on the expected arrival times of vehicles at each node so that AGVs can adjust their speed accordingly.

To streamline memory requirements, only one route itinerary per vehicle is maintained at any time. When a new itinerary is received by a vehicle, the AGV immediately commences the new trip, and any previously relayed route schedules are erased. An AGV is only routed and scheduled to its first transfer destination. After an unloading or loading activity, the vehicle requests a new itinerary to get to its next destination.

Although this scheduling framework is myopic in the sense that tasks are

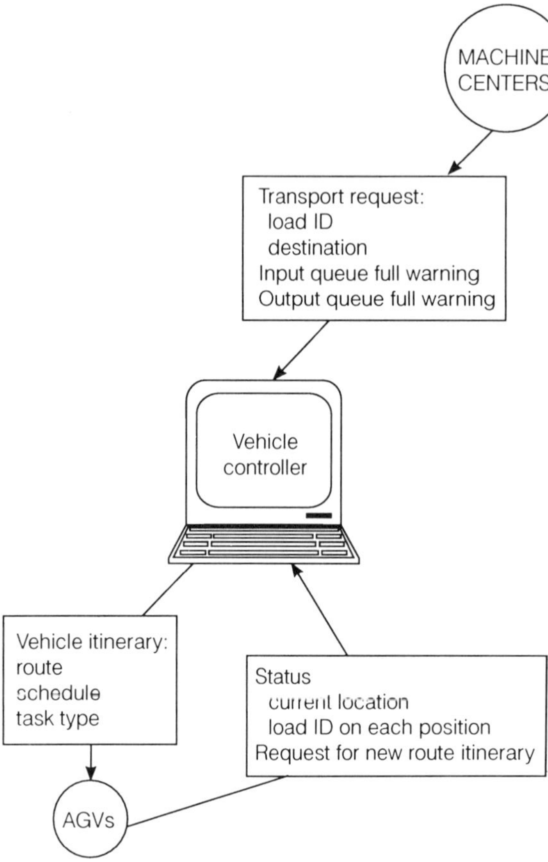

Fig. 11.4 Information flow between vehicle and local controllers.

scheduled individually, it provides advantages in terms of sensitivity to the environment. Committing an AGV to a schedule that covers all its pickup and delivery tasks in one pass is impractical because of frequently unforeseen delays at the transfer stations. Furthermore, implementation of the schedules is facilitated. A vehicle can react spontaneously to a new assignment even without complete knowledge of all its task assignments. The AGV bypasses all intermediary stations in its route indiscriminately because a transfer activity is required only at the last station in its schedule.

Figure 11.4 shows the information flow between the supervisory vehicle controller and the lower-level controllers.

(b) Event driven dispatching

The vehicle controller is invoked by any of these three events:

1. completion of a pickup activity by a vehicle;

2. completion of a delivery activity by a vehicle;
3. transmission of a transport request by a machine center.

In the first two event cases, the vehicle that has just completed a pickup or delivery calls the controller for a new itinerary. This is referred to as *vehicle-initiated-dispatching* by Egbelu and Tanchoco (1984). Before schedul-

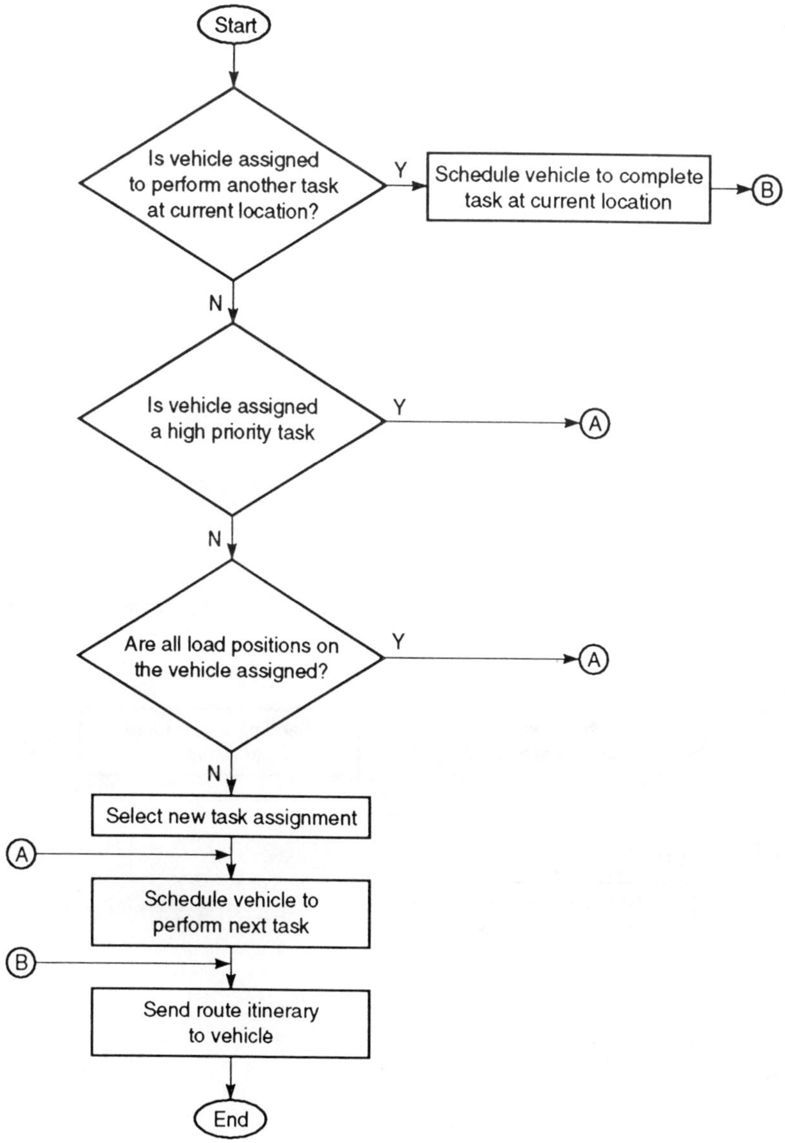

Fig. 11.5 Processing of a vehicle itinerary request after a load pickup.

ing the vehicle, the controller attempts to add another transport request to the vehicle's task list. The status of the vehicle is checked for the following:

1. if all positions on the vehicle are currently assigned to tasks;
2. if the vehicle has been assigned a high priority task; or
3. if the vehicle has another task to perform at its current location.

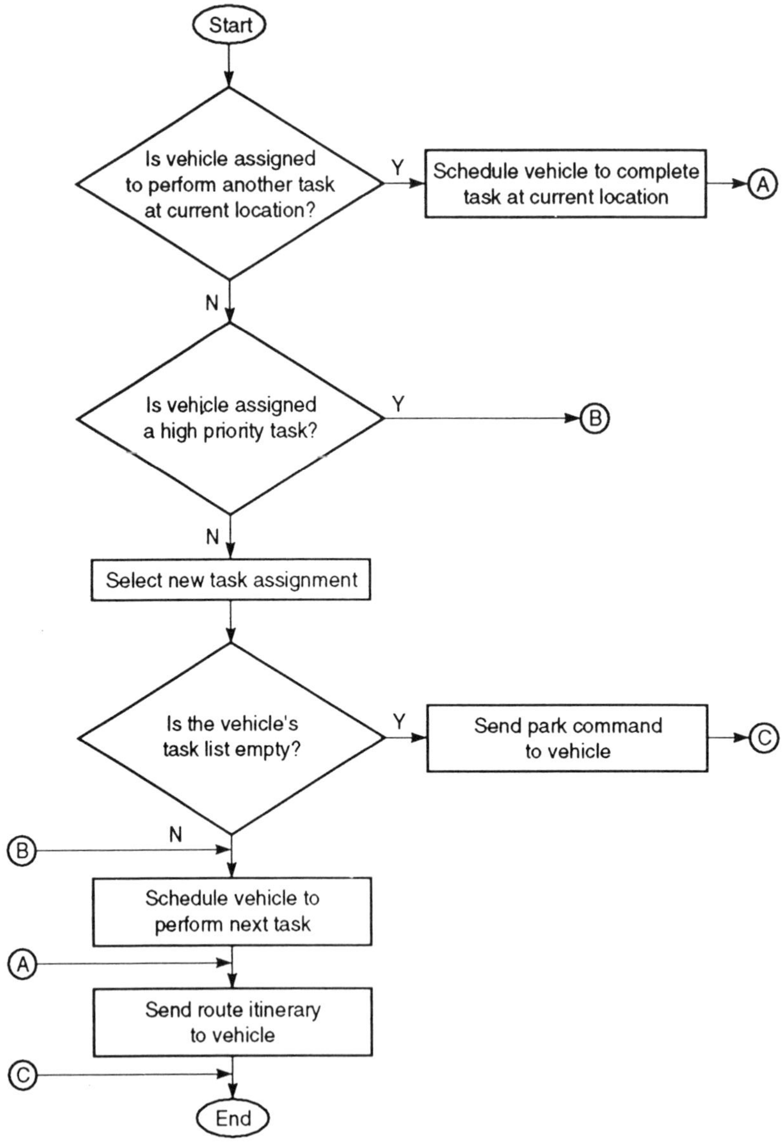

Fig. 11.6 Processing of a vehicle itinerary request after a load delivery.

If all queries are negative, the controller proceeds to assign the vehicle to a new task if possible. After the assignment procedure, the controller selects the next destination of the vehicle from the list of stops associated with the assigned positions on the vehicle. The controller then routes the vehicle to the selected destination. If the vehicle's task list is empty, then the vehicle is parked at its current location. Flowcharts for these procedures are shown in Figs. 11.5 and 11.6.

The third event type that triggers the execution of vehicle controller procedures is initiated at the machine centers. This is referred to as *workstation-initiated-dispatching* by Egbelu and Tanchoco (1984). Because the loads in the output queues are picked up sequentially, only the first load in an output queues activates a transport request to the vehicle manager. Thus, there are two instances when a transport order is sent to the controller. First, a processed load enters an empty output queue and occupies the first position in the queue. Second, a load is picked up from the output queue and the next load in line is moved downstream to the first position. This implies that the end of a loading activity could invoke a new transport request from the machine center. The vehicle controller seeks to assign the transport order to an available vehicle subject to a set of selection criteria. The controller transmits a revised itinerary if the latest assignment changes the immediate destination of the selected vehicle. If no vehicle is assigned, the transport

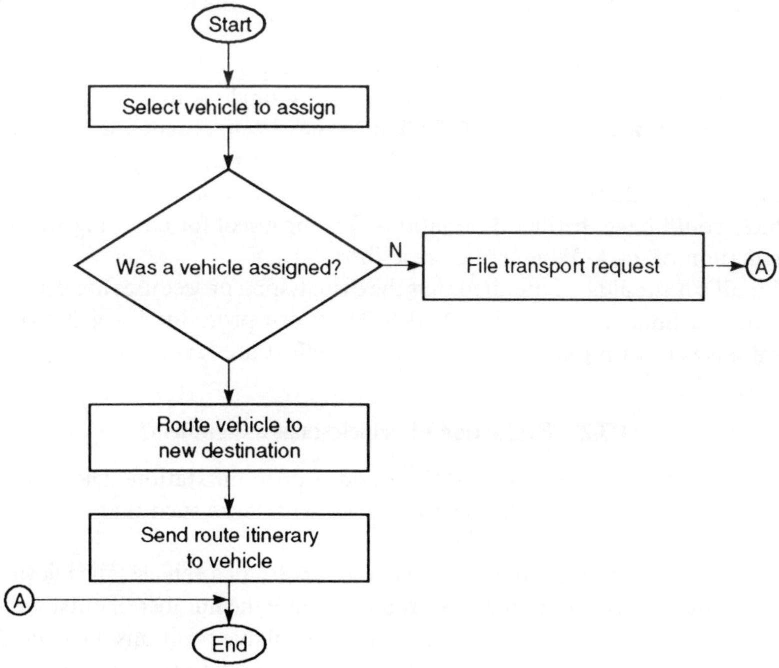

Fig. 11.7 Processing of a transport request from a machine center.

order is kept active until a vehicle becomes available. Figure 11.7 illustrates the flow of control after a transport call is received from a machine center.

11.3 VEHICLE DISPATCHING STRATEGIES

In the previous section, the main dispatching routines were simply abstracted as 'select vehicle' and 'select new task assignment' blocks in the flowcharts. This representation was deliberate because several selection criteria and procedures were considered. This section presents the details of alternative dispatching strategies.

11.3.1 Dispatching considerations for multiple-load AGVs

In AGV system where all vehicles have single load capacities, only currently unassigned AGVs are dispatched to pickup stations. With multiple load AGVs, a vehicle with pending assignments is still considered available for new tasks, provided it has at least one unassigned position. However, different considerations are involved in the dispatching of vehicles with no assignments and vehicles with current assignments.

Assigning new tasks to the latter class of AGVs may cause the preemption of a previously assigned task. The estimated transport delays associated with assigning a specific load pickup and delivery to a vehicle should reflect any such effects.

A second complication brought about by multiple load positions is the selection of the next destination of a vehicle. A single load carrier is always routed to the pickup station of the load it has been assigned to move and directed to the corresponding drop-off station after picking up the load. The routing of a multiple load AGV is not as clear cut because positions on the vehicle could have distinct destinations. The protocol for choosing the next destination of an AGV must be prescribed.

Finally, it should be reiterated that the dispatching procedures are designed for the real-time operation of an AGVS. Therefore, procedures which provide good answers with minimal computational effort are favorable.

11.3.2 Evaluation of vehicle-task assignments

A task is defined by a pickup station and a drop-off station. The primary objective of dispatching is to match vehicles with tasks such that total delay for all load movements is minimized.

Assignments are sequentially added to the task list of a vehicle. The relevance of this statement is manifested in a situation where the number of outstanding transport orders is more than the number of available positions on a vehicle. If there are x positions on the vehicle and y outstanding transport requests, then the optimal assignment list for the vehicle would be obtained by evaluating all possible combinations of x out of y tasks. The best set of x tasks are simultaneously added to the vehicle's task list. In contrast, with a sequential assignment procedure, the best assignment from the unassigned transport requests is selected at each iteration of the procedure. The corresponding transport request is added to the vehicle's task list and deleted from the list of active transport orders. At a subsequent iteration, the remaining transport requests along with any new transport requests received are evaluated. Another assignment for the vehicle is selected subject to the constraints imposed by the decision made in the previous iteration.

Once a task is assigned to an AGV, a position on the vehicle is reserved and will not be made available for other tasks until the corresponding load has been delivered. Likewise, this particular task cannot be reassigned to any other vehicle.

Evaluation of vehicle–task matching alternatives is based on the incremental delay brought about by allocating a vehicle to a task. The delay for a specific load movement refers to the expected length of time from the instant the load initiates a transport call to the time the load is dropped off at the next workstation in its process route. This estimated penalty encompasses both the time the load waits in the first position of the pickup workstation's output queue and the time the load sits on a vehicle until it is delivered. Therefore, the incremental delay is the sum of the transport delay for the load to be moved and the added delay for the other loads that have already been assigned to the vehicle.

The general form of determining the incremental delay associated with a vehicle and unassigned task pair is presented as Procedure P1.

Procedure P1

Estimating Delay of a Vehicle-Task Pair

Given:

vehicle v of capacity greater than $n+m$ with a current task list
$= \{task_1, task_2, \ldots, task_n, task_{n+1}, task_{n+2}, \ldots, task_{n+m}\}$;
unassigned load u to be picked up at station p_u and dropped off at station d_u where:

- $\{tasks\ 1, 2, \ldots, n\}$ are loads that will be picked up at stations $p_1, p_2, \ldots, p_n$ to be dropped off at stations $d_1, d_2, \ldots, d_n$;
- $\{tasks\ n+1, n+2, \ldots, n+m\}$ are loads that are already on the vehicle to be delivered to stations $d_{n+1}, d_{n+2}, \ldots, d_{n+m}$.

Step 1:
Sequence the unvisited transfer stations in the currrent task list of vehicle v:

$$\{p_1, p_2, \ldots, p_n, d_1, d_2, \ldots, d_{n+m}\}.$$

Step 2:
For each task $i, i = 1, 2, \ldots, n+m$:

Compute delay t_i^0 from the current time until the expected drop-off time of the load, where:

t_i^0 is the sum of the following terms:

- if vehicle v is currently busy at a transfer station, the remaining activity time at its current location;
- travel time from vehicle v's current location to the delivery station d_i associated with task i assuming the tour sequence established in step (1);
- loading and/or unloading time at the transfer stations that preceed station d_i in the sequence.

Step 3:
Determine total delay for the current task assignments of vehicle v:

$$y^0 = \sum_{i=1}^{n+m} t_i^0.$$

Step 4:
Sequence the transfer stations in the current task list of vehicle v along with the pickup and drop-off stations of load u:

$$\{p_1, p_2, \ldots, p_n, d_1, d_2, \ldots, d_{n+m}, p_u, d_u\}.$$

Step 5:
For each task $i, i = 1, 2, \ldots, n+m$:
Compute delay t_i' as in step (2) based on the tour sequence established in step (4).

Step 6:
Similarly, compute delay time t_u for the unassigned task u.

Step 7:
Determine total delay if u is included in vehicle v's assignment list.

$$y' = t_u + \sum_{i=1}^{n+m} t_i'.$$

Step 8:
The incremental delay associated with vehicle–task pair (v, u) is:

$$\Delta y = y' - y^0.$$

Note: The current task list of the vehicle could be empty. In which case, $\sum_{i=1}^{n+m} t_i^0$ and $\sum_{i=1}^{n+m} t_i'$ would be equal to 0.

The computed delays are based on expected values. The mean loading and unloading time is used to estimate the dwell time of the vehicle at transfer stations. Furthermore, it is assumed that the vehicle will not be blocked during its journey because the average vehicle speed is used to compute travel time. Nevertheless, the expected delay value provides a reasonable basis for ranking vehicle–task pairs in multiple-load AGV systems. The ride time of a load on a multiple-load AGV is penalized because:

1. the load is not taken along the direct path to its destination; and
2. the load has to sit in the vehicle while other loads are being transferred between the vehicle and their respective source and delivery stations.

The second reason in particular has a significant impact because current AGV technology prescribes a mean loading and unloading time of 30–45 seconds for each load to allow sufficient time for accurate positioning of the AGV and transfer of the load. With an average AGV speed of 150 feet per minute, the transfer time translates to about 75–110 feet of unrestricted run.

These major causes of delay are both due to the basic premise that a task has to be scheduled in relation to the other task assigned to a vehicle. Sequencing the transfer stations in steps (2) and (4) account for this fact.

11.3.3 Selection of the next destination of an AGV

In Procedure P1, which is executed when evaluating a potential assignment of a vehicle, the transfer stations in the AGV's task list are arranged in

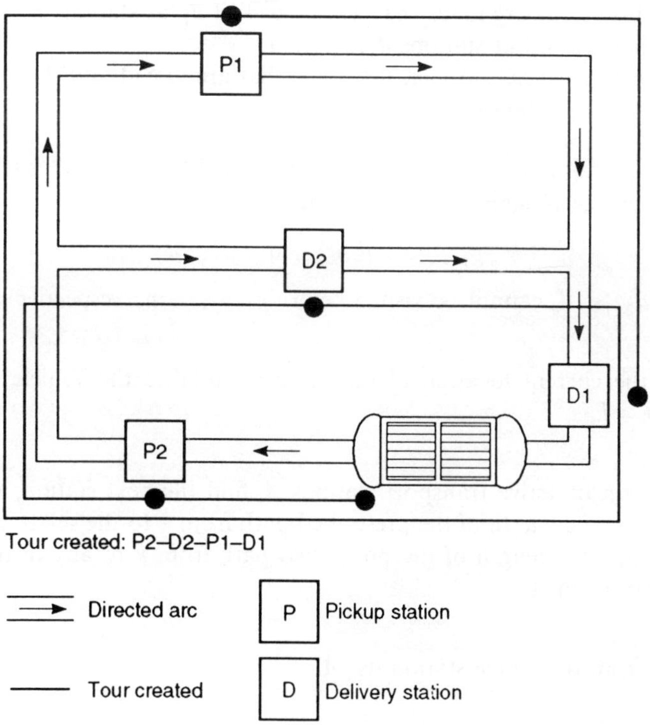

Fig. 11.8 Example of a tour formation using procedure P2.

sequence to form a tour. The sequencing routine simulates the decision process of the controller in selecting the next destination of the vehicle from its current location. In reality, the tour is seldom realized completely because the task list of the vehicle changes frequently, and consequently affects the tour sequence. For this reason, exact procedures for tour formation that minimizes travel time (Bodin *et al.*, 1983) is not recommended. Instead, a heuristic procedure for sequencing the stations is suggested.

Procedure P2 is based on the simple nearest neighbor procedure for the traveling salesman problem formalized by Rosenkrantz *et al.* (1977). However, the constraints on the precedence relationship of pickup and drop-off pairs had to be incorporated in the procedure. Figure 11.8. shows an example of a tour created using this procedure.

Procedure P2

Sequencing the Stations in a Vehicle's Task List

Given:

vehicle v's current task list $= \{task_1, task_2, \ldots, task_n, task_{n+1}, task_{n+2}, \ldots, task_{n+m}\}$

vehicle v's current location, l

where:

- $\{tasks\ 1, 2, \ldots, n\}$ are loads that will be picked up at stations $p_1, p_2, \ldots, p_n$ to be dropped off at stations $d_1, d_2, \ldots, d_n$
- $\{tasks\ n+1, n+2, \ldots, n+m\}$ are loads that are already on the vehicle to be delivered to stations $d_{n+1}, d_{n+2}, \ldots, d_{n+m}$.

Step 1:

Initialize the set of active transport stations A:

$$A = \{p_1, p_2, \ldots, p_n, d_{n+1}, d_{n+2}, \ldots, d_{n+m}\},$$

since $d_1, d_2, \ldots, d_n$ cannot be visited before $p_1, p_2, \ldots, p_n$, respectively.

Step 2:

Start at the current location of the vehicle. Initialize the vehicle location marker, $k = l$.

Step 3:

From the set of active transport stations A, find the next station, x closest to k. That is, the length of the prescribed path from k to the selected station x is less than the length of the prescribed path from k to any of the other active stations in A.

Step 4:

Delete x from the active station list A.

Step 5:

If the selected station x is a pickup station p_i,

then, add d_i to the active station list A.

Step 6:
Change $k = x$ to indicate that the station has been added to the tour.
Step 7:
If set of active stations A is not empty,
  go to step (2).
else,
    end.

Actually, in creating route itineraries it is not necessary for the controller to construct the complete tour for the AGV. The controller only needs to establish the immediate destination of the AGV. From the sequencing procedure P2, the immediate destination of the vehicle is simply the nearest active transfer station in the task list of the vehicle.

Even while a vehicle is *en route* to a specific destination, the controller could send a revised itinerary and reroute the vehicle to another destination. The dispatching policy is flexible enough to allow a vehicle to change its destination. But with the assumption mentioned earlier that vehicle–task assignments are non-transferrable, the implication is that the preempted destination will still have to be visited at a later time.

In relation to this issue, the routing scheme in this model should be clarified. Between a given pair of nodes on the network layout, a predetermined path is maintained by the controller. Although not a prerequisite in the proposed dispatching procedures, the paths prescribed in the implementation of this model are the shortest paths between nodes. Henceforth, the distance between two nodes will always be measured by the length of the prescribed path between the nodes.

The idea of rerouting the vehicle to a new destination is not equivalent to saying that the alternative paths can be generated by the controller. The controller is not intelligent enough to consider current traffic conditions when creating routes.

11.3.4 Feasibility of augmenting a vehicle's task list

The strategy adapted in adding a task to a vehicle's task list is to defer committing a position on the vehicle to a task until doing so becomes critical to routing the vehicle. That is, the task assignment is finalized only if the immediate destination of the vehicle is changed by adding the new task. As implied by the sequencing procedure P2, this would require that from the current location of the AGV, the pickup station of the new task is nearer than the current destination of the vehicle. It would otherwise be unnecessary to reserve a position on the vehicle because the load will not be picked up immediately. This strategy avoids disqualifying a vehicle when a possibly better or more important transport call is received at a later time.

Related to this idea of keeping the capacity on the vehicle free is restricting the allocation of vehicle space to a task that cannot be performed immediately because of conditions either at the pickup or delivery stations of the load. Ideally, the vehicle should not be dispatched to any station if, upon arrival at the pickup station, the transfer spur is occupied by another vehicle. Since the computational effort involved in checking the complete timetable of vehicle itineraries is huge, the feasibility query for tasks is only limited to monitoring if there is available space in the input queues of their delivery stations. Of course, this check does not preclude the possibility that another vehicle drops off loads at the delivery station and fills the input queues while a load is in transit.

An additional constraint on feasible load movements based on the current path of an AGV and the pickup station of the load may or may not be in effect. These two modes of dispatching will be referred to as **fixed-path** and **variable-path**. With a fixed-path mode of dispatching, the pickup station of a new task should lie along the path of a travelling AGV. The system operates under the variable-path mode of dispatching if an AGV is allowed to deviate from its current path to accommodate an additional task.

The fixed-path dispatching is biased towards minimizing the excess transit time of loads previously assigned to a vehicle. A direct path to transfer

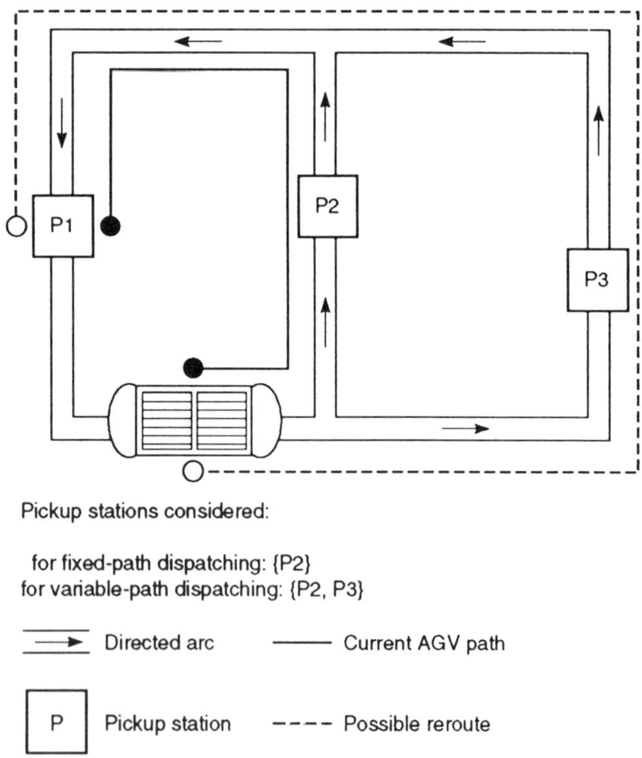

Fig. 11.9 Comparison of fixed path and variable path dispatching.

stations is taken as much as possible. On the other hand, the variable-path mode of dispatching attempts to minimize the waiting time of a load at the output queue because the AGV is scheduled to pick up loads at neighboring stations as soon as possible. Figure 11.9 illustrates the difference between the sets of feasible tasks under the two modes of dispatching.

If fixed-path dispatching is enforced, a vehicle that is traveling on a given path continues to move along the same path even if another station that could be reached sooner via another path is added to its task list. In effect, the controller sequentially transmits route itineraries corresponding to adjacent segments of the given path. The vehicle is only reoriented when it reaches the end of the original path. Thus, in the implementation of a fixed-path dispatching system, the vehicle controller stores the preempted itinerary of the AGV in memory so that the same path could be restored when the AGV resumes its trip.

Procedure P3 is a modified version of the sequencing procedure P2 for fixed-path dispatching systems. The layout and nodes in Fig. 11.10 is the

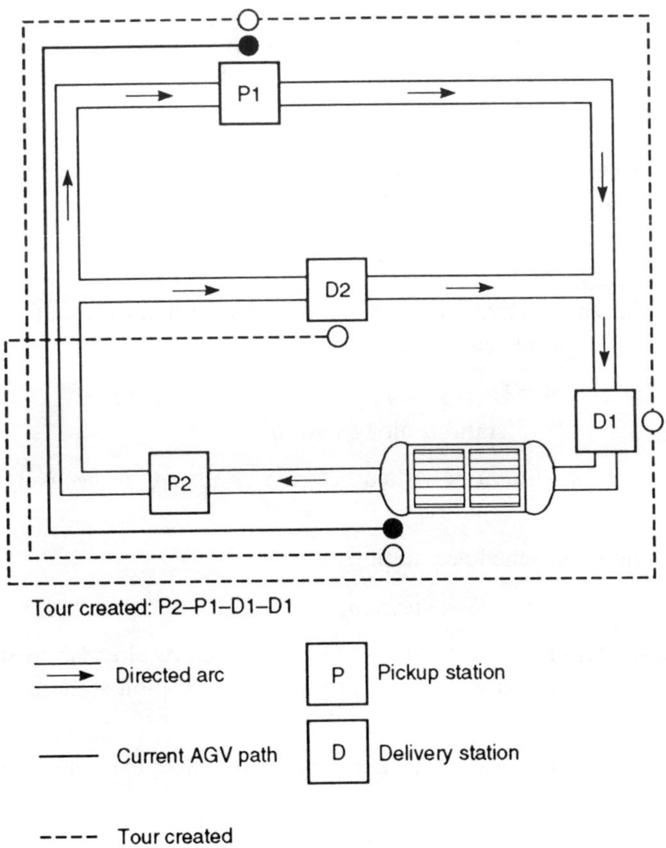

Fig. 11.10 Example of a tour formation using procedure P3.

same example shown in Fig. 11.8. However, using Procedure P3 with the additional information that the AGV is traveling towards $l' = p_1$, a different tour is created.

Procedure P3

Sequencing the Stations in a Vehicle's Task List under a Fixed-Path dispatching mode

Given:

vehicle v's current task list $= \{task_1, task_2, \ldots, task_n, task_{n+1}, task_{n+2}, \ldots, task_{n+m}\}$
vehicle v's current location, l
vehicle v's scheduled task, $task_q$, $1 \leqslant q \leqslant n+m$
where:

- $\{tasks\ 1, 2, \ldots, n\}$ are loads that will be picked up at stations $p_1, p_2, \ldots, p_n$ to be dropped off at stations $d_1, d_2, \ldots, d_n$
- $\{tasks\ n+1, n+2, \ldots, n+m\}$ are loads that are already on the vehicle to be delivered to stations $d_{n+1}, d_{n+2}, \ldots, d_{n+m}$
- $task_q$ is a load that was or will be picked up at station p_q to be dropped off at station d_q scheduled in the current route itinerary of vehicle v.

Step 1:
Set the current vehicle destination l' to:
p_q if the vehicle is on the way to pickup load q
or d_q if vehicle is on the way to drop-off load q.

Step 2:
Initialize the set of active transport stations A such that only stations along the current path $\overline{ll'}$ of the vehicle v are considered:

$$A = \{p_1, p_2, \ldots, p_n, d_{n+1}, d_{n+2}, \ldots, d_{n+m}, l'\} \cup \{\text{stations along path } \overline{ll'}\}$$

since $d_1, d_2, \ldots, d_n$ cannot be visited before $p_1, p_2, \ldots, p_n$, respectively.

Step 3:
Initialize the set of scheduled stations S:

$$S = \{p_{n+1}, p_{n+2}, \ldots, p_{n+m}\}$$

since, these stations have been visited and therefore already satisfy their precedence relationship with their corresponding drop-off stations.

Step 4:
Start at the current location of the vehicle. Initialize the vehicle location marker, $k = l$.

Step 5:
From the set of active transport stations A, find the next station, x closest to k. That is, the length of the prescribed path from k to the selected station

x is less than the length of the prescribed path from k to any of the other active stations in A.

Step 6:
Delete x from the active station list A. Add x to the set of scheduled stations S.

Step 7:
If $x = l'$,
 for $i = 1, 2, \ldots, n$:
  If p_i is not in the set of scheduled stations S,
   then add p_i to the set of active stations A.
  Else if p_i is already in the set of scheduled stations S,
   then if d_i is not in the set of scheduled stations S,
    then add d_i to the set of active stations A.
 Go to step (9).
 Note: For the condition $x = l'$ to be true, the set of active stations $A = \{\ \}$ prior to this step.

Step 8:
If the selected station x is a pickup station p_i,
 if l' is already in the set of scheduled stations S,
  add d_i to the set of active stations A.
 else if l' is not in the set of scheduled stations S,
  add d_i to the active station list A only if d_i lies along path $\overline{ll'}$.

Step 9:
Change $k = x$ to indicate that the station has been added to the tour.

Step 10:
If set of active stations A is not empty,
 go to Step (5).
else,
 end.

In summary, a transport request to move a load from pickup station p_i to drop-off station d_i can only be assigned to a vehicle currently at l traveling to l' if:

C1: The input queue at the delivery station d_i is not full.
C2: The distance between l and p_i is less than the distance between l and l'.
C3: In a fixed-path dispatching environment, p_i is along the path from l to l'.

11.3.5 Task prioritization

Since tasks are evaluated based on the location of their pickup and delivery stations, it is possible that a task becomes highly disadvantaged relative to other tasks because of the network configuration. Under the circumstance that only the first load in the output queue activates a transport request, the effects of keeping a load pickup on hold is particularly detrimental to the performance of the shop. Until the first load is picked up, the loads waiting in queue do not even exist in the transport order file of the vehicle controller. Thus, the pickup delay is propagated across all loads waiting in the output queue of the workstation.

The disadvantage of a task may be observed in two forms. First, it is possible that a task is never assigned to a vehicle at all. A second possibility is that a task is assigned to a vehicle, but because of continual additions to the task list, the accomplishment of the task is postponed indefinitely.

To safeguard the system against such possibilities, the controller classifies certain tasks as 'high-priority' tasks which should be expedited. Tasks in this list are considered in the order that they are reclassified as high-priority. The sequence in which the transport orders are placed do not have any bearing on the priority rankings.

If an unassigned task is deemed high-priority, it is immediately assigned if a vehicle is available and constraints C1, C2 and C3 are satisfied. If a task that enters the high-priority list has already been assigned, the controller informs the vehicle and reroutes the vehicle to accomplish the task at the first opportunity to do so. The nearest node policy is overridden in these occasions.

The input parameters recognized in this model to weigh the urgency of specific tasks are:

1. *max_wait_time*, the maximum length of time that a transport order is held unassigned;
2. *max_detour*, the maximum number of times that a vehicle is detoured from a direct path to a task destination; and
3. $out\_queue\_capacity_i$, the maximum number of loads in the output queue of machine station *i*.

Consequently, tasks are classified as 'high-priority' at the occurrence of the following:

1. The time since the task request was recorded by the controller exceeds *max_wait_time*.
2. $Detour\_counter_{jv}$ equals *max_detour*. $Detour\_counter_{jv}$ is incremented by one every time vehicle v is routed along a path that does not lead directly to the *task_destination* of position j on vehicle v. $task\_destination_{jv}$ is set equal to the pickup point upon assignment of a vehicle to a task, and the number of detours is accumulated from the current location of the vehicle. When the load is placed on board the vehicle, the task destination is changed to the delivery destination and $detour\_counter_{jv}$ is reset to zero.
3. The number of loads in the output queue at machine station *i* where the task is waiting is equal to $out\_queue\_capability_i$. The vehicle controller attempts to assign a vehicle to pick up a load from a full output queue when a load is placed in the last available space in the output buffer.

11.3.6 Selection of an additional task

A vehicle requests a new route itinerary after it performs all tasks required at a transfer station. At this point, the controller determines if the vehicle could accommodate an additional task. The controller first checks if there are any tasks in the 'high-priority' list. If so, the requesting vehicle is dispatched to the oldest task in the 'high-priority' list.

On the contrary, if there are no tasks that need to be expedited, the controller selects a task from the outstanding transport requests and assigns this task to a vehicle if possible. This section explains how the selection of a new task is implemented.

A special case occurs when the last container aboard a vehicle has just been unloaded and no assignments have yet been made to the vehicle. This particular situation is treated differently because (1) there are no other tasks that will be affected by the dispatching decision, and (2) in this case, there is no difference whether fixed-path or variable-path dispatching is in effect. It would therefore be practical to keep the selection procedure simple. Section 11.3.6(a) is a list of heuristic rules that could be applied in such events. Section 11.3.6(b) discusses the selection procedure for a new task that could be added to a currently non-empty AGV task list.

(a) Task selection for an unassigned vehicle

The following heuristic rules are similar to vehicle-initiated task assignment rules for single-load AGVs. This list is by no means exhaustive, but is representative of the possible criteria that could be used.

Rule A1. *Nearest_Pickup_First*
The controller assigns the load to pickup the load waiting at the station that is nearest the current location of the vehicle, thus ensuring that the empty space on the vehicle is utilized as soon as possible. This rule is equivalent to the *nearest_workstation_first* rule.

Rule A2. *Nearest_Drop-off_First*
This rule is a variation of Rule A1. Under this rule, the controller selects the task that can be delivered soonest, assuming that the vehicle is not assigned any other task. Thus, the priority of the task is evaluated by the sum of the distance traveled from the AGV's current location to the pickup point of the load, and the distance traveled from the pickup station to the destination of the load.

The rationale behind Rule A2 is analogous to the principle behind the *shortest_imminent_processing_time* rule in machine scheduling. Since allocated space on a vehicle is released from a task assignment only after the load has been delivered, the cycle time for a task ends when the AGV drops off the load.

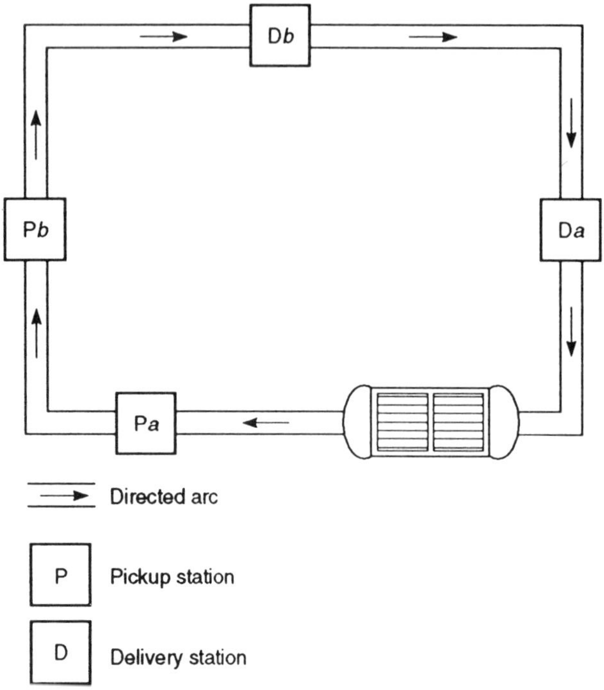

Fig. 11.11 Comparison of Nearest_Pickup_First and Nearest_Drop-off_First.

Rule A2 works well in situations where another load can be picked up close to the delivery station of the first load. However, in a situation such as the one illustrated in Fig. 11.11, Rule A1 would intuitively work better.

If no other transport calls are received and only one AGV is available, under Rule A1 the AGV will be dispatched to move load *a* first, then move *b* second. Under Rule A2 the order of dispatching is reversed, and the AGV travels a longer distance to complete both tasks. This result can again be explained in terms of machine scheduling. The vehicle's sojourn time between the delivery of one load and the pickup of the next can be interpreted like the setup time for a machine. Whereas the norm in job shop scheduling is to assume that setup time is not dependent upon the sequence in which tasks are accomplished, it is evident that with AGV dispatching the sequence in which tasks are performed affects the deadhead time between tasks.

Rule A3. *Busiest_Segment_First*
In selecting the first task for an unassigned vehicle, the effect is not just allocating one position for one task but also restricting the possible usage of the other positions on the vehicle. This rule is based on the idea that a vehicle should be dispatched to an area in the network where the extra capacity on the multiple-load vehicle will be most useful.

There are several possible indicators of how busy a segment is. The network segment of interest is the path travelled by the AGV from its current location to the drop-off point of the candidate task. Ideally, the selection would be based on the actual number of loads that are expected to be ready when the AGV reaches their respective pickup points. Again, because of the computational requirements, this method was not implemented in this model. Instead, an estimate of the average transport demand rates at pickup stations is used. One major weakness of this measure is that the present condition of the system does not in any way influence the static values.

The average transport demand rate μ_i between the current AGV location l and the drop-off point of the load d_i is computed as follows:

$$\mu_i = \frac{\sum_{p \in P'} v_p}{d_{\overline{ld_i}}},$$

where $v_p = \sum_{j \in J} \rho_j \lambda$; $J = \{\text{job types processed at station } p\}$; $\rho_j =$ probability that a job in the shop is of type j; $\lambda =$ mean arrival rate of jobs into the shop; $P' = \{\text{pickup stations along path } \overline{p_i d_i}\}$; $p_i =$ pickup station of load i; and $d_{\overline{ld_i}} =$ distance between l and d_i.

Conceptually, Rule A3 would be effective if clusters of stations are isolated in different areas of the network. An extreme case is shown in Fig. 11.12. Assumng that the average transport demand rates at the pickup stations are

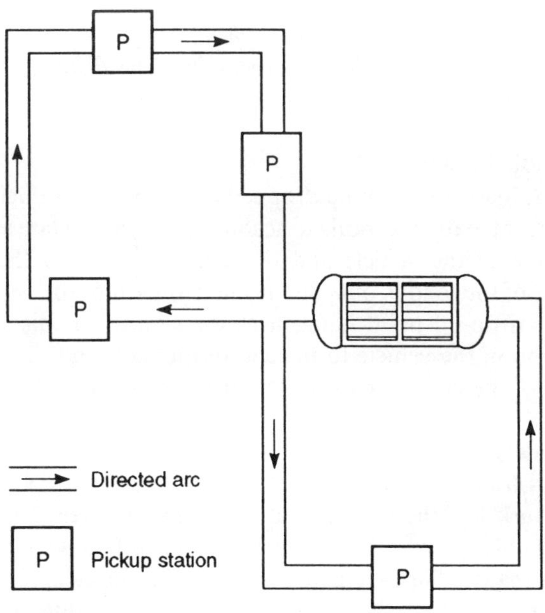

Fig. 11.12 Layout example of Busiest_Segment_First.

equal, it will be more advantageous if the AGV was dispatched to the stations in the left-hand area than to the one station in the right region.

(b) Task selection for a vehicle with pending assignments

After the vehicle loads or unloads at a transfer station, the next destination of the vehicle is tentatively set following pocedure P2 or P3 for a variable-path or a fixed-path dispatching system, respectively. The vehicle controller determines if another task can be inserted before the vehicle travels to this interim destination. The decision process is different under the two dispatching modes.

Fixed-path dispatching

Transport requests should satisfy conditions C1, C2 and C3 of section 11.3.6(a) to be considered for inclusion in the vehicle's task list. It is still possible that more than one transport order satisfies these constraints. Unlike in random dispatching, all alternatives lie along the same path. The alternatives do not differ much in terms of task sequencing.

There are two perspectives on effective vehicle dispatching that could be adopted. The first is to minimize the time a position on the vehicle is empty. The second approach is to maximize the number of loads that are moved while the vehicle travels from its current location to the end of its current path. Rules B1 and B2 are offshoots of the first and second alternative objectives, respectively.

Rule B1. *Nearest_Pickup_Within_the_Path*

This rule is equivalent to Rule A1, except that the distance to the pickup stations are traced along the current path of the vehicle.

Rule B. *Nearest_Dropoff*

As in Rule A2, loads that can be dropped off before the vehicle reaches the end of its current path are ranked according to the distance between the current location of the vehicle and the drop-off point of the load. If the drop-off point of the load is outside the current path of the vehicle, then the distance to its drop-off point is taken as the sum of the distance from the current location of the vehicle to the end of the path and the distance from the end node of the current path to the drop-off point of the load.

Random dispatching

All unassigned transport requests are evaluated. Given the vehicle's current location and task list, the incremental delay that will result from assigning each of the feasible tasks to the vehicle is computed following Procedure P1. The task which causes the least incremental delay is selected. If the constraints C1 or C2 are not violated, the minimum delay task is added to the vehicle's task list.

The controller chooses to keep the minimum delay transport request active and the available position on the vehicle free if any constraint is violated by the minimum delay task which is the most favorable assignment for the vehicle at the moment. Unless another vehicle becomes available sooner for this assignment or a better task is received, the option to assign the task to the requesting vehicle is kept open.

11.3.7 Vehicle selection for a new transport request

When a transport call is received from a workstation, the vehicle controller evaluates whether a space on a vehicle should be allocated immediately to the task. Following constraint C1, if the delivery queue specified in the transport call is full, the controller simply files the request and does not take any action to assign the task.

If C1 is not true, then all vehicles with at least one unassigned position become candidates for assignment to the new transport order. This would include busy AGVs performing a transfer activity, assigned AGVs moving to a destination and idle AGVs parked somewhere in the network.

Using Procedure P2, the incremental delay that would result from including the new transport order into each of the candidate vehicle's current task list is computed. The vehicle-task pair that corresponds to the minimum delay is determined. If constraints C2 and C3 are not violated with the assignment of the transport order to the selected vehicle, then a position on the vehicle will be reserved for the task. On the other hand, if any of the constraints are not satisfied, the request is filed and left unassigned on the pretext that when the selected vehicle becomes available, it will opt to perform the task. This event would not always be true, because more favorable transport requests may exist in the active list when the vehicle requests for a new assignment.

11.3.8 Summary of the capabilities and limitations

As a concession to the incessantly changing conditions of the environment, two features were incorporated to enhance the responsiveness of the AGV system:

1. Although vehicle–task assignments are permanent, the control system has the flexibility to overwrite an AGV's active route itinerary and reroute the vehicle to another destination.
2. The controller defers commiting vehicles to tasks until the decision is critical in the selection of the vehicle's destination.

A mechanism for expediting tasks that have been indefinitely postponed is included in the control system. This feature, along with the policy of monitoring delivery queues, are preventive measures against the situation described in Egbelu and Tanchoco (1984), where all movements in the shop are locked. This event occurs when the input and output queues of machine centers are full, and all vehicles are unavailable.

Two limitations in the dispatching control system proposed in this research should be noted. The first stems from the fact that tasks are sequentially assigned. The sequential assignment policy was adopted because real-time operation requires fast procedures. But sequential assignment procedures do not ensure optimal assignments.

The selection is always a one-to-many matching in dispatching vehicles to tasks. For instance, a vehicle requests an itinerary and a task is selected from the outstanding transport orders, or a transport order is received and a vehicle is selected from among available vehicles. A better solution could possibly be obtained if the set of all available vehicles is matched against the set of all outstanding transport orders whenever the dispatching process is invoked.

When the vehicle requests for an itinerary, the best task that could be assigned to it is added to its task list. There is no assurance that this vehicle is the best vehicle to assign to the task. A strategy that was initially considered was not to assign a task to a vehicle unless the vehicle is the best candidate for the task. Based on preliminary experiments, this approach was not effective because a vehicle could always be the second best vehicle for all tasks and end up not being utilized at all.

In the case of matching a new transport request to candidate vehicles, the task need not be immediately assigned even if a vehicle is available. Thus, the procedure provides reasonable opportunity for the task to be assigned to the best vehicle possible. However, once the assignment of a transport request has been deferred, the request is not reconsidered until it is reclassified as an urgent task, or a vehicle requests a new assignment. As with vehicles, a similar argument is true for tasks that may remain unassigned because it is always the second best alternative for all vehicles.

The second weakness of the controller is that it is not intelligent enough to release a reserved position temporarily to perform another task that could be completed before the preempted load occupies a position on the AGV. In the same token, the controller only allocates tasks to positions on the vehicles that are immediately available. A reserved position is not considered, even if it is currently being unloaded or can be unloaded on the way to picking up a ready load.

11.4 IMPLEMENTATION AND RESULTS

Thus far, the differences between single-load and multiple-load AGVs have been examined from a purely qualitative point of view. This section documents the simulation experiments that were conducted to compare single-load and multiple-load AGVs. Although the dispatching procedures discussed earlier are not bounded by vehicle capacity, only one-position and two-position AGVs were considered in these experiments.

11.4.1 Software description

In this research, a simulation package, AGVSim2 (Gaskins and Tanchoco, 1989) was used to demonstrate the performance of the proposed vehicle management system. AGVSim2 is a vehicle system simulator written in C that was specially designed for evaluating AGV control systems. Subroutines that emulate the functions of the vehicle controller were developed and linked to AGVSim2.

The simulator models the job shop environment in which the AGV system operates, and generates the events that call the vehicle controller routines. All the procedures and alternative heuristic rules discussed in the previous sections were implemented. Prioritization parameters and heuristic rules can be specified by the user. The user also has the option of selecting a fixed- or variable-path mode of dispatching.

The third element of the software is the scheduler, which maintains a timetable of non-conflicting vehicle arrivals at nodes to create vehicle route itineraries. A LISP-based controller for single-load AGVs was developed by Taghaboni and Tanchoco (1988). So as not to duplicate the previous work, vehicle scheduling routines in the LISP-based controller were extracted and modified to meet the specifications of multi-load AGV problem. Primarily, the revisions are due to the fact that ongoing vehicle trips can be interrupted and replaced with new assignments.

The event simulators and the vehicle controller, both written in C, are executed as a single process. This process runs simultaneously with the LISP scheduler. Communication between the two processes is accomplished using files during the initialization stage, and whenever a new route itinerary is required for a vehicle (Gaskins and Tanchoco, 1989).

To initialize the scheduler, information on the layout of the guide path network and vehicle parameters are required from the vehicle controller. To request route itineraries for a specific vehicle, a file containing the vehicle identification code, the destination of the vehicle, the current location of the vehicle and the expected length of time the vehicle will be required at the destination is sent to the scheduler. The scheduler returns an itinerary for the vehicle from its current location to its destination.

11.4.2 Comment on the AGV path definition scheme

As mentioned earlier, the AGVs travel between any two nodes using a prescribed path. The prescribed paths in this implementation correspond to the shortest distance path between two nodes. In practice, it may be necessary to design the paths such that traffic is distributed across the network.

There are several published algorithms to find the shortest paths between two nodes in a network. In this work, the algorithm developed by Floyd (1962) was selected. Floyd's algorithm solves a multiterminal shortest path problem with *n* nodes in *n* iterations.

The intermediate nodes that form the shortest path between nodes are tracked using a matrix procedure explained by Hu (1969). A route matrix is updated to store the first intermediate node in the shortest path between two nodes. Hu's procedure is computationally efficient in that final paths are concatenated from shorter paths.

11.4.3 Design of the simulation experiments

Preliminary runs confirmed the results reported in Tanchoco *et al.* (1987) and Ozden (1988) on the relationship of the throughput in an AGVS-based system and the number of vehicles operating in the system. Given the same number of jobs that enter the system, the throughput within a fixed interval of time behaves in a concave manner with respect to fleet size as illustrated in Fig. 11.13. As fleet size is increased, the throughput level approaches an upper bound due to the finite shop capacity. Furthermore, not all jobs that enter the system are finished at the end of the simulation because of unavoidable delays such as actual processing time and travel time. Consequently, because vehicles cease to be the constraining resource, additional vehicles do not improve the throughput above the maximum level.

The design of the simulation experiments in this research is based on this observation. The objective is to determine both the maximum throughput level of a given system, and the minimum fleet size required to achieve this throughput level.

Two alternative vehicle systems are evaluated: a system where single-position AGVs are used; and a system where two-position AGVs are used. The two vehicle systems are evaluated based on the system throughput.

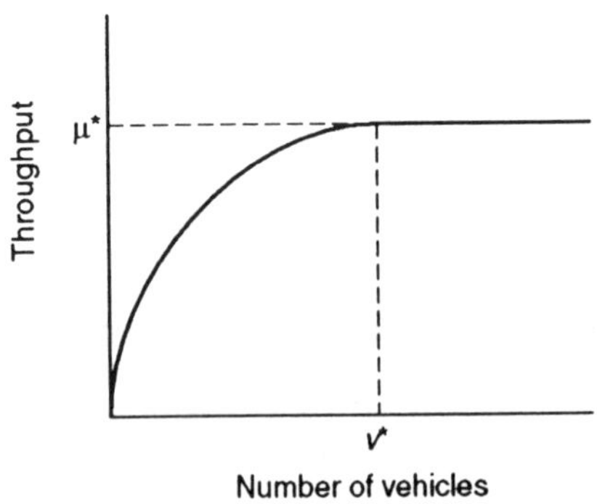

Fig. 11.13 General behavior of throughput as a function of fleet size. μ^* = maximum throughput; v^* = threshold fleet size.

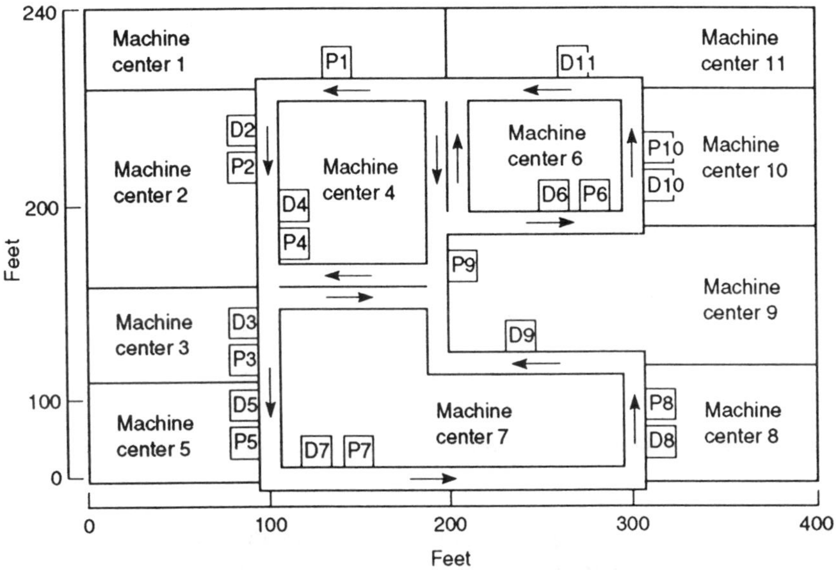

Fig. 11.14 Example job shop layout.

Throughput is defined as the number of jobs completed by the end of the simulation run. The alternative vehicle systems are compared under two cases. For both cases, the job arrival rates are simulated with all other shop parameters held constant. For each case, several replication runs (a minimum of three) were made until a 90% confidence interval is achieved.

The job shop layout used is shown in Fig. 11.14.

11.4.4 Discussion of results

The results of the simulation are shown graphically in Fig. 11.15. For each level of job arrival rate, the ratio of the number of jobs completed to the number of jobs that entered the system are plotted against the number of vehicles.

The advantage of two-load AGVs over single-load AGVs is observed in both cases. With the high rate of transport demand in case 3, there is a high probability that two loads will be ready at a pickup station when a vehicle arrives. In this case, the control system specification, that only loads at the head of the output queues, activates a call for a vehicle which presents an advantage for multiple load vehicles. If a second load moves up to the head of a queue after a vehicle picks up one load, a new transport call is transmitted to the controller. With two-position AGVs, the controller immediately assigns the same vehicle to pick up the second load if a space is available on the vehicle. With single-position AGVs, a second vehicle has to be dispatched from another location, and therefore time is lost due to deadhead travel.

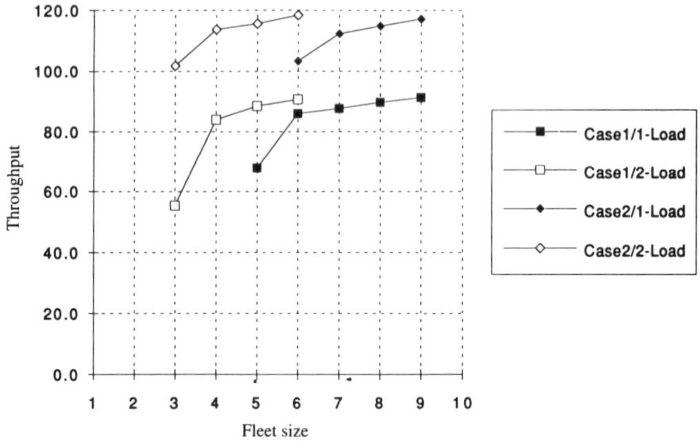

Fig. 11.15 Throughput comparison of single-load vs. two-load vehicle systems.

Furthermore, with the high transport demand rate, the chances of assigning both positions of a vehicle to tasks along the same path increases.

One last comparison on the two types of AGV systems simulated could be inferred from the observed relationship of the vehicle fleet size and system throughput. An alternative perspective can be taken with regards to the fact that at the threshold fleet number the marginal contribution of an additional two-load AGV is greater than that of a single-load AGV at its corresponding threshold fleet number. If both systems were operated with the threshold number of vehicles, the potential drop in throughput in case of a vehicle breakdown would be less if single-load AGVs are used.

REFERENCES

Bodin, L.D., Golden, B.L., Assad, A.A. and M.O. Ball (1983) Routing and scheduling of vehicles and crews. *Computers and Operations Research*, **10**, 63–211.

Co. C.G. and Tanchoco, J.M.A. (1991) A review of research on AGVS vehicle management. *Engineering Costs and Production Economics*, **21**(1), 35–42.

Egbelu, P.J. and Tanchoco, J.M.A. (1982) Operational Considerations for the Design of Automatic Guided Vehicle Based Material Handling Systems. *Technical Report No. 8201*, Virginia Polytechnic Institute & State University, Blacksburg, VA.

Egbelu, P.J. and Tanchoco, J.M.A. (1984) Characterization of automatic guided vehicle dispatching rules. *International Journal of Production Research*, **22**(3), 359–74.

Egbelu, P.J. and Tanchoco, J.M.A. (1986) Potentials for Bidirectional Guidepath for Automated Guided Vehicle Based Systems. *International Journal of Production Research*, **25**(5), 1075–99.

Egbelu, P.J. (1987) *Integration of Unit Load Selection in the Design of AGV Based Automated Manufacturing Shop.* Proceedings IXth International Conference on Production Research, Cincinnati, OH, pp. 2159–66.

Floyd, R.W. (1902) Algorithm 97 – Shortest path. *Communications of the ACM*, **5**, 345.

Gaskins, R.J. and Tanchoco J.M.A. (1989) AGVSim2 – A development tool for AGVS controller design. *International Journal of Production Research*, **27**(6), 915–26.

Hodgson, T.J., King, R.E. and Monteith, S.K. (1987) Developing control rules for an AGVS using Markov decision processes. *Material Flow*, **4**(1–2), 85–96.

Hu, T.C. (1969) *Integer Programming and Network Flows*, Addison-Wesley, Reading, MA.

Ozden, M. (1988) A simulation study of multiple-load-carrying automated guided vehicles in a flexible manufactuing system. *International Journal of Production Research*, **26**(8), 1353–66.

Rosenkrantz, D., Sterns, R. and Lewis, P. (1977) An analysis of several heuristics for the traveling salesman problem. *SIAM Journal on Computers*, **6**, 563–81.

Solberg, J.J. and McGillem, C.D. (1987) An intelligent factory transport system, in *Automated Guided Vehicle Systems*, ed. R.H. Hollier, pp. 333–44, IFS Publications/ Springer-Verlag.

Steudel, H.J. and Moon, H.K. (1987) *Selection of Unit Load Size for Automated guided Vehicles in Large Cellular Manufacturing Environments.* Proceedings IXth International Conference on Production Research, Cincinnati, OH, 2189–209.

Taghaboni, F. and Tanchoco. J.M.A. (1988) A LISP-based controller for free-ranging automated guided vehicle systems. *International Journal of Production Research*, **26**(2), 173–88.

Tanchoco, J.M.A., Wysk. R.A., Davis, R.P. and Agee, M.H. (1979) *Economic Unit Loads and Its Impact on Material Flow Systems Planning.* Fall IE Conference Proceedings, pp. 141–7.

Tanchoco, J.M.A., Egbelu, P.J. and Taghaboni, F. (1987) Determination of the total number of vehicles in an AGV-based system. *Material Flow*, **4**(1–2), 33–51.

PART FIVE

Tooling Requirements and Transport Equipment

CHAPTER 12

Tool automation in computerized manufacturing systems

L.C. Leung and S.K. Khator

12.1 INTRODUCTION

The key factor in the success of computerized manufacturing systems (CMS) is their ability to route workpieces among workstations. This integration of workstations is accomplished via a computer-controlled material handling system for the workpieces. As manufacturing systems are becoming more refined, and as material handling systems become increasingly more sophisticated, it is not uncommon to have automated routing of tools to workstations as well (Gaalman *et al.*, 1987). In essence, CMS could conceivably consist of two material handling systems: one for workpieces and another for tools. Such simultaneous routings of workpieces and tools could be a major element in the design of future CMSs. Crite, Mills and Talavage (1985) define automatic tool handling as:

> The use of a tool transport system, coupled with tool exchanges, robotics or otherwise, at the processing units, so as to allow the completely unmanned movement of tools around a system, including the tool warehouse and servicing area.

Aver'yanov *et al.* (1986) have stressed the need for tool automation:

> To increase productivity when machining in FMSs and to enable the system to operate in unmanned technology mode, it is necessary to develop automated Tool Management Systems. This should comprise: an automatic tool changer, a device for automatic tool replacement, devices for automated transportation of tools... (AGV), an automated central tool store, devices for tool monitoring..., a (tool) control system.

A system design of coordinated tool transfer aims at improving tool sharing, tooling flexibility and tool-cutting quality. With tool automation becoming a reality, problems in tool management need to be explicitly identified. There

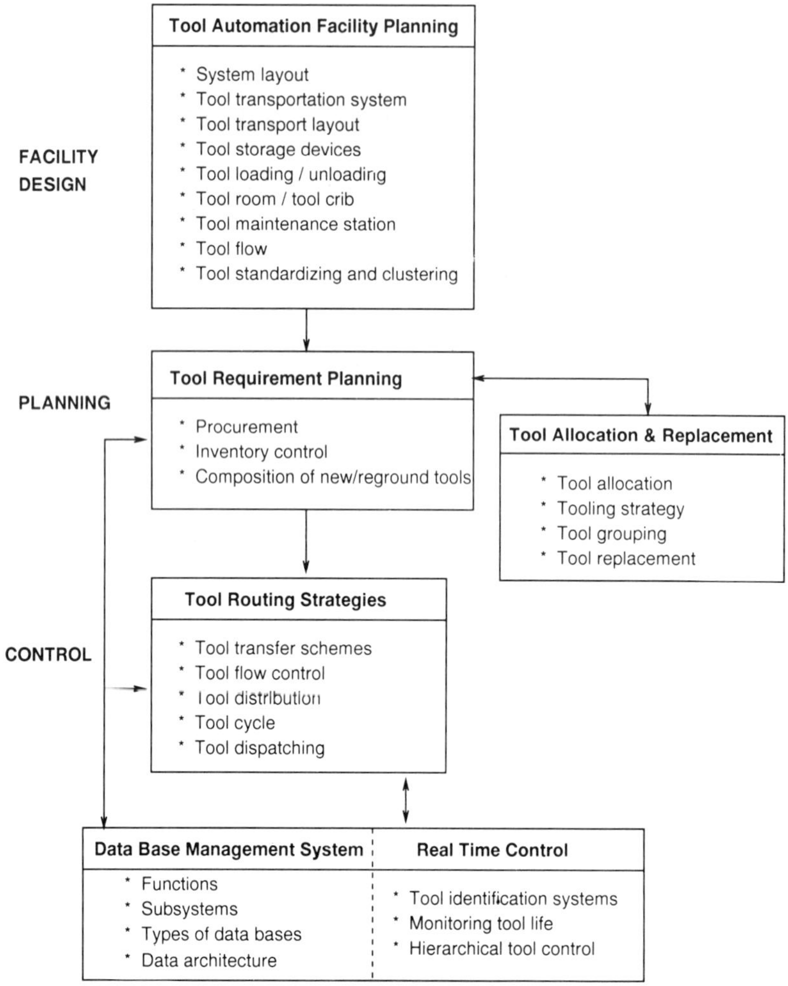

Fig. 12.1 Design, planning and control issues in tool automation.

are two basic categories of tool management problem: tool planning and tool control. The planning problems include:

- design of tool-related facilities;
- tool requirement planning; and
- tool allocation and replacement.

The control problems include:

- tool-routing strategies;
- database support; and
- real-time tool monitoring.

The interplay of these problem areas is depicted in Fig. 12.1.

12.2 TOOL AUTOMATION FACILITY PLANNING

While planning for a facility capable of handling tools in automated manufacturing systems, one needs to consider:

- layout of the machine tools in relation to the cutting tools' storage and other tool-related facilities, such as the tool grinding and tool inspection stations;
- decisions regarding the tool transporter to be used, and the layout of the tool movement path;
- movement of tools with respect to part movement
- tool standardization as well as standardization of tool holders.

Each of these issues are discussed in detail in the following sections.

12.2.1 System layout

While designing the layout of a manufacturing facility, relative locations of various departments are decided largely on the basis of the flow of parts between manufacturing and assembly departments, as well as some other departments such as receiving, storage, warehousing and shipping. Relationships, other than the flow-based ones, are kept in mind while locating some of the departments, e.g. frequency of breakdowns while locating the maintenance department. The layout of the individual departments (indicating machine and isle locations) can be created by following a similar process. The presence of a tool movement system, however, adds another dimension to the machine layout problem.

12.2.2 Components of a tool handling system

To understand the problem of tool automation facility planning, a brief description of the components of such a system is given in this section.

(a) Tool transportation system

An effective tool delivery system will be critical to the success of a modern automated manufacturing system. Either an existing workpiece transfer system or a separate dedicated system can be used for transporting tools. Some of the factors affecting this decision are (ElMaraghy, 1985):

- load on the workpiece transfer system;
- frequency of tool and workpiece transportation;
- relative size of the workpieces and tools;
- loading and unloading methods; and
- cost of a tool transportation system.

Various methods of automated tool delivery system are briefly discussed here (ElMaraghy, 1985; Kiran and Krason, 1988).

Gantry type crane

A gantry type crane has been used to transport a complete tool drum between the tool crib and the machine tools. The major drawbacks associated with this system are: capability of machine tools to be able to exchange tool drums; large number of tool drums and their storage requirements, both in the tool crib and at the machine; as well as the requirement of a gantry crane. Such a system was designed by Yamazaki (Kurimoto *et al.*, 1988).

Automated guided vehicle system

Yamazaki, in their modified design, used an AGVS to transport tools. This system has the same problems as the previous version; however, the system has higher flexibility in transporting tools. The system works as follows. A portable tool magazine is first mounted on a movable pallet. Then an AGV transports the pallet from the tool crib to the machine. Here, an interchange arm is used to exchange tools between the machine's permanent tool magazine to the portable unit on the AGV. The entire process reduces the need for a large number of permanent tool magazines. Furthermore, the exchange of tools can be done without machine stoppage.

Overhead transport carrier

An overhead transport carrier capable of delivering several tools was used by a Czechoslovakian manufacturer (Tomek and Zeleny, 1984). Machine tools are equipped with a robot arm to exchange tools between its magazine and the overhead tool transporter.

Rail-guided Cart

Tools are stored on a rack behind the machining center in a German FMS application (Fritz Werner). A rail-guided cart was used for transferring the tools to the tool rack.

Mobile robot

To automate the replacement of tools in the magazine, it is possible to use a special mobile robot which can serve several machines (Aver'yanov *et al.*, 1986).

(b) Tool transport layout

Of the above-mentioned tool transportation methods, the AGV system has been reported most frequently in the literature. Two alternative track layouts have been studied by ElMaraghy (1985). Figure 12.2 shows a straight line configuration, while Fig. 12.3 incorporates a circular track layout for tool transportation. The straight line system is simpler to operate, since no interference takes place between the part and the tool movement. However, if the existing layout of the machines is not in a straight line, a circular tool transportation system will be needed. The design of the control systems for

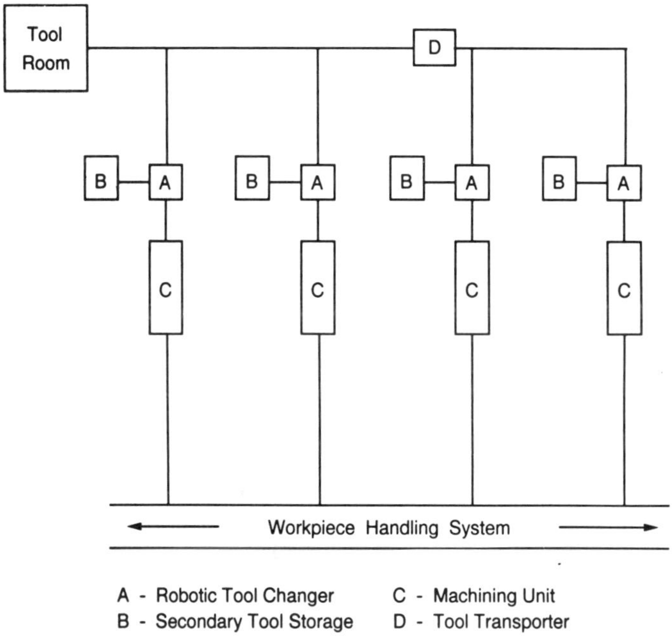

Fig. 12.2 Linear track for tool transfer: A = robotic tool changer; B = secondary tool storage; C = machining unit; D = tool transporter (reproduced with permission of SME.)

both workpiece and tool handling becomes more involved in this case due to the interaction between the two systems.

(c) Tool storage devices

ElMaraghy (1985) reported that an automated tool transporter and changing system with a secondary buffer storage reduces the need for larger tool magazines. The following are some of the commonly used tool storage devices.

- *Tool magazines.* The size of tool magazines has increased considerably, storing as many as 120 tools in a magazine. This capacity has been duplicated by building machine tools capable of supporting two magazines.
- *Racks or carousel.* Use of racks or carousel for storing large number of tools.
- *Compact tool storage unit.* Mazak designed a compact tool storage unit (called a 'tool-hive') capable of holding several hundred tools. The objective is to eliminate the need for a machine tool magazine. A robot arm transfers a tool from the tool-hive to the tool changer.
- *Block tooling system.* Sandvik designed a block tooling system where hundreds of preset block tools can be stored and automatically retrieved and loaded on the machine's turret.

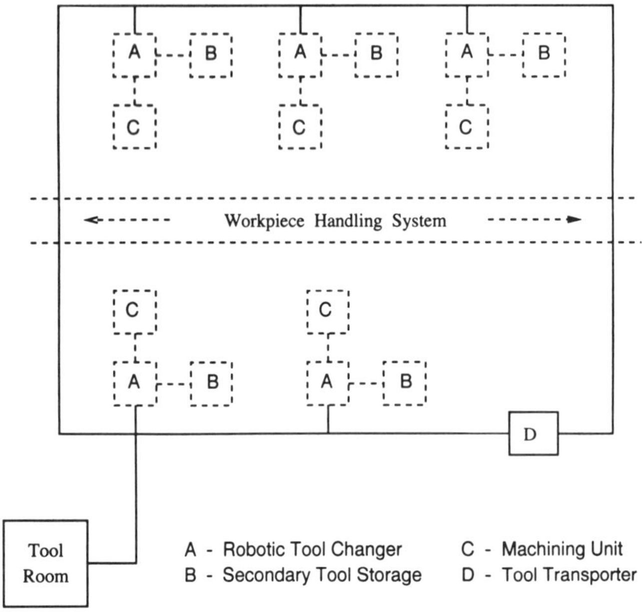

Fig. 12.3 Outside loop for automatic tool transfer: A = robotic tool changer; B = secondary tool storage; C = machining unit; D = tool transporter (reproduced with permission of SME).

(d) Tool loading/unloading mechanisms

Tools can be interchanged from a secondary system without interrupting the cutting process. However, some systems allow such an interchange only when the machine spindle is stationary, resulting in idle machine time.

(e) Tool Crib

A large number of cutting tools are needed in today's computer integrated manufacturing systems, which necessitates the need for an effective tool identification system. Knowledge of the different tools in a machine's magazine slot is also essential. Tool kitting, assembly and inspection, in addition to tool storage, are the functions performed in a tool crib.

(f) Tool maintenance station

A tool needs to be reground or replaced when it can no longer achieve the required tolerance and surface finish. When a tool becomes dull due to excessive wear, cutting forces start to increase, eventually leading to breakage (and failure) of the tool. Loktev (1983) stressed the need for a centralized tool regrinding facility as opposed to assigning it to a machine tool operator.

With centralized regrinding it is possible to introduce advanced and more efficient tool grinding methods. In the case of tool inserts, the mounting of new inserts can also be done at this station.

12.2.3 Tool flow in an automated manufacturing system

Figure 12.4 shows the flow of tools in an FMS (Rhodes, 1986). Incoming tools are first stored as components. Components are then assembled and preset. The prepared assemblies may go into the kitting process, or be stored as assemblies. Either a set of tools for specific operations or complete tool magazines are kitted for delivery to the machine tool, where it is used for an operation. After completing an operation, a tool may remain in the magazine or it may be sent to another machine if tool sharing is permitted. After completion of a batch, the tools are returned to the tool assembly area.

12.2.4 Tool standardization and clustering

Tool standardization will reduce the variety of tools to be stored, and will have several other benefits, such as smaller tool magazine size, reduced tool handling

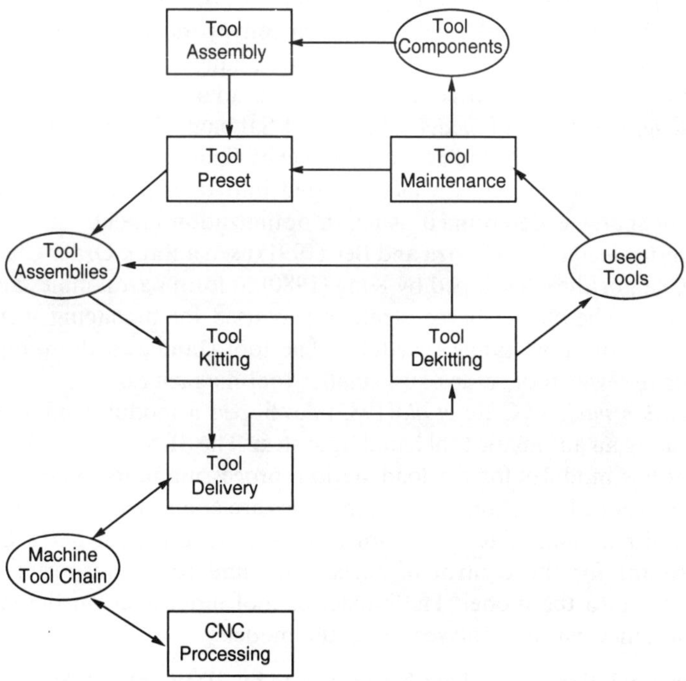

Fig. 12.4 Tool flow in a CMS (reproduced with permission of SME).

and lower tool inventories. Martin (1989) described the benefits obtained by standardizing tools at a plastic machinery plant; the company even redesigned parts to conform to the standard tooling. A cellular manufacturing system was used at this facility. Carter (1985) described a flexible tooling system which can reduce downtime associated with tool changing. DeSouza and Bell (1991) presented a tool cluster-based strategy for the management of cutting tools.

12.2.5 Models used

There has not been any major research reported in the literature dealing directly with the facility planning/layout associated with tool automation. Manual layout planning methods such as Systematic Layout Planning (SLP) and Travel Charting, or computer-aided layout planning packages such as CRAFT, CORELAP, ALDEP, PLANET, COFAD and the recently developed BLOCPLAN, can be used to determine the relative locations of tool-related facilities such as the tool crib, tool maintenance station, etc. Additionally, Petri nets, queueing theory, clustering methods and simulation techniques have been used by several authors to study tooling-related issues. Some of these studies are briefly reported here:

- *Petri net-based model.* Bruno and Biglia (1985) used a Petri net-based model in a system where conveyors handle tools. The modeling objective was to find the optimal balance between conveyors and machines. A real life problem with several machines cannot be handled using this technique, since the network size becomes too large to analyse.
- *Queueing theory-based model.* Vinod and Sabbagh (1986) modeled a tool sharing system as a closed network of queues. Spare tools and tool replacement upon tool failure are incorporated into their model. The optimal level of spares is determined using an optimization model.
- *Clustering method.* DeSouza and Bell (1991) used a Rank Order Clustering (ROC) algorithm developed by King (1980) to form part families and tool grouping. The cluster-based strategy was used for managing work and tool flow for short range schedules. The tool standardization approach results in fewer tool setups and smaller tool magazines.
- *Simulation models.* Crite *et al.* (1985) developed a modular simulator for evaluating an automatic tool handling system. The SLAM-based simulation model has modules for the load station, processing units, unload station and the central warehouse. These modules are connected by two separate material handling systems, one for tools and the other for parts. Control algorithms for the control of parts, tools and tool transporter (AGV) are built into the model. The impact of tool movement on the system's performance can be analysed using the model.

  Other simulation studies have been reported by ElMaraghy (1985), Gaalman *et al.* (1987) and Kashyap (1992).

12.3 TOOL REQUIREMENT PLANNING

Tool requirement planning (TRP) is concerned with assessing the number of tools or tool life required for the machining of workpieces during a given planning horizon. Gray *et al.* (1990) have drawn the parallel between MRP and TRP. While a conceptual framework for tool resource planning was outlined, the authors also suggested the limitation of using MRPII models:

> "Existing resource planning models like MRPII address the planning and control of material flows and machines, but are incapable of dealing simultaneously with the constraints imposed by tool requirements."

Chung (1991) also placed TRP within the context of MRPII (see Fig. 12.5), where TRP is likened to capacity requirement planning except for one major

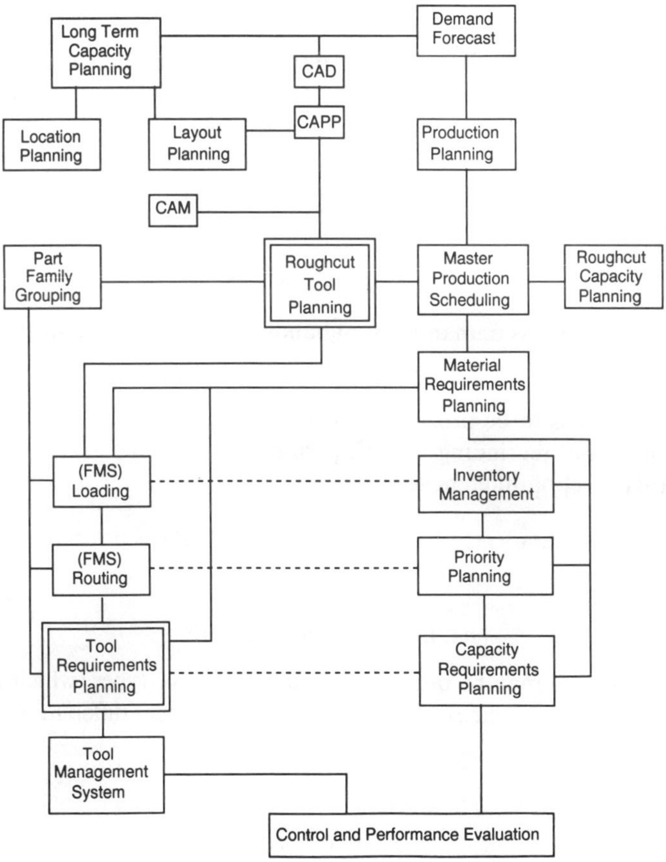

Fig. 12.5 Roughcut tool planning and tool requirements planning (reproduced with permission of SME).

difference: while production capacity is relatively fixed, tools can be purchased and reground in a relatively short period of time.

12.3.1 Relevance of TRP to efficient manufacturing

Major problems as a result of poor planning of tool availability had been observed by Mason (1986) and Chung (1991) to include:

- high level of tool inventory;
- significant system idle time due to lack of tools;
- unnecessary handling of tools;
- hampering of production flow;
- increased queues; and
- unnecessary tool duplicates.

Further, tool costs could represent as much as 25% of operating costs (Cumings, 1986). Rovito and Hankins (1986) showed that different tooling strategies significantly affect tool inventory requirement. Hammer (1984) showed that the added cost of tool automation can bring in great improvements in economic efficiency. ElMaraghy (1985) discussed various tool automation components, and identified the need to plan and control tools in terms of usage, availability, replenishment, etc.

12.3.2 Issues in TRP

For manned as well as unmanned tool handling, the major decisions in TRP are:

- number of tools to be procured in each period;
- amount of tool regrinding in each period; and
- inventory level of different tool types.

These decisions are affected by various factors, outlined below.

(a) Use of alternative tools

An operation type may be processed by different tool types, which may be loaded on different machines, and may be processed under different machining parameters.

(b) Tool sharing

Whether a tool resides on one machine or migrates to other machines during the manufacturing cycle, it can work on different parts.

(c) Tool failures

A considerable percentage of tool inventory is commonly lost due to tool failure.

(d) Tool regrinding

The life of a tool can usually be extended via an extensive regrinding process. However, a tool can only be reground a specified number of times, as prescribed by the tool manufacturer, and is to be disposed of beyond that point.

(e) Ratio of new versus reground tool life

To ensure quality machining, an appropriate mix of new and reground tool inventory is desired; this is necessary as certain operations may require new tools for a higher degree of tolerance, a higher degree of precision, a better surface finish, etc. This means that a balanced inventory of reground tool life and new tool life must be maintained.

(f) Tooling strategies and tool readiness

The level of tool inventory requirement is affected by the tooling strategy employed, as well as the need for buffer tools (due to tool failure, tools being circulated, tools being reground and new tools in preparation). Also central to the determination of tool inventory is the 'readiness' of the tool automation system: tool room support, tool transport devices and tool monitoring.

(g) Time-phasing of projected tool usage

Tool usage needs to be projected according to the manufacturing lead time of components.

12.3.3 TRP models

Evidently, there is a need to have a planning model which would address the type of tools that should be purchased, and what the inventory levels of certain tool types during a given planning horizon should be. Savoie (1988) proposed a simple method by which to estimate tool requirements. Koulamas (1991) studied tool requirements in a multi-level bills-of-materials-based machining system, but the emphasis is on the end product's completion time, with tool requirement computed based on the parts explosion concept of manufacturing resource planning (MRP). Graver and McGinnis (1989) analysed tool provisioning at the machine level, i.e. how many tools should be loaded on to a particular machine. The need to account for tools which are being circulated, as well as those that are being reground, is identified. Tool requirements are estimated using both a static approach (worst case scenario) and a simulation model.

(a) Chung (1991) developed a rough-cut tool planning model and two TRP models: one for a bulk exchange policy, and one for tool sharing during a frozen cycle policy. The different types of tooling strategies will be discussed in detail in section 12.4.2. In brief, 'bulk strategy' provides tools for each workpiece that visits the machine, and 'sharing during frozen cycle' provides tools for workpieces which visit the machine during the cycle. Chung's two TRP models are:

Bulk exchange

$$\text{Min:} \quad \sum_h \sum_j \sum_t \{C_{j1} Y_{hjt} + C_{j2} Y'_{hjt} + C'_{j1} I_{hjt}\}$$

s.t.

$$I_{hjt-1} + Y_{hjt} + Y'_{hjt} - I_{hjt} = r_{hjt} \quad \text{for all } h, j, t$$

$$Y'_{hjt} \leqslant I_{hjt-1} \quad \text{for all } h, j, t,$$

where

C_{j1} = replacement cost for each unit of tool type j;
C_{j2} = reconditioning cost for each unit of tool type j;
C'_{j1} = holding cost per period for each unit of tool type j;
Y_{hjt} = amount of tool type j (for workpiece type h) replaced during period t;
Y'_{hjt} = amount of tool type j (for workpiece type h) reconditioned during period t;
I_{hjt} = on hand inventory of tool type j;
r_{hjt} = amount of tool type j (for workpiece type h) required during period t.

Note that 'amount of tool type' can mean either copies of tools or tool life.

Sharing of tools during the frozen manufacturing cycle

$$\text{Min:} \quad \sum_j \sum_t \{C_{j1} Y_{jt} + C_{j2} Y'_{jt} + C'_{j1} I_{jt}\}$$

s.t.

$$I_{jt-1} + Y_{jt} + Y'_{jt} - I_{jt} = r_{jt} \quad \text{for all } j, t$$

$$Y'_{jt} \leqslant I_{jt-1} \quad \text{for all } j, t,$$

where

Y_{jt} = amount of tool type j replaced during period t;
Y'_{jt} = amount of tool type j reconditioned during period t;
I_{jt} = on hand inventory of tool type j;
r_{jt} = amount of tool type j required during period t.

The sharing model differs in that the requirement for a tool type is not specified for a workpiece type. The key implicit assumption is that r_{jt} is less than $\sum_h r_{hjt}$.

(b) Khator and Leung (1994) developed a tool planning model where tool sharing takes place during and across manufacturing cycles; this means that tool migration is permitted within the system, assuming the existence of an efficient tool transfer system. The model objective is minimization of the total costs of tool purchase, tool regrinding, tool inventory and operation (machining and inspection). The decision variables are the quantity of purchase, the quantity of regrinding and the assignment of tools (quantity and type) to part operations. The constraints consist of:

- part-operation assignment and demand requirements;
- recursive inventory relationships;
- tool life cost due to tool breakage;
- buffer tool inventory;
- tool regrinding reserve;
- regrinding capacity;
- tool purchase budget;
- proportion of new tool and reground tool.

12.4 TOOL ALLOCATION AND REPLACEMENT

This section discusses the static problems of tool preparation (*versus* the dynamic aspects of tool flow). Different approaches to tool allocation and tool replacement are addressed.

12.4.1 Tool allocation

Tool allocation is concerned with the availability of tool life in the magazine. The decisions are (1) what type of tools, and (2) how many copies of each tool are to be loaded on to a particular machine during a manufacturing cycle. Such determination is based on the process plan, workpiece mix and their production levels, with consideration given to the limited nature of the tool magazine as well as the tool inventory.

Since workpiece assignment (machine loading) and tool allocation are reciprocal in nature – assignment of a workpiece to a particular machine depends upon the allocation of tools to that machine, and *vice versa* – these two tasks ought to be performed concurrently. Mathematical programming models for the concurrent determination of these two decisions can be found in Sarin and Chen (1987), Han *et al.* (1989) and Leung *et al.* (1992). The effect of workpiece movement on tool allocation was explicitly addressed by Leung *et al.* (1992). However, no literature to-date has examined the effect of tool movement on tool allocation. There are certain rule-of-thumb type tool allocation approaches, referred to as **tooling strategies**.

12.4.2 Types of tooling strategies

Rovito and Hankins (1986) proposed the following tooling strategies, each of which has its own degree of tool sharing:

- *Bulk exchange.* This strategy suggests the provision of a copy of each tool needed for each job visiting the machine. Bulk exchange will undoubtedly result in a relatively high tool inventory and considerable tool handling. It is a method most suited to the production of high-volume/low-variety part mix. The implementation of this strategy is relatively simple and is easy to control.
- *Sharing tools in a frozen production window.* During the frozen production window, tools identified as having common workpiece types will not be duplicated in the tool matrix. The tool matrix is set up at the beginning of the window. This method allows the sharing of tools during each production window, and may be viewed as a kind of bulk exchange during each frozen window. To implement this strategy, a robust database and decision scheme is required.
- *Migration at completion of a workpiece type.* By allowing tools to be removed once their services are no longer required, this strategy allows the subsequent loading of other tool types required for machining workpieces at a later time. In essence, the sharing of tools is extended across production windows. A sophisticated decision support system is required to determine the set of tools that needs to be removed from the magazine.
- *Resident tools.* High usage tool types are identified and are to reside on certain machines or groups of machines. Such dedication of specific tool types to machines allows flexibility in the system to respond to changes to the routing of workpieces due to machine breakdown or changes in the production mix.

12.4.3 Tool grouping

The identification of high usage tools is but one of the many criteria for the grouping of tools. Due to the large variety of tool types found in most manufacturing systems, it is often more beneficial to group tools into categories. Zeleny (1981) proposed a group technology-based organization of tools and workpieces which reduced the number of required tools and workpiece pallets in a highly automated manufacturing system used for prismatic parts.

12.4.4 Models on tool sharing

(a) Analytical model

Vinod and Sabbagh (1986) modeled the tool sharing system as a closed queueing network, with p_{ij} defined as the probability that a job visits station j after visiting i, and π_i as the steady state distribution of tool availability.

It is assumed that when a job visits a station, it will stay there until the desired tool arrives. First, analytical results were derived for situations when there were no spares in the system – a finite source scenario. Via an approximation procedure, the system performance is measured by the increase in processing time due to the lack of tools. It was stated that "Experimental evidence suggest that the approximation is sufficiently accurate for practical applications." Then, spare tools and tool replacement upon tool failures are incorporated. The optimal level of spares is determined using an optimization model. In general, while their model is based on several simplifying assumptions, it appears to be the only analytical work on tool availability. All other models are simulation-based.

A mathematical model for predicting the system reliability of an automatic tool changing system with various carbide inserts and spares subject to Weibull failures was developed by Pan *et al.* (1986). The authors considered tooling as a 'series system with spares'.

(b) Simulation models

Gaalman *et al.* (1987) studied tool sharing and its impact on tool investment. The trade-off between tool investment and investment in tool transport devices is examined. Operational strategies for tool sharing are designed into the simulation study. Tool savings are made from the difference in costs between a complete quadruplicate tool set and the selected tool mix for a type.

Hammer (1984) discussed issues in tool automation: tool changing, replacement and monitoring. A simulation experiment is utilized to analyse the added costs of tool automation and the benefits that resulted. It was shown that great improvements can be made in economic efficiency.

Rovito and Hankins (1986) designed a simulation experiment to analyse the impact of tool migration on tool change time and setups. Based on the simulation results, the authors suggested that tool change time has a significant effect on the lost spindle time – a finding that strengthened the argument for automated tool loading and unloading.

12.4.5 Tool replacement

When a tool is returned to the tool room, an inspection process commonly takes place to determine whether the tool should be reground or disposed of. Since the pioneering work on tool economics by Taylor, there is now a voluminous literature on the economic life of tools. A bibliography of the papers dealing with tool life and cutting tool economics can be found in Gray *et al.* (1990). As our focus is on automation aspect of tools, economic tool life issues are not discussed here. Instead, we will concentrate on the implementation of tool replacement at the operational level.

Cherpakov (1983) suggested a method for replacing tools in automatic lines. Zavanella *et al.* (1990) dealt with tool failures and tool replacement policies when tool life was considered to be stochastic. An analytic procedure was suggested for calculating the reliability level of jobs. The reliability level was dependent on two factors: the true tool life, and an established tool change time. The tool replacement policies were:

1. Replacement of a tool when its machining time equals or exceeds established tool change time less maximum machining time for a job – this allows the remaining tool life to perform reliably for the longest operation.
2. Replacement of a tool when its machining time equals or exceeds established tool change time less minimum machining time for a job – this allows the remaining tool life to perform reliably for the shortest operation.
3. Replacement of a tool when its machining time equals or exceeds the established tool change time.

The actual values obtained through a simulation experiment were compared with the analytically developed reliabilities.

Sharit and Elhence (1989) proposed a framework for the computerization of tool replacement decision making in a flexible manufacturing system (FMS). A procedure that addressed the decision of whether a tool should be replaced prior to its processing of a workpiece was outlined. The proposed procedure attempted to minimize the tool-related economic losses associated with these decisions while maximizing part throughput.

12.5 TOOL-ROUTING STRATEGIES

Tool inventory-related decisions such as tool procurement and tool inventory level have already been discussed above. Various approaches to tool allocation and replacement were also described in the preceding section. Tool routing strategies are the next aspect that must be dealt with while designing an automated tool handling system. The following topics should be considered:

- tool transfer schemes;
- tool flows;
- tool distribution amongst the machine tools;
- tool cycle; and
- tool dispatching priorities.

A detailed discussion of each of these issues follows.

12.5.1 Tool transfer schemes

Tools are allocated to various machines in the system according to the respective tooling strategies. If the tools are shared amongst machines, then control methods are required to transfer tools from one machine to another.

The first issue dealing with tool transfer is associated with tool dispatching, which can be described as follows. There are one or more copies of a tool available in the system. However, not all machines have a copy of the tool. When an incoming workpiece needs a tool which is currently not resident on the machine, a request is made for this tool. Two situations may arise in this context. In the first case, the tool is available (not currently in use) on more than one machine. Which of these available tools must be brought to this machine? In the second case, all copies of the tools are in use. When this tool becomes free, there may be more than one request pending for this tool. Where should this tool be sent?

Furthermore, a policy regarding the return of the tool is needed in this situation:

- returning the tool to the machine from where it came, after processing of the current workpiece has been completed.
- no return of the tool, i.e. leaving it at the current machine and transferring it when it is needed by some other machine tool.

The impact of these two policies on the tool handler workload and machine idle time would be different.

12.5.2 Tool flows

Zhang *et al.* (1989) described the tool flow problems in highly automated turning systems. A hierarchy of levels of tool flow automation, with each level having its own focal point of tool supply, is considered. Tool flow originates at the central tool room and continues to the cell consisting of a group of machines (secondary tool buffer), then proceeds to the machine magazine, and finally, to the machine turret. A return path begins at the machine turret and ends at the central tool room.

Zeleny (1985) used the following criteria for tool flow optimization in a shop where boxed shape castings were machined:

- identical operations, especially final ones, should be performed on the same pallet and, preferably, on the same machine;
- each machine should operate for a certain time period within a chosen group of machining operations.

12.5.3 Tool distribution

The number of tools required by a manufacturing system do not increase in direct proportion to the variety of workpieces (Rhodes, 1985; Zeleny, 1985). Several researches have studied the effect of tool sharing and tool distribution in a computer integrated manufacturing system. Kashyap (1992) studied the impact of tool duplication on job flowtime and machine utilization in a tool constrained system.

12.5.4 Tool cycle

A tool cycle begins at the tool room. After kitting and assembly, a tool is transported to a machine. After completing a machining operation, tool may either remain on the machine or go to another machine (tool sharing). Eventually, a tool is returned to a tool inspection station, and it may either be sent to a tool maintenance station or to the tool room.

12.5.5 Tool dispatching

Iwata *et al.* (1982) used three different dispatching rules for the selection of an alternative machine tool and workpiece transporter. These rules were based on the processing time, early start time and early finish time. Tool selection was based on the early start time.

Han *et al.* (1989) studied a system where a part was processed on one machine only. Tools needed for various operations were taken from other machines. Various tool return policies and job dispatching rules were evaluated.

12.5.6 Models used

Attempts made in modeling tool flow strategies fall under mathematical programming approaches and simulation studies. A brief description of some of the work is given here.

(a) Mathematical programming approaches

A framework for tool management consisting of a tool automation level, a tool management strategy and a tool issue strategy was presented by Zhang *et al.* (1989). An algorithmic approach to the design, control and operation of a real life tooling system was provided. Solutions to the tool flow problem were derived from the following decision making levels:

- selection of a machine tool;
- selection of cutting tools and their machine assignments;
- selection of a tool transport device.

The user interface was provided through a conversational language to describe the manufacturing system parameters. An interactive database contained information on the machine, tools, parts and system configuration. Outputs from the model included schedules for tool transfer, tooling analysis and other cell-related performance measures.

Iwata *et al.* (1982) also modeled a tool constrained flexible manufacturing system. Similar to the previous approach, three decision making levels were selections of the machine, cutting tools and transporter. The effect of the following dispatching rules on the makespan, flow time, machine and transporter utilization was evaluated:

- shortest operation time with alternatives considered (SOTA);
- earliest starting time with alternatives considered (ESTA);
- earliest finishing time with alternatives considered (EFTA).

Results of a case study, consisting of three machine tools and a loading and unloading station, found the EFTA rule to be more desirable for both machine and transporter selection.

Han *et al.* (1989) gave a mathematical programming formulation for a manufacturing system, where different machine centers were capable of performing all the operations on a part. However, the tools needed to complete all the operations might not reside in a machine tool's magazine. A mathematical formulation for tool loading to minimize tool movement between the machines was developed. A heuristic solution method was developed for the tool return policy. The effect of a no tool return policy *versus* an immediate tool return policy was investigated on the production makespan. Several part selection rules were also evaluated. A simulation experiment was performed to see the difference between the performance of the mathematical formulation and the heuristic method.

(b) Simulation models

ElMaraghy (1985) described the use of simulation modeling in evaluating alternative automated tooling systems which helped in choosing the optimal design parameters. A GASP IV simulation language-based package, TOOLSIM, was developed. The input to the package consisted of information regarding workstations, parts, tools and tool transporter. The sharing of standard and special tools among machines was allowed via a random tool flow strategy. A tool ranking scheme attempted to fill the tool requests from the tool magazine, intermediate tool buffers of the machine itself or other machines, and finally, from the central tool room. Performance measures such as the number of parts produced by each machine and the overall production rate, tool preparation time statistics, machine and tool transporter utilization, machine queue sizes and tool breakdown statistics were reported by the program. The location of intermediate tool storage buffers, as well as tool room, could be defined as part of the system layout.

Kashyap (1992) studied the effect of tool selection rules (which of the multiple tools available to use) and request selection rules (which of the pending requests for a tool to be satisfied next) using a simulation model within an FMS context. He found that the tool transporter utilization was affected by these rules. The shortest distance rule performed the best in the case of tool selection. The impact of these rules on the makespan of the production schedule was insignificant in his study. When only one copy of a tool was available in the system, the effect of the request selection rule on the system performance was found to be significant. Tool duplication, as would be expected, improved system performance.

Other simulation studies have been reported by Crite *et al.* (1985), Carrie and Perera (1986) and Gaalman *et al.* (1987).

12.6 TOOL MANAGEMENT INFORMATION SYSTEM

A sound tool management information system is paramount to the effective planning, control and organization of tools in an automated manufacturing system. The complexity of the information system required is directly proportional to the level of tool automation in the manufacturing system. The efficient storage and retrieval of pertinent tool information enables the successful implementation of tool requirement planning, tool flow strategies, tool allocation and replacement, and tool automation in general. Tool information system is the backbone to the productive utilization of:

- tool sharing;
- tool routing flexibility;
- product design flexibility;
- machine flexibility;
- tool reliability;
- tool maintainability; and
- quality of tool-cutting.

We separate the task of designing a tool management information system into two areas: design of the database system, and design of real-time tool control.

12.6.1 Database management

Our discussion on database management will be broken down into four areas of concern: functions, subsystems, database types and data architecture. More detailed discussions on one or more of these aspects can be found in Ber and Falkenburg (1985), Rovito and Dvorsky (1986), Eversheim *et al.* (1987), Garapati and Wang (1988), Little *et al.* (1988), Kokko (1988), Amazon Computers Ltd. (1990), Ranky (1988), Mason (1986, 1992) and Bell and DeSouza (1987).

(a) Functions

The database system can provide such management planning information as:

- purchasing of tools;
- inventory control of tools;
- tool cost control;
- scheduling of tool requirements; and
- utilization of tools.

Operationally, the management system could be utilized for:

- tool kitting;
- tool flow control;
- tool maintenance; and
- tool monitoring.

(b) Subsystems

The database system can be broken down into different subsystems, with each responsible for different phases of tool management:

- *Tool crib subsystem.* This subsystem is responsible for the inspection, recondition and disposal of tools. Tool information needed at this subsystem includes size, geometry, edge preparation, tool material and grade of the insert.
- *Tool requirement subsystem.* This subsystem provides information for the analysis of tool procurement and tool inventory planning. Operational information would include such data as tools in stock, in process, in preparation, etc. Also, information on vendors, tool life as well as demand profile are included.
- *Tool monitoring subsystem.* This subsystem is responsible for the movement of tool flow, tool location, tool changes and the status of the tool recycling system.
- *Tool/parts production planning subsystem.* This subsystem provides information for process control, part programming, tool preset, tool maintenance, tool assembly, etc. Here, data is needed to relate tools, machines and parts.
- *Subsystems for unique features.* It is possible that certain systems may have a unique subsystem to handle such specialized functions as kitting and assemblies, gauges, jigs and fixtures, store room information, presetting, etc.

(c) Database types

There are three types of databases as characterized by their respective nature: static, real-time and support.

Static database

A static database comprises of data which is updated periodically on a batch basis. Primarily, this is data that will remain unchanged for at least one manufacturing period. A typical sample of such files and their data are as follows:

1. Tool classification files: tool code number, tool type, tool manufacturer, type of holder, tool geometry.
2. Tooling files by workpieces: part operations by tool types, processing time, etc.

3. Tooling files by machine: tool-machine compatibility data, machining capability, etc.

Real-time database
Real-time databases comprise of tool data which is updated on a real-time basis:

1. Tool-life status file: useful tool life, remaining tool life, regrinding history, etc.
2. Tool status file: tool location, tool breakage condition, idle or in use, tool magazine information, etc.

Support databases
Support databases are those to which the tool management information system needs to have access. These databases include: process plan, MRP database, factory control database, NC programming files, CAD system for workpiece design, scheduling and resource management, etc.

(d) Data architecture

Relational databases have been adopted for designing tool data architectures. Ranky (1988) proposed a generic tool data architecture, with a relational database, for FMSs. There are four levels of information: 'tooltype', 'element and edge', 'subassembly' and 'one tool' – with each level representing a tool data collection situation. These levels are logically linked together. Little *et al.* (1988) utilized a Structured System Analysis and Design Method (SSADM) to design a relational database management system which can be either machine- or system-centered.

12.6.2 Real-time control

For the efficient operation of an automated manufacturing system, it is important to know the location and status of various tools in the system. Both hardware and software support are essential for high productivity of such a system. The following are some of the issues that relate to the real-time control of tools in a computer integrated manufacturing system:

- tool identification systems;
- monitoring tool life;
- hierarchical tool control.

(a) Tool identification systems

Since a large number of tools are needed in today's computer integrated manufacturing systems, an effective tool identification system is a must for its efficient operation. At the machine level, the types of tools loaded in the magazines and the status of these tools must be known.

The tool management computer stores the following data for each tool:

- number and name;
- length and diameter;
- life and accumulated usage time;
- tool offsets;
- tool supplier code.

The following methods for tool identification have been reported:

- bar coding;
- memory chips;
- vision system.

Bar coding

Bar coding is the most commonly used tool identification system. The bar code label can either be printed by the tool crib attendant manually entering all the tool information, or by inserting the tool into a presetting machine which measures the tool dimensions and transmits it to the printer. The printed label is then put on the tool holder. A laser reader mounted on the tool magazine relays the tool information, including the slot number, to the tool management computer.

Memory chips

A read/write chip inserted into the tool holder is programmed to store all the necessary tool information mentioned earlier. The tool crib attendant inserts the tool into a tool presetting machine, and once all the tool measurements have been performed, it is written on the chip. The machine tool magazine is equipped with a chip reader, which reads this information while indexing around the tool. Kurimoto *et al.* (1988) have described an advanced intelligent tool holder with an IC chip with 32 bytes of memory capable of storing a large amount of tool data.

Each time a tool is returned to the tool crib it is inspected, and the updated information is printed on a new bar code label or written to the chip.

Vision system

A vision system consisting of a small camera and CRT display that identifies a cutting tool has been developed by Giddings and Lewis (Kiran and Krason, 1988). The system uses General Electric's optimization vision system and Gidding and Lewis' software to identify a cutting tool by type and dimension.

(b) Monitoring tool life

The tool life of a cutting tool depends on several factors, such as the tool material, work material, speed, feed and depth of cut, etc. Most computer integrated manufacturing systems (Albert, 1987; Martin, 1989) work from a fixed tool life determined conservatively (to avoid tool breakage and probable loss of the job) by the user based on his previous experience of the tool and other cutting conditions (e.g. condition of the machine). Tool manufacturers also indicate the recommended tool life for a given work material and cutting conditions. The tool life is commonly measured in time units. However, there are problems associated with determining tool life prior to the use of the tool. First, the actual cutting conditions may vary. Second, a tool may be used on more than one workpiece with different work materials and other cutting conditions, thereby invalidating the previously determined tool life. A more efficient approach, which will result in optimal tool utilization and reduction in wasted tool life, will be to detect the tool wear or possible tool breakage by monitoring the tool's condition. ElMaraghy (1985), Rhodes (1985), Kiran and Krason (1988) and Jacobs *et al.* (1988) have described the use of probes and sensors in monitoring tool life. The use of adaptive control which not only monitors the cutting conditions but also provides feedback to the machine tool is also recommended.

Use of a probe

An N/C part programmer indicates the checking of a tool for wear or breakage. A fixed probe permanently mounted on the machine tool comes into contact with the cutting tool. The probe software verifies the tool length, and a change in the tool length indicates either a worn or broken tool.

Use of sensors

Various sensors have been developed which measure:

- cutting force;
- power input at the machine spindle;
- temperature at the cutting edge;
- vibration frequency;
- electrical resistance;
- acoustic (sound) frequencies.

An increase/decrease in the value of these parameters is an indication of worn or broken tool. Force and power sensors are the most commonly used.

Adaptive control systems

Sensors are used to measure spindle force and/or spindle horse-power. The measured values are compared with the predetermined value to adjust the spindle speed (rpm) and the feed rate to avoid tool breakage and loss in machining accuracy. Work in this area is still going on to provide a reliable, cost-efficient method. Kurimoto *et al.* (1988) have described a Mazatrol CNC system where the predicted tool life and actual cutting time of the tool are monitored. When the accumulated cutting hours exceeds those predicted, the controller issues instructions for it to be replaced. The individual tool life is determined by utilizing the tool data obtained from past operational experience. The Mazatrol controllers also have the ability to monitor the cutting power being consumed by each power unit. The maximum power available to the specific tool can be individually predetermined. When the cutting load exceeds the predetermined maximum load, the feed rate is automatically reduced to maintain the cutting load to preset values. This adaptive feed rate control (AFC) maintains optimum, but reliable, cutting loads, even allowing for casting stock variations, hardness and for tool wear.

(c) Hierarchical tool control

A tool management system which tracks the tool at different levels in an automated manufacturing system has been recommended by various researchers. Aggregate information regarding the tool availability at the component level may be sufficient for tool procurement and inventory control. Knowledge of the production schedule and the tool data are required at the assembly and kitting level. At the lower level of a tool control system, the tracking of individual tools, including their locations on a machine tool, and the status of the tools is needed for a proper tool management system. A general view of the different levels of tool management, as given by Rhodes (1985), is shown in Fig. 12.6.

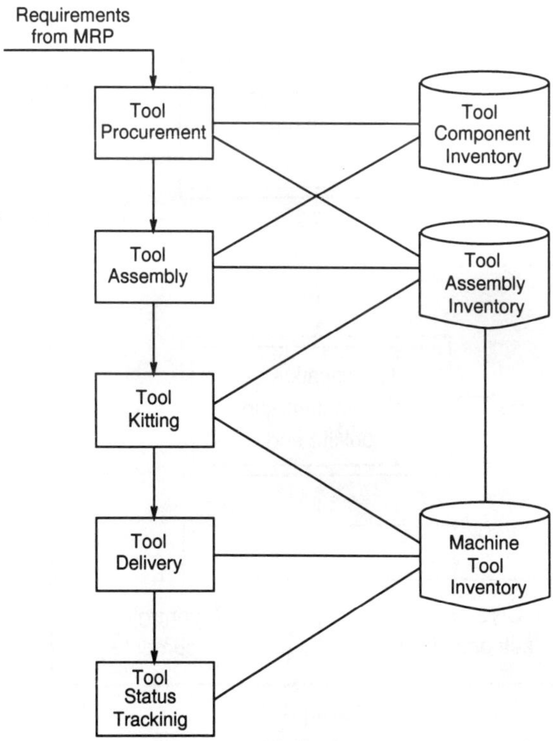

Fig. 12.6 Levels of tool management (reproduced with permission of SME).

(d) Development of control schemes

Recently, attempts have also been made in developing appropriate schemes for adaptive control. Some of the research work done in this area is reported below.

Process monitoring

Jacob *et al.* (1988) describe work done at The Dresden University of Technology, where tool monitoring was the focus of process monitoring. Force signals for

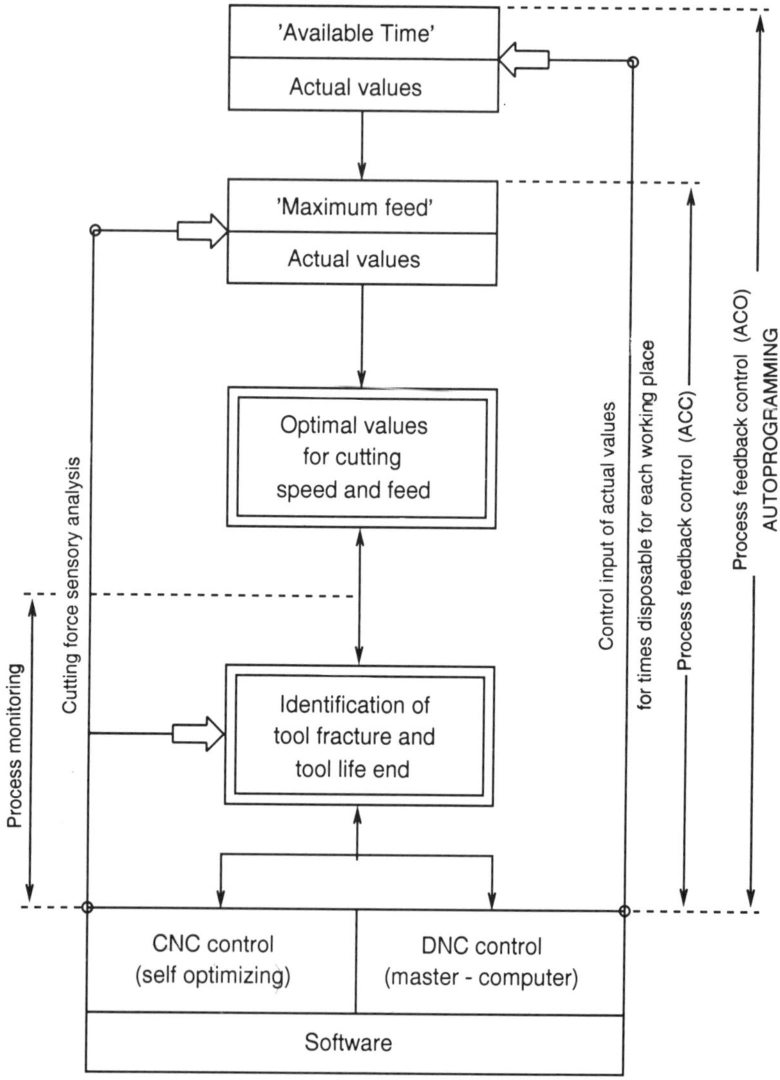

Fig. 12.7 Conceptual framework for self-optimizing control (reproduced with permission of Taylor & Francis Ltd).

different process states such as tool breakage, hard spot, etc., are recognized, and optimization models for an adaptive control constraint (ACC) system and for adaptive control optimization (ACO) developed. Their conceptual framework for self-optimizing control is shown in Fig. 12.7.

Expert systems
ElMaraghy (1985) has identified the need for developing:

- effective, economical and reliable sensors to monitor machine operation and signal malfunction;
- efficient adaptive control algorithms to ensure a quick response by the machine as a result of feedback from the sensors.

Use of artificial intelligence technology such as the expert systems approach to monitor operation of the machine tool is indicated as being very promising.

12.7 FUTURE RESEARCH ISSUES

In this section, future research issues are outlined. These issues are organized under each tool planning or control category.

12.7.1 Tool automation facility planning

While planning and designing a manufacturing system with automated tool movement, several issues need to be addressed. These range from the relative location of tool-related entities to the selection and economic justification of the tool transfer system.

(a) System layout incorporating simultaneous movement of tools and workpieces

Guidelines need to be developed regarding the spatial planning of tool-related facilities such as the tool crib, tool grinding station and tool transporter path layout. These would assist a manufacturing facility designer in developing an effective system layout.

(b) Selection of tool handling system

Research needs to be directed towards the modeling of decisions regarding the use of a common work and tool handling system or separate handling systems. The impact of such a decision on the throughput rate, WIP levels and tool inventory needed for efficient system operation should be assessed. Such work should also include the selection of an appropriate tool handling system and its cost justification.

(c) Modeling tool handling capacity

Determination of the tool handling capacity required at different stages of the tool cycle needs to be made. Tool handling loads at various stages should be balanced for smooth system operation.

12.7.2 Tool Requirement Planning

(a) Modeling TRP with tool handling consideration

None of the TRP models to-date incorporate any consideration of the system's tool handling capability. The extent to which the sharing of tools can take place in a manufacturing system is constrained by the level of tool automation, which is dependent on the capability of the tool handling system. To accurately determine the tool requirement, a TRP model ought to address the issue of tool handling.

(b) Simulation experiments to test the performance of TRP models

TRP models are planning models, and hence they do not address the operational aspects of machining. To truly test the efficacy of these models, the model results need to be tested based on experiments which simulate the operational aspects of the machining environment.

(c) Sensitivity analysis of relevant TRP parameters

To gain insights on the economic impact of TRP parameters such as demand, tool life, inventory cost, etc., a comprehensive sensitivity analysis of TRP parameters should be performed.

(d) Economic lot size for tools

Analogous to the EOQ concept of products, it may be possible to analyse tool inventory under similar conditions. One benefit of using such an approach (as opposed to some mathematical programming approach) is that a lot sizing perspective is much more easy to implement.

(e) Probabilistic aspects of TRP

Existing TRP models are deterministic formulations, yet many aspects of TRP are probabilistic in nature (demand, tool life, tool failure, etc.). It would appear that efforts directed at the formulation of a probabilistic TRP model should be worthwhile.

12.7.3 Tool allocation and replacement

(a) Concurrent workpiece assignment and tool allocation with simultaneous routing of workpieces and tools

The simultaneous routing of tools and workpieces presents a new way of looking at workpiece assignment and tool allocation. A normative model for the assignment of workpieces and allocation of tools should incorporate the trade-offs between workpiece and tool transportation. This problem may also be examined from a queueing perspective. In essence, the system may be viewed as having two queueing situations – one for the workpiece (the customers) and another for the tools (the servers). Similarly, simulation experiments could be performed to explore the relationship between workpiece assignment and tool allocation under such an environment.

(b) Modeling strategies for resident tools with tool handling consideration

One strategy for the allocation of tools is to have certain tool types resident on a particular machine or particular groups of machines. With the in-process movement of tools, such a strategy takes on a new dimension, and research should be continued to investigate the variety of tools which should be resident.

(c) Preparation of tools in a tool automated environment

At the tool preparatory stage there should be quite a number of new research problems as a result of tool automation. With a high degree of tool sharing, the kitting of tools could have a different orientation. Also, the organization of tools in the tool crib may need to be modified so as to improve tool handling. In fact previously, tools have been placed into different wear-and-tear categories for the purposes of tool monitoring and replacement.

12.7.4 Tool routing strategies

(a) Workpiece release based on tool availability

Tool automation has affected the availability of tools, which in turn has changed the nature of workpiece release in a manufacturing system. Rules based on workpiece attributes such as the number of operations, total processing time, due date, etc., are no longer adequate. Determination of the availability of tools at a machine before releasing a workpiece could help in reducing parts waiting for tools. Similarly, when a machine becomes free it could request a workpiece whose large number of operations can be completed from the currently loaded tools in the machine magazine.

(b) Relationship between tool and workpiece movement

Studies have been conducted where a performance comparison has been made between the following two approaches: (1) no tool movement – a workpiece moves from one machine to another until all its operations are completed; or (2) no workpiece movement – a part waits on a machine until all its operations are completed. Hybrid strategies, where both part and tool movements are allowed, may be worth researching. The effect of these strategies on part flow time, WIP, tool inventory, machine setup time and material (both workpiece and tool) handling should be investigated.

(c) Tool routing

In a manufacturing system where tools are shared among machines, the routing of tools may affect the system's performance. An effective tool scheduling strategy needs to be developed which will minimize any delays associated with the non-availability of tools.

12.7.5 Real-time tool control

(a) Tool life determination

Due to tool automation, a tool has to experience changing cutting conditions, such as different machines and workpieces. A large amount of research has been done in estimating tool life. However, models are still not fully developed in accurately predicting the life of a tool which is dependent on varying cutting conditions.

(b) Adaptive control systems

Work is still continuing in designing cost-effective adaptive control systems capable of varying the cutting speed, feed and depth of cut as the tool wears and/or cutting conditions (such as workpiece hardness) vary. Tool automation can be more effective with an advanced control system.

(c) Tool monitoring systems

Efficient tool monitoring is essential to the implementation of tool automation. Artificial intelligence technologies such as expert systems hold promise for developing an effective tool monitoring system based on the signals received from machine sensors.

12.8 SUMMARY

In the past decade or so, the flexibility of manufacturing systems has significantly improved due to the automation of workpiece handling. Now, with the automation of tool handling, such flexibility can be further enhanced. One might speculate that tool handling is in general more cost efficient than workpiece handling, as tools are more suited to standardization, as well as being more portable. Whether a system is purely tool automated or workpiece automated, or a hybrid, the analysis of trade-offs between workpiece travel and tool travel will need to be properly analysed.

In this chapter, we have discussed various issues in the planning and control of tools within a tool automated environment. We broke the issues down into five general categories: tool automation facility planning; tool requirement planning; tool allocation and replacement; tool routing strategies; and tool management information system. For each of these general categories, we described the problems as well as the problem environments. Various approaches to these problems were delineated, and discussions on the corresponding literature were provided throughout. We have also attempted to identify some of the potential research issues which would advance the state-of-the-art of tool planning and control. In sum, we believe that the automation of tool handling will be a major issue in designing future manufacturing systems.

REFERENCES

Albert, M. (ed) (1987) Tool management for a ten-year-old FMS. *Modern Machine Shop*, **59**, March, 54–64.

Amazon Computers Ltd. (1990) *Computerised Tool Management System: Amazon Technology*, March.

Aver'yanov, O.I. (1986) Tool management systems for flexible manufacturing systems. *Stanki i Instrument*, **57**(8), 2–4.

Ber, A. and Falkenburg, D.R. (1985) Tool management for FMS. *Annals of CIRP*, **34**.

Bell, R. and DeSouza, R.B.R. (1987) *The Management of Tool Flows in Highly Automated Flexible Machining Installations*. Second International Conference on Computer-Aided Production Engineering, Edinburgh, UK, April.

Bruno, G. and Biglia P. (1985) *Performance Evaluation and Validation of Tool Handling in Flexible Manufacturing Systems using Petri Nets*. Proceedings International Workshop on Timed Petri Nets, Torino, Italy, pp. 64–71.

Carrie, A.S. and Perera D.T.S. (1986) Work scheduling in FMS under tool availability constraints. *International Journal of Production Research*, **24**(6), 1299–308.

Carter, N. (1985) The application of a flexible tooling system in a flexible manufacturing system. *Robotica*, **3**, 221–8.

Cherpakov, B.I. (1983) Group replacement of tools in automatic lines. *Stanki i Instrument*, **54**(10), 6–7.

Chung, C. (1991) Planning tool requirements for flexible manufacturing systems. *Journal of Manufacturing Systems*, **10**(6), 476–83.

Crite, G.D., Mills, R.I. and Talavage, J.J. (1985) PATHSIM, a modular simulator for an automatic tool handling system evaluation in FMS. *Journal of Manufacturing Systems*, **4**(1).

Cumings, S. (1986) *Developing Integrated Tooling Systems: A Case Study at Garret Turbine Engine Company*. Proceedings Fall Industrial Engineering Conference, Boston, MA.

DeSouza, R.B.R. and Bell, R. (1991) A tool cluster based strategy for the management of cutting tools in flexible manufacturing systems. *Journal of Operations Management*, **10**(1).

ElMaraghy, H.A. (1985) Automated tool management in flexible manufacturing. *Journal of Manufacturing Systems*, **4**(1).

Eversheim, W., Jacobs, S. and Weinand, L. (1987) Structure and application of a universal company-independent data bank or tools. *Annals of CIRP*, **36**(1).

Gaalman, G.S., Nawijn, W.M. and Platzer, L.W. (1987) Tool sharing in an FMS – a feasibility study. *Engineering Cost and Production Economics*, **12**.

Garapati, S. and Wang, H.P. (1988) *Decision Support System for Tool Management in FMS*. Proceedings IEEE International Conference on Computer Integrated Manufacturing, Troy, New York.

Graver, T.W. and McGinnis, L.F. (1989) A tool provisioning problem in FMS. *International Journal of Flexible Manufacturing Systems*, **1**(3), 239–54.

Gray, A.E., Seidmann, A. and Stecke, K.E. (1993) A synthesis of tool-management issues and decision problems in automated manufacturing. *Management Science*, **29**(5), 549–67.

Hammer, H. (1984) *Improving Economic Efficiencies through Flexible Automation in Milling and Boring*. Proceedings Second Biennial International Machine Tool Technical Conference, National Machine Tool Builders' Association, Mclean, VA.

Han, M., Na, Y.K. and Hogg, G.L. (1989) Real-time tool control and job dispatching in flexible manufacturing systems. *International Journal of Production Research*, **27**(8), 1257–67.

Iwata, K., Murotsu, A., Oba, F. and Yasuda, K. (1982) Production scheduling of flexible manufacturing systems. *Annals of CIRP*, **31**, 319–22.

Jacob, H.J., Hentschel, B. and Stange, B. (1988) Intelligent tool monitoring for machining. *International Journal of Production Research*, **26**(10), 1579–92.

Kashyap, A. (1992) Control rules for tool sharing in flexible manufacturing systems. *Unpublished Master Thesis*, Department of Industrial and Management Systems Engineering, University of South Florida Tampa, FL, USA.

Khator, S.K. and Leung, L.C. (1994) Intermediate tool requirement planning for FMS. *Journal of Manufacturing Systems*, **13**(1), 9–19.

King, J.R. (1980) Machine component grouping in production flow analysis: An approach using a rank order clustering algorithm. *International Journal of Production Research*, **18**(2), 213–32.

Kiran, A.S. and Krason, R.J. (1988) Automating tooling in a flexible manufacturing system. *Industrial Engineering*, April 52–7.

Kokko, T. (1988) *Total Tooling Management System (TMS)*. Proceedings 7th International Conference on Flexible Manufacturing Systems, September, pp. 69–77.

Koulamas, C.P. (1991) Total tool requirements in multi-level machining systems. *International Journal of Production Research*, **29**(2), 417–37.

Kurimoto, A., So, K. and Hill, D. (1988) *Fully Integrated Tool Management System for CIM*. Proceedings 7th International Conference on Flexible Manufacturing Systems, September, pp. 25–38.

Leung, L.C., Maheshwari, S.K. and Miller, W.A. (1992) Concurrent part assignment and tool allocation in FMS with material handling considerations. *International Journal of Production Research*, **31**(1), 117–38.

Little, D., Kehoe, D.F. and Al-Maliki, I. (1988) *Tool Management: The Key Issues for Integrated Manufacture*. Proceedings 7th International Conference on Flexible Manufacturing Systems, Stuttgart, Germany, September, pp. 53–68.

Loktev, A.D. (1983) More efficient use of cutting tools. *Stanki i Instrument*, **54**(7), 14–16.

Martin, J.M. (1989) Managing tools makes the cell click. *Manufacturing Engineering*, April, 59–62.

Mason, F. (1986) Computerized cutting-tool management. *American Machinist & Automated Manufacturing*, May, 105–20.

Mason, F. (1992) Computerized tool management system. *SME Blue Book*, August.

Pan, J.N., Kolarik, W.J. and Lambert, B.K. (1986) Mathematical model to predict the system reliability of tooling for automated machining system. *International Journal of Production Research*, **24**(3), 493–501.

Ranky, P.G. (1988) A generic tool management system architecture for flexible manufacturing systems (FMS). *Robotica*, **6**, 221–34.

Rhodes, J.S. (1985) *The Development of Integrated Control Systems for Flexible Manufacturing Systems.* Proceedings CASA/SME Autofact Conference, Detroit, MI.

Rhodes, J.S. (1986) *FMS Tool Management Systems.* Proceedings CASA/SME Flexible Manufacturing Systems Conference, SME, Deorborn, MI, March.

Rovito, V.P. and Dvorsky, R.E. (1986) *Planning Models for Designing FMS.* Proceedings Advanced Manufacturing Systems Exposition and Conference, June, Chicago, IL.

Rovito, V.P. and Hankins, L. (1986) *The Impact of Tooling in Flexible Manufacturing Systems.* Proceedings Advanced Manufacturing Systems Exposition and Conference, pp. 249–71, June, Chicago, II.

Sarin, S.C. and Chen, C.S. (1987) The machine loading and tool allocation problem in a flexible manufacturing system. *International Journal of Production Research*, **25**(4), 1081–94.

Savoie, R.M. (1988) *Tool requirements planning: A new approach.* Proceedings American Production and Inventory Control Society, pp. 195–7.

Sharit, J. and Elhence, S. (1989) Computerization of tool-replacement decision making in flexible manufacturing systems: a human-systems perspective. *International Journal of Production Research*, **27**(12), 2027–40.

Tomek, P. and Zeleny, J. (1984) FMS in Czechoslovakia. *FMS Magazine*, 35–40.

Vinod, B. and Sabbagh, M. (1986) Optimal performance analysis of manufacturing systems subject to tool availability. *European Journal of Operations Research*, **24**(3), 398–409.

Zavanella, L., Maccarini, G.C. and Bugini, A. (1990) FMS tool supply in a stochastic environment: Strategies and related reliabilities. *International Journal of Machine Tools Manufacturing*, **30**(3), 389–402.

Zeleny, J. (1981) Flexible manufacturing systems with automatic transport of tools. *Annals of CIRP*, **30**.

Zeleny, J. (1985) Unmanned technology and control strategies in flexible manufacturing systems with random flow of parts and tools. *Annals of CIRP*, **34**(1), 391–5.

Zhang, P., DeSouza, R. and Bell, R. (1989) Tool Management for Highly Automated Turning Systems. *Loughborough University of Technology*, UK.

CHAPTER 13

Guidance and navigation techniques for guided and autonomous vehicles

C.B. Besant

13.1 INTRODUCTION

Automated Guided Vehicles (AGVs) have a wide range of applications, particularly in the field of industrial automation and general materials handling, such as in warehousing. There are also an increasing number of specialized applications for AGVs such as for fire-fighting, underwater maintenance of offshore structures and handling hazardous materials under accident conditions. The concepts of vehicles used in all these applications have many principles that have some commonality, such as guidance and sensor systems. The level of guidance techniques and sensor systems on an AGV depends on the nature of the application. AGV applications range from vehicles operating in highly structured environments such as a factory workshop, to autonomous vehicles that may have to work in an unstructured environment such as in fire-fighting applications.

In most AGV applications, the most important aspect is the accuracy and reliability of the guidance system for the AGV. This essentially consists of the determination of the AGV's position and heading relative to some reference point and direction. The AGVs currently in use in industrial and office environments have these tasks simplified, since the paths for the AGVs are defined by rails, buried wires or painted lines. The AGVs that follow buried wires or painted lines have simple, on-board inductive or optical sensors to continuously track the fixed floor paths, thus achieving vehicle guidance.

AGV systems based on fixed paths suffer from the disadvantage that path-network changes are difficult to achieve. A more flexible approach is to define the path-network in software on each AGV, so that changes can easily be made. However, such a scheme greatly magnifies the problem of vehicle guidance. Various methods have been used to achieve this, using on-board and a combination of on- and off-board sensors. The idea of an autonomous or self-reliant AGV is attractive where flexibility of use is a priority.

A wide range of AGV guidance and navigation concepts will now be discussed in some detail.

13.2 VEHICLES GUIDED BY OFF-BOARD FIXED PATHS

Many current AGV systems fall into the category of fixed path systems. There are three widely used guidance techniques for such systems: rail, inductive and optical guidance.

13.2.1 Rail guidance

This is a trivial guidance technique compared to others, but it does offer reliability and robustness at the expense of flexibility. In this case, the AGV is constrained to move in one direction using rails. Kerney *et al.* (1983) used such a system as a feeder to machine tools. This system had the advantage that mains power was supplied to the vehicle via a third rail which obviated the need for batteries. The AGV's position was determined by reading magnetically coded identification plates fixed to the floor and adjacent to the centre rail. Rail-based AGVs are useful if large capacity mobile robots are required, which can consume significant amounts of power. Such systems are being used by BMW (Anon, 1980).

13.2.2 Inductive guidance

Inductive guidance has been widely used by manufacturers of many AGV systems. This method uses a network of AC signal-carrying buried wires which are arranged in the form of closed loops, as shown in Fig. 13.1. The wires are laid in channels 6–10 mm wide and 15–20 mm deep cut into the floor. The wire is then sealed in position with a synthetic resin. Each guide-wire loop in the network is supplied with an electric current of a particular frequency, known as the 'steering frequency', from a quartz crystal stabilized oscillator with a power stage and transmitter. The steering frequencies are in the range 5–35 kHz. The guide-wires contain currents of 100–500 mA at voltages of 20–50 V.

The principle of the inductive-guidance is based on two concentric coils mounted on the vehicle just above the floor level. The steering frequency of the current flowing in the guide-wire is picked up by the search coils on the AGV via electromagnetic induction. Each steering coil signal is then amplified and compared to produce an error signal which is proportional to the lateral deviation of the vehicle from the guide-wire. This error signal is then used by the AGV steering servo system to steer the vehicle such that it is always above the guide-wire. The vehicles themselves are either the 'tricycle type',

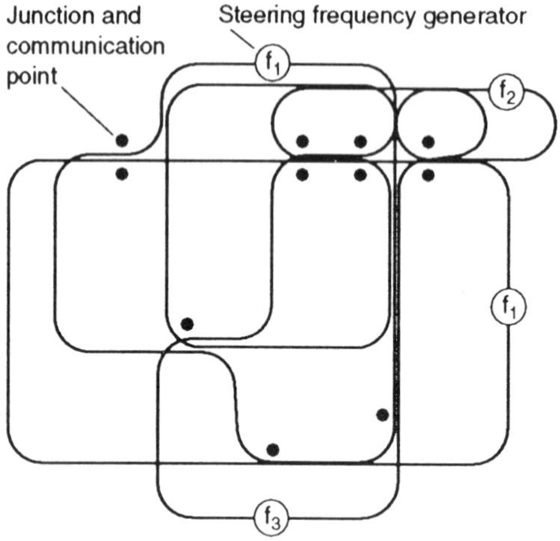

Fig. 13.1 Typical wire-guided AGV path network.

where the single front wheel is used to steer the vehicle, or the type with two wheels located at either side of the vehicle on its central axis, with castors at the front and rear. In the latter type, steering is effected by differentially driving the central wheels.

If the guide-wire network is driven by a single steering frequency signal, then active switching is used at each network junction to direct a vehicle along a required path by switching off the signal to all other paths. Such a network is termed 'active'. If multiple frequencies are used in the network, usually harmonics of a base frequency which differ from each other by a factor of 1.2, then a 'passive network' results. In a passive network the guide-wire loops are arranged in such a way that at network junctions a left turn or a right turn is characterised by a steering frequency which is different from the frequency the AGV has been previously tracking. Thus, active switching is completely avoided in passive networks, which inherently improves system reliability. The guide-wire network also contains buried communication cables which link buried inductive communication points, as is shown in Fig. 13.1, to the guide-wire network controller.

One of the central tasks of the guide-wire network is that of traffic control. This task is facilitated by dividing the network up into areas called 'regions', which are subdivided into areas called 'zones' that are subdivided into 'blocks'. A block is normally a few metres of guide-wire, and represents the smallest controlled portion of the guide-wire network. All network features such as regions, zones, block sections, branch communication points, stopping points and position referencing points are uniquely coded by the network controller when the system is installed to ensure unambiguous identification. The coded network, which is already stored on the network controller, is then stored on-board each vehicle.

When routing a vehicle, the network controller simply gives the vehicle a destination address, after which the AGV's on-board route controller navigates it to the desired place. While an AGV is moving it uses either 'incremental' or 'absolute digital' floor fixed position recognition systems to determine its own position. The incremental method determines relative positions by counting the number of floor-fixed permanent magnets that have been encountered by the vehicle, whilst in the absolute digital method, the vehicle inductively reads floor-fixed codes stating absolute positions. The absolute method is superior to the incremental method because in the latter method errors can occur when counting becomes frequent, as in highly branched networks. However, the basic problem with both methods is that the techniques are short-range in nature and require a considerable amount of expensive buried magnetic devices.

When an AGV approaches the start of a block section permanent magnets, located underneath the vehicle, trip switches buried in the floor. The network controller is thus made aware of the presence of an AGV. The network controller only allows a vehicle to proceed into the block section if it is not already occupied by another vehicle.

The overall control of a wire-guided AGV system requires:

- AGV scheduling and AGV operations management;
- control of the buried guide-wire network, including all communication points and blocking control; and
- control of all load and unload stations.

The size and complexity of the overall AGV control system depends on the number of vehicles and the complexity of the guide-wire network. Large systems are hierarchical in nature to distribute the various control tasks, so making the computations more manageable.

Inductive guidance is simple and reliable for most AGV systems, but it does have the following disadvantages.

- the system lacks flexibility in that paths cannot easily be changed because of the buried guide-wire;
- the cost of laying the guide-wire is significant, and it is disruptive if laying takes place in an operational plant due to the dust created by cutting the concrete floor;
- if the guide-wire breaks due to floor stresses, or is in close proximity to ferrous metals, the magnetic guidance field becomes distorted or disappears; and
- inductive guidance is subject to electromagnetic interference.

13.2.3 Optical guidance

Optical guidance systems were originally developed for office or light engineering environments. They have similarities to wire-guide systems except for their guidance technique. The system is based on the use of lines, painted with visible or invisible fluorescent paint, on the floor which act as paths for AGVs to follow. Two photosensors are fixed to the underside of the vehicle and in a similar manner to the steering coils on a wire-guided AGV, detect the intensity of the fluorescence. The signals from these photosensors are amplified and compared to produce an error signal that is used to steer the vehicle along the desired path. Stops, turns and other vehicle commands are made using magnetic plates in the floor in a similar manner to that used in wire-guided systems.

In a system with a large number of vehicles, the supervisory computer is used for vehicle dispatching, routing, traffic control and collision avoidance. Communication between the supervisory computer and vehicles is via a FM radio link.

The advantage of the optical guidance method over wire guidance is that routes can easily be changed by removing the lines and painting new ones. However, the optical system can have problems if lines are erased through wear, or if lines are obscured by objects.

13.3 VEHICLES GUIDED BY ON-BOARD SOFTWARE PROGRAMMABLE PATHS

It has already been shown that fixed-path AGV systems suffer from a lack of flexibility, as well as other disadvantages such as the cost of installation. There is a need, therefore, for systems that are free-ranging in concept and that operate in a manner analogous to humans in terms of navigation. In a factory environment a supervisory system will timetable routes for vehicles to perform certain duties. Appropriate information is then sent to the vehicles, which then have to fulfil each requested transport order based on its guidance and navigation system. The guidance system then has to ensure that the vehicle goes to the requisite points in the environment, is at those points at a specified time, and follows programmed paths safely and reliably.

There are various ways of considering the guidance problem for free-ranging vehicles. Harmon (1987) classified the capability of the sensors required into three horizons. The closest is just under the vehicle, giving positioning and orientation information. The second horizon identifies the capability of the vehicle to locate and recognize nearby markers and obstacles. The final horizon defines the ability of the vehicle to characterize the distant terrain.

Isik (1988) uses a principle called 'nested hierarchical control', where the navigation is divided to three planning and control levels. The three levels are: Planner, Navigator and Pilot. The Planner is used to develop the map so as to determine goals to reach an assigned task. The Navigator works on a map of a higher resolution to provide detailed paths. The final, Pilot level provides optimal motion control, obstacle avoidance and maintains continuity of motion along a path.

Table 13.1 Problem solving layers of the CMU mobile robot projects

8. Supervisor
- Global supervision of system behavior
- User interface

7. Global Planning
- Task-level planning to provide sequences of sensory, actuator and processing (software) actions
- Simulation
- Error-recovery and replanning in case of failure or unexpected events

6. Control
- Scheduling of activities
- Integration of plan-driven with data-driven activities

5. Navigation
- Navigation modules provide services such as path planning and obstacle avoidance

4. Real-World Modelling
- Integration of local pieces of correlated information into a global real-world model that describes the robot's environment of operation
- Matching acquired information against stored maps
- Object identification
- Landmark recognition

3. Sensor Integration
- Information provided by different sensors is correlated and abstracted
- Common representations and compatible frames of reference are used

2. Sensor Interpretation
- Acquisition of sensor data (vision, sonar, laser rangefinder)
- Interpretation of sonar data

1. Robot Control
- Set of primitives for robot operation
- Actuator control (e.g. locomotion)
- Sensor control
- Internal sensor monitoring

The CMU Mobile Robot Project (1987) uses a layered architecture to define software processing levels, as shown in Table 13.1. Sensor information is integrated to create or update the real world map, which is used by the navigation layer for activities such as path planning and obstacle avoidance.

Free-ranging AGVs do not have physical guide paths which they are forced to follow and, therefore, it is usually necessary for such vehicles to have more than one guidance method to ensure robustness of the vehicle's movements. Various guidance techniques are now described.

13.3.1 Dead reckoning based on odometry

Navigation is a process of directing movement of the vehicle from one position to another. To accomplish this, the present position and velocity must be known relative to a fixed datum. The combination of distance travelled from a known position with an estimate of the direction of the path being followed is the basis of the dead reckoning method, and it is one of the earliest methods of navigation. In this method, wheel rotations on the vehicle are measured (odometry) using optical encoders, and the control system on the vehicle is able to determine the position (x, y) and heading (θ) of the vehicle.

Odometric calculations for two different vehicle configurations are now given.

(a) Differential drive vehicles

Two wheels at each side of the centre of the vehicle are driven by electric motors, with wheel rotations measured using optical encoders. The vehicle can be driven along a curved path by driving the two motors at different speeds to each other. The centre of rotation of the vehicle is at the centre (CR), between the two wheels, as shown in Fig. 13.2.

Consider the vehicle moving from position (x_1, y_1) to (x_2, y_2) in time ΔT. The instantaneous velocity of the vehicle along the path x' is:

$$v = (v_r + v_l)/2,$$

where v_r = velocity of right hand wheel (ω_r/R); v_l = velocity of left-hand wheel (ω_l/R); ω and R are wheel angular speed and radius, respectively; $v_r - (x_{r1} - x_{r2})/\Delta t$; and $v_l = (x_{l1} - x_{l2})/\Delta t$.

The rate of change of the vehicle's heading is:

$$\dot{\theta} = (v_r - v_l)/\omega.$$

The new coordinates of the vehicle are:

$$\theta_2 = \theta_1 + \theta,$$
$$x_2 = x_1 + v \cos\theta_2 \cdot \Delta t,$$
$$y_2 = y_1 + v \sin\theta_2 \cdot \Delta t.$$

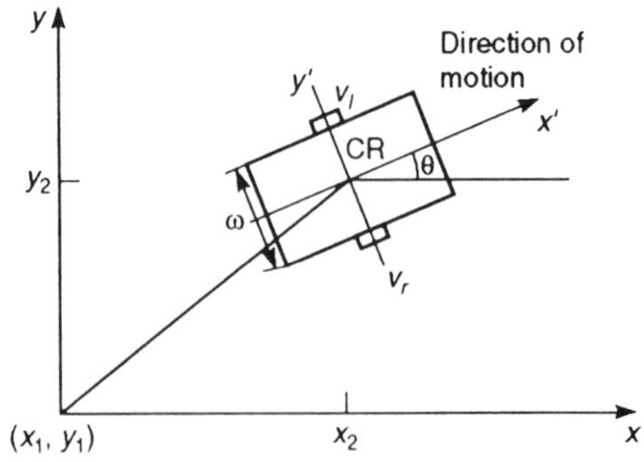

Fig. 13.2 Differential steering.

With the driving wheels at the centre of the vehicle, it can move equally well forwards and backwards.

(b) Steered wheel tricycle vehicle

The tricycle vehicle is controlled by the front wheel steering angle α and the drive wheel angular velocity ω, as shown in Fig. 13.3. With reference to the centre of rotation CR of the vehicle, the vehicle's speed v and position (x, y) are related to the steering and drive angular coordinates (α, ω) by the following kinematic equations:

$$v = R\omega \cos \alpha,$$

where R = radius of the drive wheel, and

$$\dot{\theta} = (R/b)\omega \sin \alpha,$$

where b = distance from the centre of rotation CR to the steering wheel,

$$\dot{x} = v \cos \theta,$$

$$\dot{y} = v \sin \theta,$$

where θ = heading of the vehicle.

Most AGVs now in use are steered by a single wheel or wheel pair, or are steered by differentially driven wheels. Both systems suffer from steering discontinuities for ideal path following along an arc. The steering wheel configuration has discontinuities in the desired steering angle, whereas the differential drive configuration has discontinuities in the desired velocities of the drive wheel. Nelson (1989) shows that by proper choice of a guidepoint (GP in Fig. 13.3), which is the point on the vehicle that is designated to follow the reference path, these discontinuities can be eliminated. The choice of guidepoint affects the space swept by the vehicle, and its kinematics and dynamics. It also affects the desired steering and drive functions (inverse kinematics) for path following.

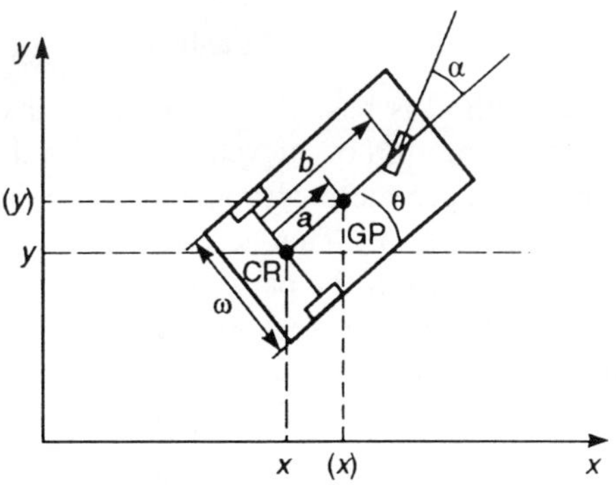

Fig. 13.3 Tricycle cart configuration.

(c) Odometric errors

An application of possible errors in a guidance technique is very important, especially with free-ranging AGVs. The frame of reference for dead reckoning is on the vehicle. Any error in the parameters of the odometric equations will result in an incorrect position for the vehicle with respect to the environment (McGillern *et al.*, 1988).

The odometric equations for a differential drive vehicle are:

$$\Delta\theta = 4\pi R\{(e_{r2} - e_{r1}) - (e_{l2} - e_{l1})\}/\omega,$$
$$\Delta x = \pi R \cos\theta\{(e_{r2} - e_{r1}) + (e_{l2} - e_{l1})\},$$
$$\Delta y = \pi R \sin\theta\{(e_{r2} - e_{r1}) + (e_{l2} - e_{l1})\}.$$

For straight line motion, e_r should equal e_l and the heading zero. If there is a slight error in one of the encoder readings, then not only is the distance travelled in error, but, more seriously, the heading of the vehicle is in error. This results in a cumulative effect resulting in a serious overall error of position. The errors can be minimized through calibration, but errors can still result when the vehicle is in operation due to the following causes:

- changes in wheel radius through wear or weight of load being carried by the vehicle;
- wheel slippage through patches of grease or liquid on the floor;
- uncertainty in separation of wheel contact points (ω) since the floor contact with the wheel extends over the width of the wheel; when cornering, this uncertainty can be magnified, producing a further uncertainty in vehicle heading;
- small obstacles on the floor under a wheel can affect a wheel measurement.

Odometry does, therefore, have problems of accuracy when in practical use. It is, however, useful in providing closed-loop control on the drive system of the vehicle, but a secondary guidance system is necessary to provide security in vehicle guidance.

13.3.2 Inertial guidance

The inertial guidance method is based on aligning the axis of a gyroscope parallel to the direction of motion of the vehicle. If the vehicle deviates from this path, an acceleration is produced orthogonal to the direction of motion, and is detected by the gyroscope. This acceleration is integrated twice to give the positional deviation from the path, which can then be corrected by the vehicle's servo mechanism. The method is extensively used where there is no external frame of reference (guided missiles, ships, etc.). There are four conceptual elements involved in inertial navigation:

1. A vector accelerometer that measures a vector non-gravitational acceleration.

2. An altitude reference when two gyros are mounted in gimbals to maintain a predetermined spatial orientation as the vehicle moves.
3. A computer for solving the acceleration equation and performing integrations for velocity and position.
4. A clock as a datum for monitoring vehicle movements.

Inertial guidance tends to be expensive, and the axis of the gyroscope tends to drift, giving rise to errors.

13.3.3 Imaging systems

Imaging-based sensors that can locate objects and markers in the environment of the vehicle are becoming increasingly important to free-ranging or autonomous vehicles. There are now a large number of techniques based on such systems that are in current use on AGVs. The techniques range from one-dimensional systems using photosensors that make use of a collimated beam of light, to two-dimensional systems that use CCD cameras to collect information, such as markers on the floor. Finally, there are full three-dimensional systems which make use of stereoscopic vision to detect objects.

Image techniques can be computationally intensive, but parallel processing computers are allowing such techniques to be increasingly used in AGV applications. A number of imaging techniques will now be presented.

(a) One-dimensional imaging

Ren and Walker (1986) used a collimated light beam and two rows of photosensors, one row situated in the front of the vehicle and the other at the rear, as shown in Fig. 13.4. The beam angle is known with respect to world coordinates. When the vehicle crosses the beam, it is detected by the front row at d_1 and the rear row at d_2. From these two values, it is possible to calculate the heading ϕ, and the lateral deviation of the vehicle with respect

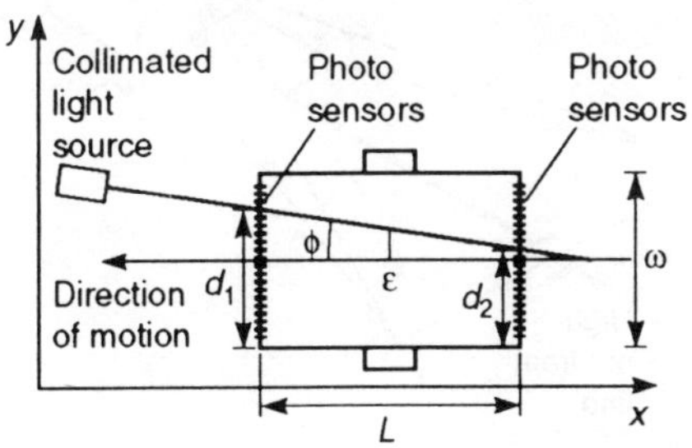

Fig. 13.4 Collimated light source and photosensors.

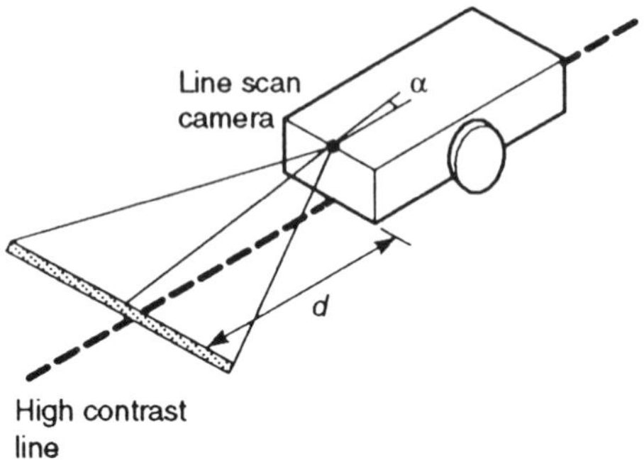

Fig. 13.5 Line scan camera and high contrast line. α = angle of inclination of the line scan camera; d = floor distance of lens aperture of camera to scanned area.

to the collimated beam as follows:

$$\phi = \tan^{-1}(d_1 - d_2)/L,$$
$$\varepsilon = 0.5 \cos \phi(\omega - d_1 - d_2).$$

The lateral deviation is determined with respect to the centre of the vehicle.

This technique was implemented on the ICAGV (Widden, 1987).

A second method is based on a line scan camera focused one metre in front of the vehicle on a painted line, as shown in Fig. 13.5. Since the velocity of the vehicle can be determined via an odometric system, the lateral and heading errors can be found.

Since both methods are one-dimensional, distance markers cannot be computed.

(b) Two-dimensional imaging

A typical two-dimensional system is shown in Fig. 13.6 where a CCD camera is mounted on-board the vehicle and looks at an area just in front of it.

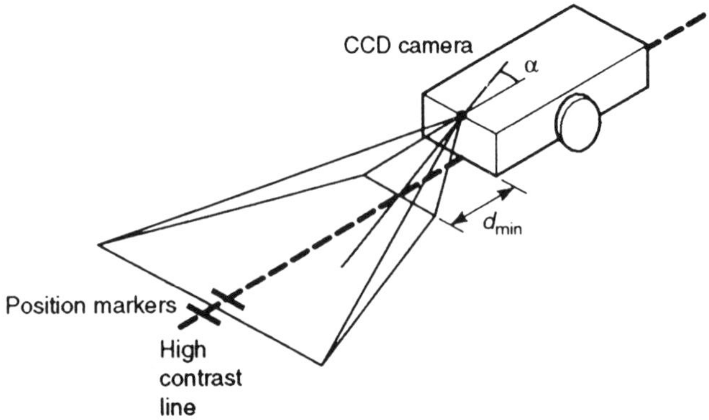

Fig. 13.6 Camera pointed at 2D floor. α = angle of inclination of the CCD camera; d_{min} = minimum floor distance between lens aperture of camera to scanned area.

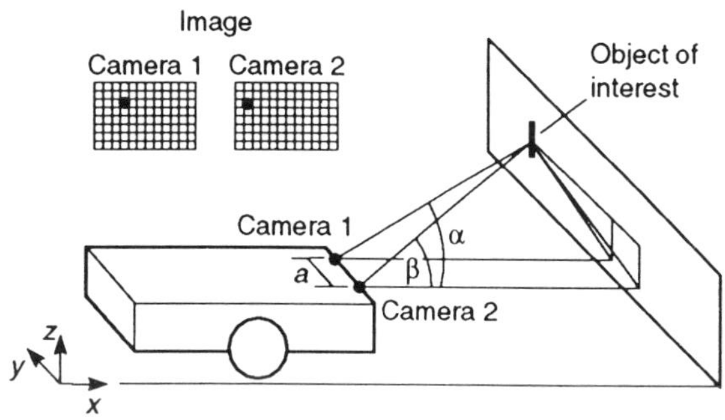

Fig. 13.7 Stereoscopic vision.

Knowing the coordinates of the camera with respect to the vehicle, an image can be mapped to the vehicle coordinate system. Pathway markers can be introduced at strategic points; for example, a datum location for precise docking or for correcting vehicle heading errors at corners (Sen *et al.*, 1991a).

(c) Three-dimensional imaging

Stereoscopic vision systems are getting much attention for autonomous vehicle applications. Two cameras are mounted on the vehicle at a known distance apart, and the scene is analysed using calculations based on triangulation (Fig. 13.7).

If only two views are used, the matching of the features between the views must be correct (Fig. 13.7). It is not easy to achieve image feature extraction with only two views, especially in real-time.

Camera calibration techniques can greatly enhance a vision system when used in 3D object recognition (Tsai, 1987). This is essential for eliminating systematic errors associated with cameras such as with the optics. Regular calibration is necessary to check for problems of drift or system failure.

Much research has also been done on geometric databases for sensor image recognition, such as that used for the Oxford University Autonomous Guided Vehicle (Cameron, 1988).

13.3.4 Ranging sensors

Ranging information can be obtained using two basic methods.

- *Triangulation* If the coordinates of two fixed points remote from the vehicle are known, then with a suitable sensor, it is possible to determine the position of a vehicle by triangulation from those points.
- *Time-of-flight* If a signal of known velocity is emitted from the vehicle to a fixed known object, then it is possible to determine the distance

between the vehicle and object by timing the emission of the signal and its return echo from the object.

The use of stereo vision on its own results in inherently high computational costs and calibration problems. Good range sensors have the potential for eliminating some of these problems. It has been shown (Herbert *et al.*, 1990) that a range sensor based on a time-of-flight laser, developed at the Environmental Research Institute of Michigan (ERIM), can be a very effective sensor for autonomous vehicles. It can provide range data without a high computational overhead, and this is important for time critical applications such as obstacle detection. It is also insensitive to illumination, and is therefore not affected by shadows.

The principle of the laser range finder is to measure the phase difference between a laser beam and its reflection from the scene. A two-mirror scanning system allows the beam to be directed anywhere within a $30° \times 80°$ field of view. The data produced by the ERIM sensor is a 64×256 range image, and the range is coded in 8-bits. The position of the vehicle in a given coordinate system can easily be derived from the measured range and the direction of the beam as follows:

$$x = R \sin \theta,$$

$$y = R \cos \phi \cos \theta,$$

$$z = R \sin \phi \cos \theta,$$

where $R =$ the measure range; $\phi =$ vertical scanning angle of the beam direction; and $\theta =$ horizontal scanning angle of the beam direction.

Although the data from such a sensor is largely independent of the environmental illumination, the reflectance image is an image of the energy reflected by the laser beam. This energy is dependent on the measured surface and its distance from the sensor. The reflectance energy can readily be calibrated and the calibration stored in a look-up table.

(a) *Laser ranging*

Laser ranging is based on the concept of a rotating laser, mounted on the vehicle, that looks for beacons on markers in the surrounding environment. If the position of the beacons or markers are known, then the position and heading of the vehicle can be determined by triangulation calculations. This basic method was used by Premi (1985), who developed active beacons on which to triangulate. A commercial system based on bar-coded markers was adopted in the Caterpillar-GEC Self-Guided Vehicle System (SGVS) (Premi, 1985; Evans, 1988).

The routes of the SGV are stored by computers both on the SGV and on the supervisory system. The laser situated on the SGV (Fig. 13.8) scans the markers and uses combinations of them to triangulate for vehicle position and heading.

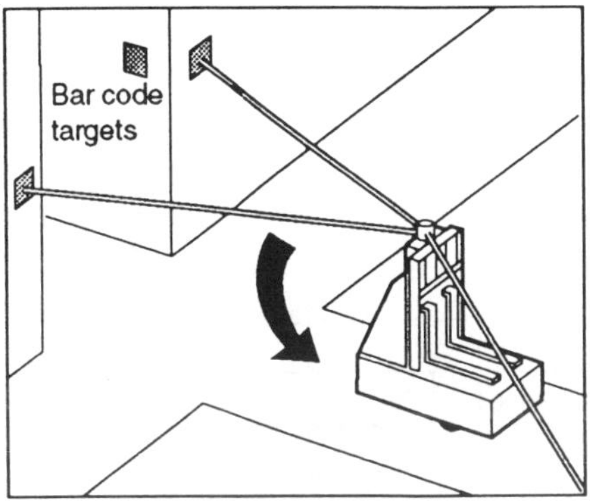

Fig. 13.8 Caterpillar self-guided vehicle system.

The SGV uses odometry as a primary guidance system, and the laser ranging system provides support in terms of correcting for cumulative errors that are inherent with odometry.

(b) Ultrasonic ranging

Ultrasonic systems have been widely used in autonomous vehicle applications, particularly for obstacle detection and avoidance. The ultrasonic ranging sensor is simple, low in cost and distances to objects are directly provided (Zelindsky, 1988).

Ultrasonic measuring systems suffer from a number of drawbacks which limit their usefulness in mapping, or in any task requiring high accuracy. These drawbacks are inherent in the principle of ultrasonic range finders at their commonly used wavelengths (Borenstein and Koren, 1990; Flynn, 1988). They are:

- Poor directionality, since most of the sound is reflected perpendicularly to the surface of an object and detected by the ultrasonic transceiver. If the surface of the object is at an angle to the incident beam, then the sonic waves are scattered and only a small amount of sonic energy may be captured by the receiver. An angle of incidence greater than 25° to the normal from the surface of an obstacle results in a weak return signal which may be undetectable.
- Range uncertainty, since the velocity of sound varies with ambient temperature. This variation is in proportion to $\sqrt{T}$, where T is the absolute temperature of the environment.

13.4 SENSOR INTEGRATION FOR FREE-RANGING AGVs

Various guidance techniques for AGV systems have been presented, and these are summarized in Fig. 13.9. It has already been mentioned that AGVs

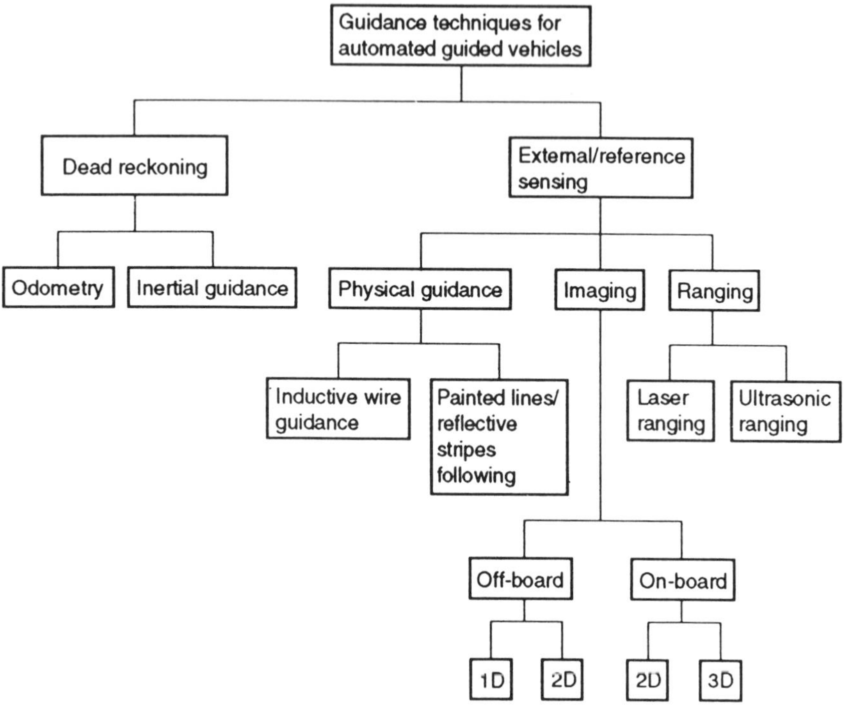

Fig. 13.9 Different guidance methods for AGVs.

often have more than one guidance system, and this is of particular importance in considering free-ranging or autonomous vehicles. The key to success for a practical free-ranging vehicle is in the integration of sensors that will give the vehicle guidance the requisite robustness for use in real working environments where safety is an important factor. A number of free-ranging or autonomous AGV systems will now be discussed to illustrate the sensor integration for vehicle guidance.

13.4.1 ICAGV system

The Imperial College free-ranging AGV system (Sen *et al.*, 1991a,b) is aimed at operating in a structured environment such as a factory shop floor. The concept sought to provide maximum flexibility in terms of the use of vehicles, and also the ease of installation. The system is based on a supervisory computer that can communicate with a small fleet of free-ranging vehicles via an FM radio link. The supervisory computer performs the usual functions of vehicle scheduling, including dynamic scheduling and dispatching. In addition, the supervisory computer is used in the system installation phase to create maps of useful environmental information such as the painted lines used to identify walkways on the factory floor which are required by factory regulations in

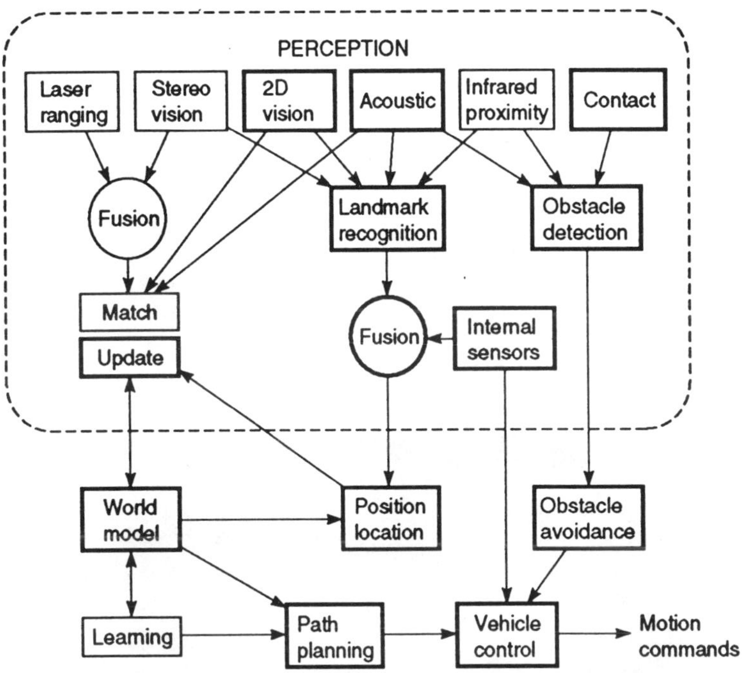

Fig. 13.10 Sensors fusion and integration for a mobile robot.

most countries. Positions of machines can also be mapped. A standard drafting system such as AUTOCAD is all that is required to create such maps. This information can then be stored in the computer on-board each vehicle, and used by sensors on the vehicle to enhance the navigation process. The supervisory computer is also used to create the vehicle network, which consists of nodes for docking stations and network junctions and intersections. This information is used in the scheduling and operational planning phase. Some additional information is added in the installation phase such as position markers to aid precision docking. The use of markers is limited in an installation to places where extra positional precision is required.

The range of sensors available to AGVs is shown in Fig. 13.10. The ICAGV uses the sensors in the shaded blocks. Dead reckoning based on odometry is used for primary guidance in following routes, and in vehicle control. The main secondary guidance is provided by a vision system consisting of a CCD camera mounted on the vehicle, as described in section 13.3.3(b). This camera is used to detect the following environmental indicators:

- Path or lane markers to check for heading or lateral drift during long straight journeys. The lines defining walkway boundaries are regularly sampled, and vehicle position and heading checked. The integrity of the walkway boundary lines are not critical as for AGVs that follow painted lines.

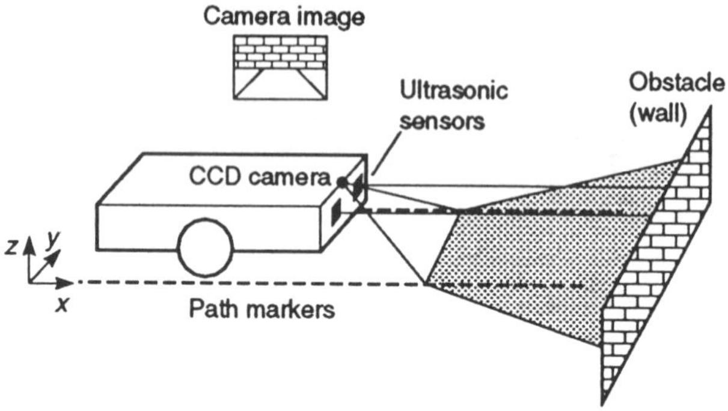

Fig. 13.11 ICAGV-II guidance system.

Fig. 13.12 Imperial College Automated Guided Vehicle (ICAGV-II).

- Position markers for real world updates.
- Docking or datum markers for precise motion.
- Initial calibration of odometry system.

The ICAGV also contains ultrasonic sensors for obstacle detection. Figure 13.11 gives a summary of the secondary guidance systems, and Fig. 13.12 shows the actual ICAGV.

13.4.2 The IPAMAR system

The IPAMAR system was developed at the Fraunhofer Institut IPA in Germany (Forster, 1989). It is based on a vehicle with a number of sensors that are integrated at a task planning level in the control architecture for determining the action for a vehicle. At the highest level in the architecture, a CAD model of the environment together with task specification and knowledge are fed into a software module for route and sensor action planning. Odometry is used on the vehicle as the primary guidance system, with secondary support guidance by an ultrasonic ranging system. The sensor planner informs the vehicle's sensor control system when to use the ultrasonic range finder at specific points in the working environment as a check on odometry.

A sequencing of motion and sensor commands are sent to the next module, which is for task execution. This module takes in information from collision-detecting ultrasonic sensors. Error recovery and local replanning takes place at this level.

Single motion and sensor commands are passed to the lowest level in the hierarchy, which is the motion control module. In this module, trajectory generation, either fixed or sensor controlled, takes place together with path control, coordinate transformation and motor control.

There are conceptual analogies between the IPAMAR and ICAGV systems in that both make extensive use of CAD design information at a planning level to identify vehicle routes and, most importantly, the provision of information on when various sensors should be used. For example, in the IPAMAR system, ultrasonic range finders are selected at points for updating vehicle positions. This is obtained by selection of the most suitable locations for referencing. Consequently, only parts of the real world have to be exactly known for operation. Normally, parallel or perpendicular walls are chosen as references to obtain clear conditions while sensing. The ICAGV system also makes use of environmental information that can be precisely located such as the lines defining the boundaries of pathways.

The on-line motion guidance for both systems is used for obstacle avoidance strategies initiated by local replanning. The sensory controlled trajectory generation functions do not require model information. It is worth commenting on the use of odometry, since many free-ranging vehicles place a good deal of reliance on it. Both the IPAMAR and ICAGV vehicles have been extensively tested for performance on odometry. In both cases, floor surface together with rates of acceleration, braking or cornering were shown to be important considerations for loss of accuracy. In the case of the IPAMAR system, the distance travelled error using odometry was reduced from 1–0.1% by careful calibration. The compensation for slippage was based on knowledge of the instantaneous driving torque to the wheels and the acceleration of the vehicle, in a model containing the relevant vehicle constants.

13.4.3 The NAVLAB system

The NAVLAB autonomous vehicle (Thorpe *et al.*, 1988) was developed at Carnegie-Mellon University, and shows the importance of fusion integration.

A mobile robot needs more than just navigation capabilities, since it must be able to extract semantic descriptions from its sensors. A vehicle sensory system should not only be able to build a geometric representation of an object, but also to relate it to a stored model. A range sensor, like that already described, is not capable of extracting semantic information from a scene. A video camera can capture semantic data and Herbert *et al.*, have used a colour camera to extract colour data from objects. Data from both range and video sensors are merged by estimating the relative position of the data which is achieved through calibration. This results in a set of pixels in a video image to be directly related to that in a range image.

The result of this sensor fusion is an enhanced feature extraction system enabling the mobile robot to not only detect objects, but also to recognize them. This increases the positional integrity of the vehicle which is important in most AGV application.

REFERENCES

Anon (1980) BMW buys KUKA portable robots. *The Industrial Robot*, **7**(4), 259.

Borenstein, J. and Koren, Y. (1990) Task-level tour plan generation for mobile robots. *IEEE Journal of Robotics and Automation*, **20**(4), 938–43.

Cameron, S. (1988) *A Geometric Database for the Oxford Autonomous Guided Vehicle.* Proceedings NATO Conference on CAD Based Programming for Sensory Robots (ed. R.R. Bahram), pp. 511–26. Springer-Verlag (NATO ASI Series (F) *Computer Systems Sciences.* **5**(511–26).

Elfes, A. (1987) Sonar based real world mapping and navigation. *IEEE Journal of Robotics and Automation*, **3**(3), 249–65.

Evans, D.F. (1988) *Non Wire Guidance: System Flexibility is what Counts.* Proceedings 6th International Conference on Automated Guided Vehicle Systems, pp. 255–62.

Flynn, A.M. (1988) Combining sonar and infrared sensors for mobile robots. *International Journal of Robotics Research*, **7**(6), 5–14.

Forster, S. (1989) *Mobile Autonomous Robot performs free-range AGV Operation.* Proceedings 7th International Conference on Automated Guided Vehicle Systems, pp. 57–68. June, Berlin.

Harmon, S.Y. (1987) The grand surveillance robot. GSR: An autonomous vehicle designed to transmit unknown terrain. *IEEE Journal of Robotics and Automation*, **3**(3).

Herbert, M., Kweon, I. and Kanade, T. (1990) 3D vision techniques for Autonomous vehicles, in *Vision and Navigation*, ed C.E. Thorpe, Kluwer Academic, New York.

Isik, C. and Meystel, A.M. (1988) Pilot level of a hierarchical controller for an unmanned mobile robot. *IEEE Journal of Robotics and Automation*, **4**(3).

Kerney, Trecker and Moswin (1983) *Guided Vehicles for the Small Manufacturing System.* Proceedings 2nd International Conference on Automated Guided Vehicle Systems, Stuttgart, Germany, 7–9 June.

McGillern, C.D., Callison, J.D. and Rapport, T.S. (1988) *Autonomous Roving Vehicles, Navigation, Control and Communication.* Proceedings 6th International Conference on Automated Guided Vehicle Systems, pp. 179–88, October.

Nelson, W.L. (1989) Continuous steering-function control of robot carts. *IEEE Transaction on Industrial Electronics*, **36**(3).

Premi, S.K. (1985) The design of a free-ranging automated guided vehicle system. *PhD Thesis*, Department of Mechanical Engineering, Imperial College of Science, Technology and Medicine, London.

Ren, Y.T. and Walker, S.P. (1986) Position determination of Automated guided vehicles. *International Journal of Advanced Manufacturing Technology*, **1**(5), 45–53.

Sen, A., Wang, C.S., Ristic, M. and Besant, C.B. (1991a) *The Imperial College Free-Ranging Automated Guided Vehicle System.* Proceedings 7th Conference on Computer Aided Production Engineering (ICAPE). 12–15 August. Tenn.

Sen, A., Wang, C.S., Ristic, M. and Besant, C.B. (1991b) *The Supervision System of the Imperial College Free-Ranging Automated Guided Vehicle.* Project 0-7803-0233-8/91. IEEE Conference on System, Manufacturing and Cybernetics. October. Charlottesville, Va.

Thorpe, C.E., Herbert, M., Kanade, T. and Shafer, S.A. (1988) Vision and navigation for the Carnegie-Mellon Navlab. *IEEE Transactions on Pattern Recognition and Machine Intelligence*, **10**(3).

Tsai, R.Y. (1987) A versatile camera calibration technique for high-accuracy 3D machine vision metrology using off-the-shelf TV cameras and lenses. *IEEE Journal of Robotics and Automation*, **3**(4).

Widden, D.E. (1987) The design and implementation of a docking system for the ICAGV. *MSc Thesis*, Department of Mechanical Engineering, Imperial College of Science Technology and Medicine, London.

Zelindsky, A. (1988) Environmental mapping with a mobile robot using sonar. Proceedings AI 88 *2nd Australian Joint Artificial Intelligence Conference.* November, Adelaide, S.A. *Journal of Artificial Intelligence*, **2**(6), 363–78.

Index

Page numbers appearing in **bold** refer to figures and page numbers appearing in *italic* refer to tables.